Earthworm Assisted Remediation of Effluents and Wastes

Sartaj Ahmad Bhat • Adarsh Pal Vig •
Fusheng Li • Balasubramani Ravindran
Editors

Earthworm Assisted Remediation of Effluents and Wastes

Springer

Editors
Sartaj Ahmad Bhat
River Basin Research Center
Gifu University
Gifu, Japan

Fusheng Li
River Basin Research Center
Gifu University
Gifu, Japan

Adarsh Pal Vig
Botanical and Environmental Sciences
Guru Nanak Dev University
Punjab, India

Balasubramani Ravindran
Department of Environmental Energy
and Engineering
Kyonggi University
Suwon, South Korea

ISBN 978-981-15-4521-4 ISBN 978-981-15-4522-1 (eBook)
https://doi.org/10.1007/978-981-15-4522-1

This Springer imprint is published by the registered company Springer Nature Singapore Pte Ltd.
The registered company address is: 152 Beach Road, #21-01/04 Gateway East, Singapore 189721, Singapore

Preface

Water is one of the essential requirements for all oxygen-dependent living organisms because water can regulate physical and chemical parameters. Approximately, 71% of the planet is covered by water and oceans contain 96.5% of earth's water. The main resource of water includes rainwater, wells, streams, natural springs, ocean, and rivers. In the last few decades, there is a rapid development of human populations and industrial revolutions. Accordingly, various industries are released wastewater/effluent which generates serious environmental problems, especially water pollution. On the other hand, indiscriminate usage of synthetic fertilizers for crop production, during rainy days can migrate into the water bodies which also cause water pollution. In addition, water pollution can affect living organisms and alter the overall food chain.

Nowadays, a huge amount of wastewater sludge/solid wastes are produced by various industries and human beings. These sludge/solid wastes contain a significant amount of hazardous materials that generate soil pollution. In soil, hazardous pollutants are potentially toxic to living organisms, and they alter the chemical and biological reactions. Currently, various peoples have been using numerous methods (like physical, chemical, and biological methods) to combat water and soil pollution, but these methods contain several disadvantages. Therefore, there is urgent requirement of cost-effective and environment-friendly techniques to remediate pollutants. This book "*Earthworm Assisted Remediation of Effluents and Wastes*" introduces various remediation strategies. For example, vermifiltration of wastewater/effluent employing earthworms is a recently established "*novel*" technology. This term filtration technology is based on the capability of worms to consume and break down various organic waste materials and heavy metals from effluent, and their capacity to remove different pollutants from effluent by absorption via body walls of the earthworms. Vermifiltration is an effective and environment-friendly technology for wastewater/effluent treatment. In addition, earthworms can eliminate toxic hazardous materials from solid wastes and also enhance the microbial populations which stimulate crop production. Internal body of the earthworms has metallothioneins, protein that can bind with heavy metal ions, and also an earthworm detoxifies the various soil pollutants. This book contributed by an interdisciplinary group of water and soil scientists which provides new knowledge in the field of environmental pollution. We wish to thank all of the referees, who generously contributed their time and talent to maintain the high quality of this volume. We also express our thanks to the springer nature for their invaluable support and cooperation in the publication of the book.

Gifu, Japan — Sartaj Ahmad Bhat
Amritsar, Punjab, India — Adarsh Pal Vig
Gifu, Japan — Fusheng Li
Suwon, South Korea — Balasubramani Ravindran

Contents

Part V Soil

Part V Soil

About the Editors

Sartaj Ahmad Bhat is working as Post-doctoral Researcher in the River Basin Research Center, Gifu University, Japan. He received his Ph.D. for Environmental Sciences from Guru Nanak Dev University, India in 2017. He is efficient in waste management techniques, especially towards the vermicomposting and substrate compatibility, nutrient enrichment and heavy metal accumulation dynamics. So far, Dr. Bhat has authored more than 30 research publications in peer-reviewed international journals. He is also an editor, editorial board member and reviewer of many international reputed journals published by PLOS, De Gruyter, Springer, SAGE, MDPI, Elsevier and Taylor and Francis. He has been recently awarded as a Top Peer Reviewer 2019 in Environment and Ecology by Publons Web of Science Group and has more than 200 verified reviews to his credit.

Adarsh Pal Vig is a Professor and former Head, Department of Botanical and Environmental Sciences, Guru Nanak Dev University, Amritsar, India. Dr. Vig is having teaching and research experience of more than 26 years, about 105 publications in National/International journals with a h-index of 21 and had supervised 10 Ph.D. students. His research mainly focuses on biological treatment technologies for agricultural and industrial wastes. He has been awarded Distinguished Teachers Award, 2012 and Environmentalist of the Year, 2016 & 2018. Presently also working as Director of UGC—Human Resource Development Centre and Project coordinator, FDC and NRC under Pandit Madan Mohan Malviya National Mission on Teachers and Teaching, MHRD, Govt. of India, at Guru Nanak Dev University.

Fusheng Li is a Professor in the Division of Water System Safety and Security Studies and the Graduate School of Engineering at Gifu University, Japan. He received his BS degree for environmental engineering from Lanzhou Jiaotong University of China in 1986, MS degree from Kitami Institute of Technology of Japan in 1994 and PhD degree from Gifu University of Japan in 1998. Dr. Li is directing the Division of Water Quality Studies that covers the fields from water quality to water and wastewater treatment, and recently to resource and energy recovery from organic waste. The ongoing research projects in his lab include adsorption; membrane filtration, enhanced coagulation, disinfection; biological water and wastewater treatment; vermicomposting treatment of vegetable waste and activated sludge; microbial fuel cell; physicochemical water quality assessment; biological water quality assessment. He has over 350 scholarly publications, including more than 160 in peer reviewed journal papers. As principal supervisor, he has guided so far 41 masters and 13 doctorate graduate students to the completion of their degrees. Dr. Li is the recipient of awards from several academic societies and associations for his research work on water treatment and water quality dynamics studies.

Balasubramani Ravindran is currently working as an Assistant Professor in Department of Environmental Energy & Engineering, Kyonggi University, South Korea. He obtained his doctorate from the Environmental Science and Engineering Division, Council of Scientific & Industrial Research (CSIR) Central Leather Research Institute (CLRI), which is an affiliated with the University of Madras, Tamil Nadu, India, by 2013. His primary research focuses on development and evaluation of treatment technologies for solid waste and wastewater from domestic and industrial outlets. He has published more than fifty research papers in peer-reviewed journals, few book chapters and three patents to his credit. He is as a potential reviewer in top international journals and also received "Outstanding reviewer award" from Elsevier and Springer Journals. He has also received prestigious "Best Researcher—IBET 2017" award (in waste management research) for Exceptional Performance and Contributions to 'International Bio-energy technology/eco-protection/organic food/green business' given by the International Centre for Biogas & Bio-energy Technology, India.

Part I
Wastewater Alone

Chapter 1
Applicability of Vermifiltration for Wastewater Treatment and Recycling

Bhavini, Kavita Kanaujia, Amber Trivedi, and Subrata Hait

Abstract With the rapid population growth and wastewater generation due to anthropogenic activities, availability of freshwater is decreasing annually. Untreated wastewater discharged from the municipal and industrial sectors reaches to the local surface water bodies and degrades water quality. Conventional wastewater treatment systems possessing high carbon footprint require mechanistic operations and need to be made affordable with ease of operation. To overcome the impediments associated with the conventional treatment systems, vermifiltration technique employing earthworms in a filter bed has emerged as an alternative for wastewater treatment and recycling. Further, the potential of macrophyte has also been explored by integrating with the vermifiltration system for wastewater treatment. This chapter presents the applicability of vermifiltration technique with various filter design configurations and mechanisms involved for the treatment and recycling of both sewage and industrial effluents. Further, the influence of different operational parameters like hydraulic retention time (HRT), organic loading rate (OLR), hydraulic loading rate (HLR), filter media bed design, earthworm density and flow mode on organic, nutrient and pathogen removals from domestic and industrial wastewater is discussed concisely. Moreover, future perspectives have been provided towards the improvement of the efficacy of the vermifiltration system for wastewater treatment and recycling.

Keywords Vermifiltration · Earthworms · Macrophytes · Integrated system · Wastewater treatment and recycling

Bhavini · K. Kanaujia · A. Trivedi · S. Hait (✉)
Department of Civil and Environmental Engineering, Indian Institute of Technology Patna, Patna, Bihar, India
e-mail: shait@iitp.ac.in

S. A. Bhat et al. (eds.), *Earthworm Assisted Remediation of Effluents and Wastes*,
https://doi.org/10.1007/978-981-15-4522-1_1

1.1 Introduction

Increase in global population, urbanization and industrialization has resulted in environmental pollution and degradation including diminished water quality (Verma et al. 2012). Disposal of untreated sewage and industrial effluents into the surface water bodies leads to water pollution (Goel 2006). Wastewater carrying organics like biochemical oxygen demand (BOD), chemical oxygen demand (COD) and nutrients like nitrogen and phosphorus results in the problems like depletion of dissolved oxygen (DO) and eutrophication (Metcalf et al. 1991; Zheng et al. 2013). In addition, exposure to the water contaminated by the release of pathogens from sewage into the surface water leads to water-borne diseases (Reddy and Smith 1987). Thus, deterioration of river ecology along with the loss of freshwater sources creates an unhealthy environment for humans (Wang et al. 2012). Furthermore, the per capita available water is becoming less with an increase in the population pertaining to the limitation of freshwater sources (Pimentel et al. 2004). Therefore, it becomes necessary to reuse wastewater generated from households and other places after giving a certain level of treatment. Owing to the water scarcity and water pollution due to anthropogenic activities, there is an urgent need to treat and reuse the treated effluent in industrial, agricultural and non-potable purposes.

For wastewater treatment, anaerobic and aerobic processes are being used worldwide (Speece 1983). In the anaerobic process, microbes convert organic matters into methane and carbon dioxide, whereas in the aerobic process, aerobic microbes convert organic matters into biomass and carbon dioxide (Metcalf et al. 1991). The anaerobic process is effective for high COD wastewater, requires less energy, and produces less sludge in comparison to aerobic process. However, it has been documented that the aerobic process is comparatively better than the anaerobic process in terms of acclimatizing the variation in pH, temperature and organic loading rates (OLR) (Degremont 1991). Further, the aerobic process requires less time to restart and can work between a range of temperature from 25 to 35 °C as compared to the optimum temperature for the anaerobic process is 30 °C (Singh et al. 2019b). However, both conventional wastewater treatment techniques required high capital cost, recurring expenditures, skilled manpower, more time to restart after complete shutdown and mechanized and energy-intensive operations (Noumsi et al. 2005). In addition, the sludge generated from conventional processes requires further treatment before getting disposed into the environment. Other than the biological treatment process, physical and chemical processes are also being used in some part of the world (Adin and Asano 1998). However, physical and chemical processes are not efficient organic and nutrient removal from wastewater (Ra et al. 2000). Thus, in the present scenario, an economical and sustainable process is required to treat wastewater with less capital and operation and maintenance cost and ease of operation process.

Integration of earthworms in wastewater filtration process has evolved as an eco-friendly and economical alternative to conventional wastewater process, collectively known as vermifiltration (Tomar and Suthar 2011). Wastewater passing through the initial layer, where the organic matter is converted into humus by earthworm, is followed by the filtration through filter media which supports micro-organism's growth and subsequently secondary treatment occurs. Recent studies have shown that the vermifiltration technique can emerge as a suitable and sustainable alternative for wastewater treatment and recycling. Thus, the chapter presents an overview of the vermifiltration technique with various filter design configurations, applicability and performance evaluation of the technique for the treatment and recycling of sewage and industrial effluents, explaining the mechanisms involved. Additionally, the performance of an integrated macrophyte-vermifiltration system for wastewater treatment and recycling has also been presented. Further, the effects of different filter design and operational parameters on the system performance have been summarized. Moreover, future research perspectives have been provided towards the improvement of the efficacy of the system for wastewater treatment and recycling.

1.2 Overview of Vermifiltration Technique

Vermifiltration system comprises an earthworm active zone along with filter media bed which supports microbial community for domestic and industrial wastewater treatment. The species of earthworms employed in vermifiltration include *Eisenia fetida*, *Lumbricus rubellus*, *Eudrilus eugeniae* and *Eisenia andrei* with filter bed consisting of soil, compost and cow dung which are available for pollutant degradation in earthworm active zone (Singh et al. 2019b; Xing et al. 2011). In filter media design, different materials like sand, gravel, cobblestone and quartz sand are commonly used (Singh et al. 2019b). In vermifiltration system, wastewater is firstly passed through earthworm active zone followed by filter media bed. Depending on the wastewater flow direction, vermifiltration system, in general, can be of two types: horizontal flow system (HFS) and vertical flow system (VFS). In HFS, wastewater flows horizontally through the bed while in VFS wastewater is fed vertically through the bed as shown in Figs. 1.1 and 1.2, respectively. A hybrid system combining both horizontal and vertical flow systems in the sequence is used for the treatment of wastewater. The flow of wastewater in the hybrid system is either from a horizontal

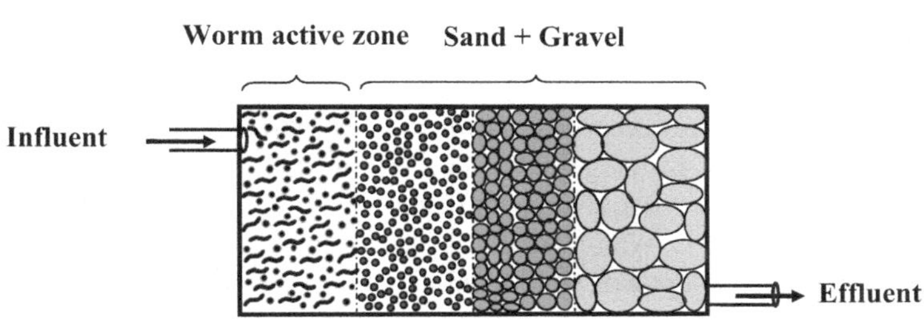

Fig. 1.1 Schematic of a typical horizontal flow vermifiltration system

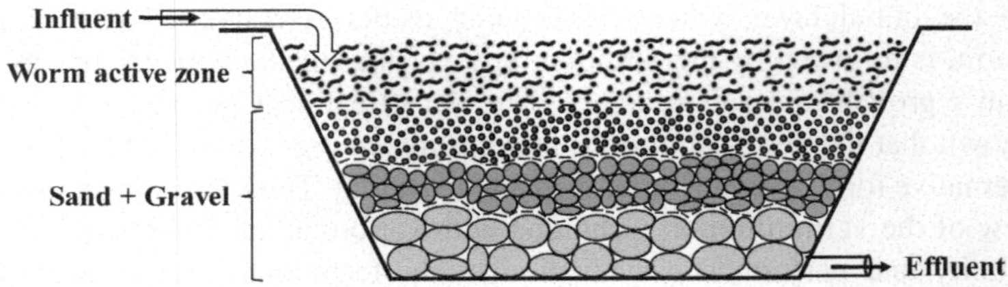

Fig. 1.2 Schematic of a typical vertical flow vermifiltration system

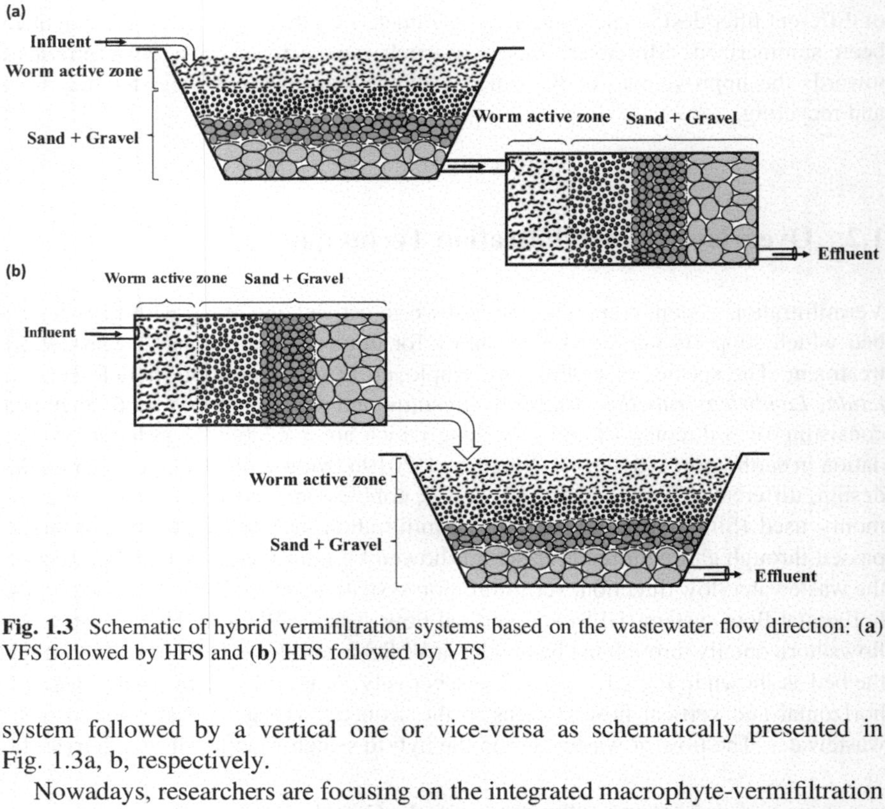

Fig. 1.3 Schematic of hybrid vermifiltration systems based on the wastewater flow direction: (**a**) VFS followed by HFS and (**b**) HFS followed by VFS

system followed by a vertical one or vice-versa as schematically presented in Fig. 1.3a, b, respectively.

Nowadays, researchers are focusing on the integrated macrophyte-vermifiltration system to improve the wastewater treatment efficiency. In macrophyte-assisted vermifiltration system, the concept of wetlands using different plant species like *Canna indica*, *Phragmites australis*, *Typha angustifolia*, *Saccharum spontaneum*, *Cyperus rotundus*, etc. is integrated with vermifiltration system for wastewater treatment (Chen et al. 2016; Nuengjamnong et al. 2011; Samal et al. 2017a; Tomar and Suthar 2011; Wang et al. 2010b). Removal from wastewater takes place when macrophyte uptakes significant amount of nutrients for their growth. A macrophyte-assisted vermifiltration system has been schematically shown in Fig. 1.4. The root or rhizospheric zone of plants helps to provide a favourable

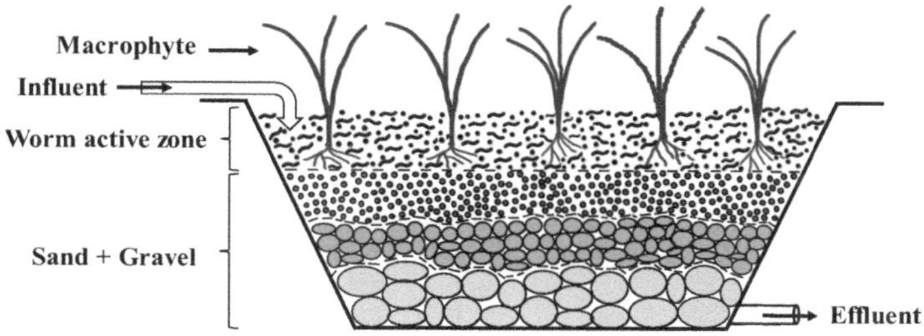

Fig. 1.4 Schematic of a macrophyte-assisted vermifiltration system

environment for the growth of the diverse microbial community to degrade organic contaminants (Bezbaruah and Zhang 2005). Further, researchers have found that macrophyte transfers oxygen from the atmosphere to the rhizosphere which is further consumed by the microbial community (Bezbaruah and Zhang 2005; Brix 1994). Increased oxygen is responsible for maintaining aerobic condition for the microbes as well as for earthworms which is useful to accelerate the degradation of organic contaminants.

1.3 Performance Evaluation of Vermifiltration System

1.3.1 Applicability of Vermifiltration for Sewage Treatment

It has been reported that the vermifiltration technique is an efficient and eco-friendly process to treat wastewater originating from households (Kumar et al. 2016; Li et al. 2009). Vermifiltration technique has been applied to domestic wastewater treatment and has shown a significant reduction of COD and NH_3^+-N (Sinha et al. 2008; Wang et al. 2010a; Xing et al. 2011). Applicability of vermifiltration technique with associated process parameters for wastewater treatment is summarized in Table 1.1. Earthworms consume retained suspended particles in the filter during ingestion and significantly reduce BOD by more than 90% and COD in the range of 80–90% and a significant reduction in nutrients concentration (Li et al. 2009; Wang et al. 2011). According to Kumar et al. (2016), application of vermifiltration employing earthworm species *Eisenia fetida* and *Eudrilus eugeniae* to treat wastewater generated from domestic activities has shown the reduction of about 88% and 70% BOD, 78% and 67% TSS and 75% and 66% TDS, respectively, whereas a laboratory-scale study has revealed the removal of contaminants like BOD_5, COD and TSS from domestic wastewater in the range of 55–66%, 47–65% and 57–78%, respectively (Xing et al. 2010). The earthworm species *Eisenia fetida* is one of the most common species employed to treat domestic wastewater (Gunadi et al. 2002; Hughes et al. 2009; Sinha et al. 2008). In another study on domestic wastewater treatment, employing *Eisenia fetida* as an earthworm species has shown removal of

Table 1.1 Application of vermifiltration technique with associated process parameters for the treatment of sewage and industrial effluents

Wastewater	Flow direction	Earthworm species	Earthworm density	Macrophyte (if any)	Dimensions (L × W × H) (cm)	Filter bed: Bedding material (top to bottom) and thickness (in cm)	HRT (h)	OLR (kg COD/ m^3/d)	HLR (m^3/ m^2/d)	Duration (d)	Organic removal (%)	Nutrient removal (%)	Pathogen removal (%)	References
Synthetic wastewater spiked with sewage	Vertical	*Eisenia fetida*	10,000 earthworms/m^3	–	30 × 25 × 60	Gravel and vermicompost (30); sand (10); and coarse gravel (15)	6	–	1.3	70	COD: 74; BOD: 85	–	TC; FC; FS; *E. coli*: 99	Arora et al. (2014)
Pharmaceutical wastewater	Vertical	*Eudrilus eugeniae*		–	–	Sand; vermicast and fine soil	24–96	0.8–3.2	–	–	BOD: 86–97; COD: 84–97	–	–	Dhadse et al. (2010)
Human faeces	Vertical	*Eisenia fetida*	4 kg/m^2	–	37 × 27 × 25.5	Coir; woodchip; mixture of coir and woodchip; mixture of coir; woodchip and vermicompost	–	–	–	360	COD: 87–89	TP: 56–59	*E. coli*: 99	Furlong et al. (2014)
Sewage	Vertical	*Eisenia fetida*	5000–6000 earthworms/m^3	–	100 × 100 × 150	Fine gravel (20); worm active zone (20); sand and compost (50); fine gravel (40); coarse gravel (20)	–	–	2	45	COD: 80	Nitrate: 60	–	Ghasemi et al. (2019)
Gelatine industry wastewater	–	*Lumbricus rubellus*	–	–	900 × 700 × 100	Sawdust; cow dung; *Leucaena leucocephala* foliage; bovine urine	–	–	–	180	COD: 90; BOD: 89	–	–	Ghatnekar et al. (2010)
Sewage	Vertical	*Eisenia fetida,*	10,000 earthworms/m^3	–	25 × 20 × 30	Vermicompost (5)	–	–	2.5	90	BOD: 88; TOC: 85	NH_3-N: 86	FC: 99	Kumar et al. (2016)
		Eudrilus eugeniae				River bed material (20)					BOD: 70; TOC: 62	NH_3-N: 74	FC: 90	
Sewage	Vertical	*Eisenia andrei*	3000 earthworms/m^2	–	400 × 250 × 200	Chaff; fine wood flour and turf (30); coarse wood flour and chaff (40); coarse quartz sand (10) and fine quartz sand (20)	–	–	1	365	BOD_5: 89, COD: 84	TN: 35; TP: 24	–	Li et al. (2009)

Sewage	Vertical	*Eisenia fetida*	0.008 g/m^3	–	–	Ceramsite	–	–	4.2	510	BOD_5: 78, COD: 68	NH_4^+-N: 92	–	Liu et al. (2013)
Sewage	Vertical	*Eisenia fetida*	5000–10000 $earthworm/m^2$	–	–	Soil (15); fine gravel and sand (10.25); gravel (40)	2	–	–	–	BOD_5: 98, COD: 70	–	–	Manyuchi et al. (2013)
Cheese whey wastewater		*Eisenia fetida*	–	–	–	Coarse compost (30); fine compost (100); stone (15)	7	0.3–3	–	460	BOD: 76; COD: 82	TN: 60; TP: 77	–	Merlin and Cottin (2009)
Synthetic dairy wastewater	Vertical	*Eisenia fetida*	10,000 earth-worms/m^3	*Canna indica*	–	Vermicompost and soil (20); sand (20); fine gravel (20); coarse gravel (20)	–	–	0.65	70	BOD: 81; COD: 76	TN: 24–42	–	Samal et al. (2017b)
Dairy wastewater	Vertical	*Eisenia fetida*	10,000 earth-worms/m^3	*Canna indica* (R_1), *Saccharum spontaneum* (R_2), *Typha augustifolia* (R_3)	H: 90; Dia.: 19.8	Soil and vermicompost (30); sand (10); soil (15); coarse gravel (15)	11	–	0.6	90	BOD_5 (R_1): 88, BOD_5 (R_2): 80, BOD_5 (R_3): 84 COD (R_1): 83, COD (R_2): 76, COD (R_3): 79	TN (R_1): 62; TN (R_2): 53; TN (R_3): 56; TP (R_1): 60; TP (R_2): 55; TP (R_3): 58	–	Samal et al. (2018b)
					60 × 18 × 30	Garden soil and vermicompost; lat-erite soil	10	–	–					
Hospital wastewater	Vertical	*Eisenia fetida*	10,000 earth-worms/m^3	–	40 × 40 × 120	Soil and earthworm bed (30); sand (30); detritus (30); cob-blestone (20)	–	–	1	122	COD: 90; BOD_5: 82–90	–	–	Shokouhi et al. (2020)
Synthetic wastewater	Horizontal	*Eisenia fetida*	10,000 earth-worms/m^3	–	80 × 20 × 20	Garden soil and compost (64); dolochar (16)	26.66	3.38 kg. COD/m^3. d	1.8	60	COD: 96	TN: 22; NH_4^+-N: 86; TP: 43; PO_4^{3-}-P: 61	–	Singh et al. (2019a)
Dairy wastewater	Vertical	*Eisenia fetida*	16,000 earth-worms/m^3	–	–	Soil (10); sand and gravel (20); gravel (50)	6–10	–	–	–	BOD_5: 99; COD: 80–90	–	–	Sinha et al. (2007)
Sewage	Vertical	*Eisenia fetida*	20,000 earth-worms/m^3	–	–	Soil (10); sand and gravel (20), gravel (50)	1–2	–	–	–	BOD_5: 98; COD: 45	–	–	Sinha et al. (2008)

(continued)

Table 1.1 (continued)

Wastewater	Flow direction	Earthworm species	Earthworm density	Macrophyte (if any)	Dimensions (L × W × H) (cm)	Filter bed: Bedding material (top to bottom) and thickness (in cm)	HRT (h)	OLR (kg COD/m^3/d)	HLR (m^3/m^2/d)	Duration (d)	Organic removal (%)	Nutrient removal (%)	Pathogen removal (%)	References
Petroleum industry wastewater	Vertical	*Eisenia fetida*	–	–	–	Soil (10); sand and gravel (20); gravel (50)	1–2	–	–	–	C10–C14: 99; C15–C28: 99; C26–C36: 99	–	–	Sinha et al. (2012)
Sewage	Vertical	*Perionyx sansibaricus*	0.022–0.0245 g/m^3	*Cyperus rotundus*	80 L	Soil with small stones and pebbles (25.4); leaves (5.08); sawdust (5.08); small stones and gravels (5.08); large stones (12.7)	1	–	–	–	COD: 90	NO_3^-N: 93; PO_4^{3-}:98	–	Tomar and Suthar (2011)
					59.69 × 45.72 × 38.1	Small pebbles and sand (15.24); large pebbles (25.4)								
Sewage	Vertical	*Eisenia foetida*	1000 earthworms/m^3	*Phragmites australis*		Peat (40); sand (60); gravel (30)	6	0.192	1	420	COD: 90	NH_4^+- N: 92; phosphorus: 91	–	Wang et al. (2010b)
Synthetic wastewater	Vertical	*Eisenia fetida*	0, 0.0045, 0.0085, 0.0125, 0.0165 g/m^3	–	30 × 30 × 75	Cobblestone (5); detritus (15); silver sand (15); earthworm packing bed (artificial soil and earthworm) (35)	1	–	0.2	60	COD: 68–77	NH_3-N: 72–78; TN: 63–66; TP: 80–82	–	Wang et al. (2013)
Sewage	Vertical	*Eisenia foetida*	21,000 earthworms/m^2	–	–	Ceramsite and quartz sand (20); quartz sand (10)	18.3; 9.2; 7.3; 6.6	–	2.4, 4.8, 6, 6.7	120	BOD_5: 55–66; COD: 47–65	TN: 8–15; NH_4-N: 21-62	–	Xing et al. (2010)
Domestic wastewater sludge	Vertical	*Eisenia foetida*	32 g/L	–	Dia.: 30 and H: 60	Ceramic pellets (50)	–	–	3	200	TCOD: 49–54		–	Zhao et al. (2010)
Synthetic wastewater	Vertical	*Eisenia fetida*	–	*Acorus calamus*	100 × 80 × 80 (main frame)	Slag (25); gravel (20)	–	–	–	365	COD: 87	TN: 86; TP: 83	–	Zhao et al. (2014)
					80 × 70 × 80 (vermifilters)	Artificial soil (peat soil and wood chips) (30); mixed sand (5); ceramsite (15); gravel (5)								

78% BOD_5, 68% COD and 90% TSS (Liu et al. 2013). A study has been conducted by Zhao et al. (2014) to treat synthetic wastewater through macrophyte-assisted vermifiltration using different combinations of vertical sub-surface flow constructed wetlands platned with macrophyte *Acorus calamus* and earthworm *Eisenia fetida*. Results of the study revealed the removal of up to 87% COD, 86% total nitrogen (TN) and 83% total phosphorus (TP).

Nitrogen removal from wastewater is mainly responsible for the nitrifiers and denitrifiers microbes present in the earthworm's intestinal guts (Ihssen et al. 2003). Earthworms are able to aerate the system through its borrowing action which enhances the nitrification process and creates a favourable microenvironment for the growth of aerobic nitrobacteria (Samal et al. 2017a). Wang et al. (2010b) have combined macrophyte *Phragmites australis* and earthworm species *Eisenia fetida* to treat domestic sewage with an OLR of approximately 192 g/m^2/d and hydraulic loading rate (HLR) of 1 m^3/m^2/d, and the results showed an average reduction of about 90% COD, 93% SS and 92% NH_4^+-N. Wang et al. (2013) reported 63–66% removal efficiency of TN and 72–78% removal of NH_3^-N from synthetic domestic wastewater. Liu et al. (2013) also reported about 92% NH_4^+-N removal from domestic wastewater. Further, the removal of phosphorus depends upon the sorption capacity, surface area and size of vermifilter bed material along with chemical reaction like ligand exchange reaction, complexation and precipitation (Samal et al. 2017a). Vermifiltration system combined with macrophytes *Perionyx sansibaricus* and *Cyperus rotundus* reported the reduction of wastewater pollutants like COD, total suspended solids (TSS), total dissolved solids (TDS) and NO_3^- by more than 85% (Tomar and Suthar 2011). Wang et al. (2013) reported 80–82% removal of TP using bedding material which consists of cobblestone, detritus, silver sand and earthworm bed while removal of 87% of TP using cobblestone, soil and sawdust. Furlong et al. (2014) obtained a removal efficiency of TP in the range of 56–59% in human faeces.

The most crucial parameter in the sewage treatment from the human health point of view is pathogen removal. In this context, a comprehensive review of available literature by Swati and Hait (2018) underscores that earthworms are capable of pathogen reduction from various wastes. Arora et al. (2014) reported around 99% removal of *Escherichia coli* (*E. coli*), total coliform (TC), faecal coliform (FC) and faecal streptococci (FS) from synthetic wastewater spiked with sewage in a vermifiltration system. Further, Kumar et al. (2016) have treated domestic wastewater with vermifiltration and achieved a reduction of FC by 99%. An experimental run of 365 days of vermifiltration showed the reduction of COD by more than 87 and 99% thermotolerant coliforms using domestic wastewater (Furlong et al. 2014).

1.3.2 Applicability of Vermifiltration for Industrial Effluents

Initially limited to the treatment of the domestic wastewater, the vermifiltration technique has gradually evolved to be studied for the treatment of the industrial effluents. However, very few studies (Table 1.1) are being carried out on the vermifiltration of industrial wastewater because of the sensitive nature of earthworms towards parameters like pH, heavy metals, pesticides and salinity. Regardless of this, vermifiltration applied to industrial effluent from the food and beverage sector has shown encouraging pollutant removal efficiency and can pave way for application for many other industrial effluents that have low or no toxicity (Singh et al. 2019a). Additionally, vermifiltration system has been applied to other industrial effluents, such as petroleum industry and pharmaceutical industry (Dhadse et al. 2010; Sinha et al. 2012). Sinha et al. (2007) have successfully applied vermiltration system to treat effluent from dairy industries which mainly consist of organics like proteins, carbohydrates and fats. According to the study, earthworm species *Eisenia fetida* has resulted in the removal of about 99% BOD_5 and COD in the range of 80–90%. It also leads to the removal of TDS and TSS in the range of 90–92% and 90–95%, respectively. Another study conducted by Sinha et al. (2012) on petroleum industry wastewater has shown 99% removal of C10–C14, C15–C28 and C26–C36. Further, cheese whey waste has been treated by using vermifiltration and achieved about 76% BOD, 82% COD and 77% TSS removal efficiency (Merlin and Cottin 2009). Ghatnekar et al. (2010) reported the removal of COD and BOD by 89 and 90%, respectively, from gelatine industry wastewater employing earthworm species *Lumbricus rubellus*. Dhadse et al. (2010) studied application of vermifiltration on herbal pharmaceutical effluents using earthworm *Eudrilus eugeniae* at different organic loading rates (OLR) of 0.8, 1.6, 2.4 and 3.2 kg COD/m^3/d with 3.2 kg COD/m^3/d as the optimum with COD and BOD removal efficiencies in the range of 85–94% and 90–96%, respectively. Macrophyte-assisted vermifiltration was applied to treat synthetic dairy wastewater by employing macrophyte species *Canna indica* and reported removal of BOD, COD, TSS, TDS and TN by 81%, 76%, 85%, 23% and 43%, respectively (Samal et al. 2017b).

1.4 Mechanisms of Vermifiltration Technique

Vermifiltration technique works in combination of earthworms and microbes. Evolving from the basic system, macrophyte-assisted vermifiltration has emerged as an eco-friendly alternative for wastewater treatment and recycling. In order to unravel the treatment mechanisms, the roles of various layers and components of a typical macrophyte-assisted vermifiltration system have been schematically presented in Fig. 1.5. The solids retained on the filter bed are consumed by the earthworms and converted into the humus (Sinha et al. 2008; Singh et al. 2017). A microbial layer formed on the filter bed also contributes to the degradation of the contaminants retained on the filter bed. Generally, vermifiltration system consists of components, i.e. earthworms and filter bed. Filter bed supports the earthworm growth by providing food source by sorption mechanism from the wastewater, and a microbial layer is formed because of low porosity (Liu et al. 2013; Singh et al. 2017; Wang et al. 2010a, b). Further, the earthworm active zone is also known as aerobic zone while

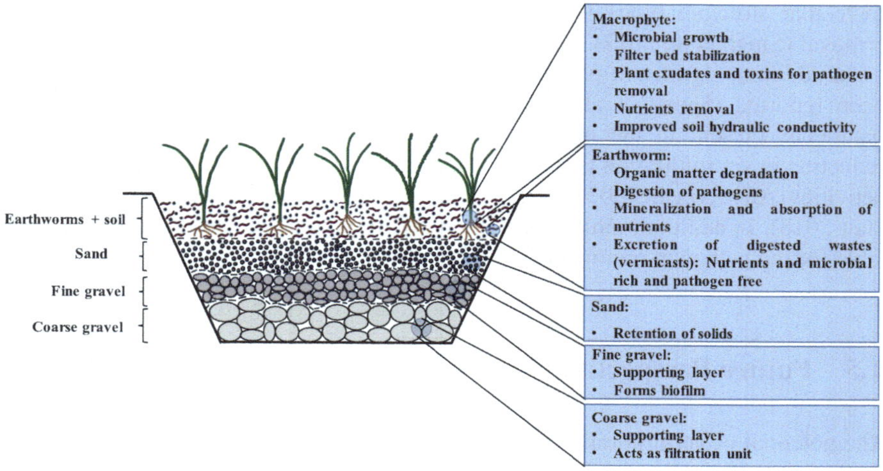

Fig. 1.5 Schematic representation of the role of different layers and components in macrophyte-assisted vermifiltration system

filter bed is called anoxic zone in vermifiltration (Samal et al. 2018a). Oxygen level is increased in filter bed by the borrowing action of earthworms. Further, the increase in the surface area of soil particles with an increase in vermibed porosity to retain more organic pollutants and suspended solids facilitates further decomposition by earthworms (Jiang et al. 2016; Sinha et al. 2008; Singh et al. 2018). Earthworms process wastes by actions like ingestion, grinding, digestion and excretion, and these actions have several physical, chemical and biological effects on the internal ecosystem of earthworm active zone (Singh et al. 2017). The ingestion and grinding actions by earthworm result in conversion of feed waste material into small particles (2–4 microns) followed by the digestion due to symbiotic action of microbes and enzymes in intestine (Kumar et al. 2015; Sinha et al. 2010; Singh et al. 2017; Wang et al. 2011). Numerous enzymes like protease, lipase, amylase, cellulase and chitinase are secreted in the gizzard and intestine of the earthworms which lead to biochemical conversion of the cellulosic and the proteinaceous materials present in the wastewater (Sinha et al. 2010). Since earthworm gut hosts diverse microbial communities, ingested food materials are excreted as vermicast into the soil with nutrients. Microbes present in the biofilm for their population growth further degrade nutrients retained on it, and the nutrients present in the vermicast (Sinha et al. 2008). Earthworms secrete mucus (slimy fluid) from their body which is composed of various metabolites to keep their body surface humid, which also helps in absorbing oxygen (Singh et al. 2017). Earthworms are able to convert large organic matter into complex amorphous solids which contains phenolic compounds and this process is called ‘humification’. These humic substances present in vermibed help in metal adsorption and contain those organic compounds which have complex molecular structure as aromatic rings, carbonyl groups, phenolic and alcoholic hydroxyl. This molecular structure binds with different metal ions and thereby helps in metal removal (Singh et al. 2017).

In addition, significant pathogen reduction by the vermifiltration technique has been reported (Samal et al. 2017a). Earthworms have the capacity to cull the pathogens present in the ingested materials (Sinha et al. 2010). The pathogen reduction in vermifiltration is caused mainly because of the enzymatic and microbial activities (Alberts et al. 2002; Hartenstein 1978; Monroy et al. 2008, 2009; Swati and Hait 2018). In addition, inhibition in humates in the guts of earthworms is responsible for the pathogen removal (Brown and Mitchell 1981; Hartenstein 1978).

1.5 Future Perspectives

The potential of vermifiltration to treat domestic as well as industrial wastewater is well documented. An insight of vermifiltration based on the experimental results, design configurations and treatment mechanism has been provided. Vermifiltration integrated with macrophyte is an emerging technique for the wastewater treatment. Most of the studies have demonstrated vertical vermifilter at laboratory-scale level only for synthetic wastewater treatment. For this purpose, vermifiltration studies with real sewage and industrial effluents will be useful to assess the organic, nutrient and pathogen removal efficiency. However, research is warranted to explore the different vermifiltration system design configuration for wastewater treatment. Various process parameters such as earthworm stocking density, flow rate, hydraulic retention time (HRT), OLR and filter bed configuration need to be optimized for scaling-up the process. In addition, most of the studies have employed epigeic earthworm species *Eisenia fetida* only. In this context, it is pertinent to explore the various other earthworm species as pure and mixed cultures as diverse earthworm species co-exist in nature. Studies are required to be conducted to explore the effect of symbiotic relationships or mixed earthworm species on the removal of contaminants from wastewater.

1.6 Conclusions

The applicability of vermifiltration technique for the treatment of both sewage and industrial effluents along with the treatment mechanisms involved has been extensively discussed. Additionally, the potential of macrophytes has also been discussed in an integrated vermifiltration system for wastewater treatment. Further, the influence of different filter design and operational parameters on the system performance has been presented. The combined effect of earthworm active zone and filter media in the vermifiltration system has been reported for the effective removal of pollutants from the wastewater. Maximum organic and nutrient removal efficiencies of 99% BOD, 96% COD, 86% nitrogen and 83% phosphorus have been reported in the vermifiltration of wastewater. Pathogen removal of 90–99% for FC and 99% for TC, faecal streptococci and *E. coli* by the vermifiltration technique has also been reported. Further, the nutrient removal in an integrated macrophyte-vermifiltration system is mainly because of uptake by macrophytes for their growth. Removal of pollutant is highly selective on the components of vermifilter like filter media composition, earthworm species and macrophyte employed in the process. Moreover, it is necessary to assess vermifiltration system for wastewater treatment employing mixed cultures as diverse earthworm species co-exist in nature. The effect of various process parameters like HLR, OLR and earthworm density during vermifiltration is not quite clear. Extensive research is warranted to optimize different process parameters along with an optimized vermifilter design for efficient wastewater treatment and recycling.

References

Adin A, Asano T (1998) The role of physical-chemical treatment in wastewater reclamation and reuse. Water Sci Technol 37(10):79–90

Alberts B, Johnson A, Lewis J (2002) Molecular biology of the cell, 4th edn. Garland Sci, New York

Arora S, Rajpal A, Kumar T, Bhargava R, Kazmi AA (2014) Pathogen removal during wastewater treatment by vermifiltration. Environ Technol 35(19):2493–2499

Bezbaruah AN, Zhang TC (2005) Quantification of oxygen release by bulrush (*Scirpus validus*) roots in a constructed treatment wetland. Biotechnol Bioeng 89(3):308–318

Brix H (1994) Functions of macrophytes in constructed wetlands. Water Sci Technol 29(4):71–78

Brown BA, Mitchell MJ (1981) Role of the earthworm, *Eisenia foetida*, in affecting survival of *Salmonella enteritidis* ser. *typhimurium*. Pedobiologia 22:434–438

Chen Z, Hu S, Hu C, Huang L, Liu H, Vymazal J (2016) Preliminary investigation on the effect of earthworm and vegetation for sludge treatment in sludge treatment reed beds system. Environ Sci Pollut Res 23(12):11957–11963

Degremont G (1991) Water technical handbook. Lavoisier Publishing, Paris

Dhadse S, Satyanarayan S, Chaudhari PR, Wate SR (2010) Vermifilters: a tool for aerobic biological treatment of herbal pharmaceutical wastewater. Water Sci Technol 61(9):2375–2380

Furlong C, Templeton MR, Gibson WT (2014) Processing of human faeces by wet vermifiltration for improved on-site sanitation. J Water Sanit Hyg De 4(2):231–239

Ghasemi S, Mirzaie M, Hasan-Zadeh A, Ashrafnejad M, Hashemian SJ, Shahnemati SR (2019) Design, operation, performance evaluation and mathematical optimization of a vermifiltration pilot plan for domestic wastewater treatment. J Environ Chem Eng 103587

Ghatnekar SD, Kavian MF, Sharma SM, Ghatnekar SS, Ghatnekar GS, Ghatnekar AV (2010) Application of vermi-filter-based effluent treatment plant (pilot scale) for biomanagement of liquid effluents from the gelatine industry. Dynam Soil Dynam Plant 4(1):83–88

Goel PK (2006) Water pollution: causes, effects and control. New Age International

Gunadi B, Blount C, Edwards CA (2002) The growth and fecundity of *Eisenia fetida* (Savigny) in cattle solids pre-composted for different periods. Pedobiologia 46(1):15–23

Hartenstein R (1978) Utilization of soil organisms in sludge management. In: Conference proceedings. US department of commerce, national technical information services, PB286932. Springfield, Virginia, p 171

Hughes RJ, Nair J, Ho G (2009) The risk of sodium toxicity from bed accumulation to key species in the vermifiltration wastewater treatment process. Bioresour Technol 100(16):3815–3819

Ihssen J, Horn MA, Matthies C, Gößner A, Schramm A, Drake HL (2003) N_2O-producing microorganisms in the gut of the earthworm *Aporrectodea caliginosa* are indicative of ingested soil bacteria. Appl Environ Microbiol 69(3):1655–1661

Jiang L, Liu Y, Hu X, Zeng G, Wang H, Zhou L, Liu S (2016) The use of microbial-earthworm ecofilters for wastewater treatment with special attention to influencing factors in performance: a review. Bioresour Technol 200:999–1007

Kumar T, Bhargava R, Prasad KH, Pruthi V (2015) Evaluation of vermifiltration process using natural ingredients for effective wastewater treatment. Ecol Eng 75:370–377

Kumar T, Rajpal A, Arora S, Bhargava R, Hari Prasad KS, Kazmi AA (2016) A comparative study on vermifiltration using epigeic earthworm *Eisenia fetida* and *Eudrilus eugeniae*. Desalin Water Treat 57(14):6347–6354

Li YS, Xiao YQ, Qiu JP, Dai YQ, Robin P (2009) Continuous village sewage treatment by vermifiltration and activated sludge process. Water Sci Technol 60(11):3001–3010

Liu J, Lu Z, Zhang J, Xing M, Yang J (2013) Phylogenetic characterization of microbial communities in a full-scale vermifilter treating rural domestic sewage. Ecol Eng 61:100–109

Manyuchi L, Kadzungura L, Boka S (2013) Vermifiltration of sewage wastewater for potential use in irrigation purposes using *Eisenia fetida* earthworms. World Acad Sci Eng Technol 78:538–542

Merlin G, Cottin N (2009) Performance of a compost biofilter containing earthworms to treat cheese whey. Environ Technol 30(10):995–1002

Metcalf L, Eddy HP, Tchobanoglous G (1991) Wastewater engineering: treatment, disposal, and reuse, 3rd edn. McGraw-Hill, New York

Monroy F, Aira M, Domínguez J (2008) Changes in density of nematodes, protozoa and total coliforms after transit through the gut of four epigeic earthworms (Oligochaeta). Appl Soil Ecol 39(2):127–132

Monroy F, Aira M, Domínguez J (2009) Reduction of total coliform numbers during vermicomposting is caused by short-term direct effects of earthworms on microorganisms and depends on the dose of application of pig slurry. Sci Total Environ 407(20):5411–5416

Noumsi IM, Akoa A, Eteme RA, Nya J, Ngniado P, Fonkou T, Brissaud F (2005) Mosquito development and biological control in a macrophyte-based wastewater treatment plant. Water Sci Technol 51(12):201–204

Nuengjamnong C, Chiarawatchai N, Polprasert C, Otterpohl R (2011) Treating swine wastewater by integrating earthworms into constructed wetlands. J Environ Sci Health A 46(7):800–804

Pimentel D, Berger B, Filiberto D, Newton M, Wolfe B, Karabinakis E, Clark S, Poon E, Abbett E, Nandagopal S (2004) Water resources: agricultural and environmental issues. Bioscience 54 (10):909–918

Ra CS, Lo KV, Shin JS, Oh JS, Hong BJ (2000) Biological nutrient removal with an internal organic carbon source in piggery wastewater treatment. Water Res 34(3):965–973

Reddy, K. R., & Smith, W. H. (1987). Aquatic plants for water treatment and resource recovery. In: Conference on research and applications of aquatic plants for water treatment and resource recovery, Magnolia Publishing Inc, Orlando, Fla. (USA), 1986

Samal K, Dash RR, Bhunia P (2017a) Treatment of wastewater by vermifiltration integrated with macrophyte filter: a review. J Environ Chem Eng 5(3):2274–2289

Samal K, Dash RR, Bhunia P (2017b) Performance assessment of a *Canna indica* assisted vermifilter for synthetic dairy wastewater treatment. Process Saf Environ Prot 111:363–374

Samal K, Dash RR, Bhunia P (2018a) Effect of hydraulic loading rate and pollutants degradation kinetics in two stage hybrid macrophyte assisted vermifiltration system. Biochem Eng J 132:47–59

Samal K, Dash RR, Bhunia P (2018b) A comparative study of macrophytes influence on performance of hybrid vermifilter for dairy wastewater treatment. J Environ Chem Eng 6 (4):4714–4726

Shokouhi R, Ghobadi N, Godini K, Hadi M, Atashzaban Z (2020) Antibiotic detection in a hospital wastewater and comparison of their removal rate by activated sludge and earthworm-based vermifilteration: environmental risk assessment. Process Saf Environ Prot 134:169–177

Singh R, Bhunia P, Dash RR (2017) A mechanistic review on vermifiltration of wastewater: design, operation and performance. J Environ Manag 197:656–672

Singh R, Bhunia P, Dash RR (2018) Understanding intricacies of clogging and its alleviation by introducing earthworms in soil biofilters. Sci Total Environ 633:145–156

Singh R, Bhunia P, Dash RR (2019a) Optimization of organics removal and understanding the impact of HRT on vermifiltration of brewery wastewater. Sci Total Environ 651:1283–1293

Singh R, Samal K, Dash RR, Bhunia P (2019b) Vermifiltration as a sustainable natural treatment technology for the treatment and reuse of wastewater: a review. J Environ Manag 247:140–151

Sinha RK, Bharambe G, Bapat P (2007) Removal of high BOD and COD loadings of primary liquid waste products from dairy industry by vermifiltration technology using earthworms. Indian J Environ Prot 27(6):486–501

Sinha RK, Bharambe G, Chaudhari U (2008) Sewage treatment by vermifiltration with synchronous treatment of sludge by earthworms: a low-cost sustainable technology over conventional systems with potential for decentralization. Environmentalist 28(4):409–420

Sinha RK, Agarwal S, Chauhan K, Chandran V, Soni BK (2010) Vermiculture technology: reviving the dreams of sir Charles Darwin for scientific use of earthworms in sustainable development programs. Technol Invest 1(03):155–172

Sinha RK, Chandran V, Soni BK, Patel U, Ghosh A (2012) Earthworms: nature's chemical managers and detoxifying agents in the environment: an innovative study on treatment of toxic wastewaters from the petroleum industry by vermifiltration technology. Environmentalist 32(4):445–452

Speece RE (1983) Anaerobic biotechnology for industrial wastewater treatment. Environ Sci Technol 17(9):416A–427A

Swati A, Hait S (2018) A comprehensive review of the fate of pathogens during vermicomposting of organic wastes. J Environ Qual 47(1):16–29

Tomar P, Suthar S (2011) Urban wastewater treatment using vermi-biofiltration system. Desalination 282:95–103

Verma AK, Dash RR, Bhunia P (2012) A review on chemical coagulation/flocculation technologies for removal of colour from textile wastewaters. J Environ Manag 93(1):154–168

Wang S, Yang J, Lou SJ (2010a) Wastewater treatment performance of a vermifilter enhancement by a converter slag–coal cinder filter. Ecol Eng 36(4):489–494

Wang DB, Zhang ZY, Li XM, Zheng W, Yang Q, Ding Y, Zeng GM (2010b) A full-scale treatment of freeway toll-gate domestic sewage using ecology filter integrated constructed rapid infiltration. Ecol Eng 36(6):827–831

Wang L, Guo F, Zheng Z, Luo X, Zhang J (2011) Enhancement of rural domestic sewage treatment performance, and assessment of microbial community diversity and structure using tower vermifiltration. Bioresour Technol 102(20):9462–9470

Wang J, Liu XD, Lu J (2012) Urban river pollution control and remediation. Procedia Environ Sci 13:1856–1862

Wang LM, Luo XZ, Zhang YM, Lian JJ, Gao YX, Zheng Z (2013) Effect of earthworm loads on organic matter and nutrient removal efficiencies in synthetic domestic wastewater, and on bacterial community structure and diversity in vermifiltration. Water Sci Technol 68(1):43–49

Xing M, Li X, Yang J (2010) Treatment performance of small-scale vermifilter for domestic wastewater and its relationship to earthworm growth, reproduction and enzymatic activity. Afr J Biotechnol 9(44):7513–7520

Xing M, Wang Y, Liu J, Yu F (2011) A comparative study of synchronous treatment of sewage and sludge by two vermifiltrations using an epigeic earthworm *Eisenia fetida*. J Hazard Mater 185 (2–3):881–888

Zhao L, Wang Y, Yang J, Xing M, Li X, Yi D, Deng D (2010) Earthworm–microorganism interactions: a strategy to stabilize domestic wastewater sludge. Water Res 44(8):2572–2582

Zhao Y, Zhang Y, Ge Z, Hu C, Zhang H (2014) Effects of influent C/N ratios on wastewater nutrient removal and simultaneous greenhouse gas emission from the combinations of vertical subsurface flow constructed wetlands and earthworm eco-filters for treating synthetic wastewater. Environ Sci Processes Impacts 16(3):567–575

Zheng C, Zhao L, Zhou X, Fu Z, Li A (2013) Treatment technologies for organic wastewater. Water Treat:249–286

Chapter 2
Vermifiltration for Rural Wastewater Treatment

Meena Khwairakpam

Abstract To improve the neglected sanitation services particularly in rural areas, proper management of wastewater is the need of the hour. About 70% of India's population lives in the villages, and they are deprived of improved sanitation. The rural areas are mostly un-sewered mainly due to inadequate water supply required for efficient functioning of water carriage system and scattered population. In such un-sewered rural areas, major problem lies in collection, removal and disposal of night soil, wastewater and garbage. The wastewater from kitchens, baths, etc are let directly into streets, resulting to breeding of flies and mosquitoes which may be a major cause for start of an epidemic. There is an urgency to invest, both in sewers and in the treatment of sewage. Possible risk of disease from coming in direct contact with sewage having pathogens can be reduced if sewage treatment plants are installed. However, most of the developing nations are unable to install such treatment process owing to lack of fund and infrastructure. Another problem for starting sewage treatment plants in the rural areas is the scattered population due to which a centralized systems cannot be adopted. The only option available is to provide on-site facility, and one of the on-site treatment techniques is the vermifiltration. Vermifiltration is a process that adapts traditional vermicomposting for treating wastewater. Such system allows water to flow by gravity vertically/horizontally through a filter media like sand and gravel of different sizes. Wastewater is purified in this process as it percolates through the vermicompost by physical as well as microbial degradation and the organics passes through the gut of the earthworm which comes out as value-added end product. There is no sludge generation in the process; instead generation of vermicompost is there which will be helpful in generation of income. This is also an odourless process, and the resulting vermifiltered water is suitable for farm irrigation and in parks and gardens. Application of the vermifiltration technology in wastewater treatment is easily adaptable in developing countries due to its simplicity and treats water to acceptable standards.

M. Khwairakpam (✉)
Centre for Rural Technology, Indian Institute of Technology, Guwahati, Assam, India
e-mail: meena.kh@iitg.ac.in

S. A. Bhat et al. (eds.), *Earthworm Assisted Remediation of Effluents and Wastes*,
https://doi.org/10.1007/978-981-15-4522-1_2

Keywords Vermifiltration · Vermifilter · Wastewater · Earthworms · Rural · On-site

2.1 Introduction

Wastewater generated from rural areas is disposed of as it is with no proper treatment. In most of the underdeveloped/developing countries, it is usually disposed of into roads, nearby water sources, fields and open spaces near the residence. This leads to nuisance in the surrounding environment and may lead to health issues to the residents as sewage carries disease-causing pathogens in addition to high organic loads. Before its disposal, sewage has to be treated otherwise its high organic contents may lead to depletion of the DO values in the discharging water bodies which would have a negative effect on the survival of all aquatic organisms in the water bodies. Since 100 years, different technologies for sewage treatment have been well developed, however, there is a need to focus for innovative and cost-effective on-site treatment processes (Leach and Enfield 1983; Kruzic and Schroeder 1990). Traditional methods used for sewage treatment in the rural areas include oxidation ponds, lagoons, stabilization ponds, activated sludge process, upflow anaerobic sludge beds, sequencing biological reactors, land treatment, etc. Many developing nations cannot afford to construct and maintain large and costly sewage treatment plants (STPs), and even in developed nations, to have a sustainable wastewater management in future, one needs to focus on decentralized systems. Moreover, most countries prefer sewage treatment processes which can provide effluent standard at minimal cost. The major expenses in centralised facilities like activated sludge process, trickling filter, lagoon, ozone oxidation; floatation, sedimentation, and wetland system are capital cost, operation and maintenance (O&M) costs, and the procurement of land. It is also difficult to operate especially in areas with low population densities and dispersed households especially in rural and hilly areas. Above all the technical difficulties to construct, operate and manage such centralized facilities in such areas, there is also lack of funds. In such cases having decentralized facilities for individual households or a cluster of homes to treat their domestic wastewater on-site will reduce the organic loads (BOD and COD) on the discharging water bodies as well as easier to monitor. The decentralized approach for wastewater treatment which employs a combination of on-site and/or cluster systems is gaining more attention. The new trend is towards decentralized and on-site treatment systems where there is scope for flexibility in management and simple in operation. Adopting decentralized system can bring a long-term solution for rural areas/small communities and is reliable as well as cost-effective. On-site treatment of wastewater is a low-cost technology with low energy consumption, simple and reliable that even private owners with little skill for operation can afford (Schudell and Boller 1989). In the context of developing countries particularly in rural areas, vermifiltration could be an ideal technology for the treatment of domestic effluents, owing to its cost-effective and ecologically sustainable characteristics. When compared to prevailing biological treatment options, vermifiltration is much more environment-friendly and economically viable (Shao et al. 2014; Chyan et al. 2013; Arias et al. 2005; Carballeira et al. 2017; Kumar et al. 2016). Vermifiltration is the bioconversion of liquid/wastewater, while vermicomposting is the biological conversion of solid waste to a value-added end product. In other words, introduction of earthworms in filtration system with suitable bedding materials to breakdown organic pollutants is called vermifiltration (Tomar and Suthar 2011). In vermifiltration degradation of organic pollutants in wastewater happens with the joint action of earthworms and microorganisms (Zhao et al. 2010; Wang et al. 2016). In 1992, Prof. Jose Toha of University of Chile initially advocated the use of vermifiltration as an alternate technology as it is a fast, odourless process producing an end product which is stable, disinfected, detoxified and highly nutritive effluent (Wang et al. 2010; Xing et al. 2010). It is an economically and environmentally preferred decentralized technology compared to other biological process.

2.2 Wastewater Treatment

With the surfacing of problems associated with centralized wastewater treatment facilities, decentralized wastewater treatment options and water reuse are gaining importance at a fast pace. Wastewater management is normally done through end of pipe system which is the conventional method of wastewater treatment. These systems are treating huge quantities of waste and are becoming unmanageable leading to severe water pollution problems. Decentralized wastewater treatment options play an important role in managing and improving rural environments in the long run. Vermifiltration can be one of the promising decentralized wastewater treatment techniques which provide treatment of wastewater by filtering through a vermicomposting mass. The treatment of raw domestic wastewater through a filtration process has been investigated before (Lens et al. 1994). Xing et al. (2005) carried out a pilot-scale study on vermifiltration of sewage at Shanghai Quyang Wastewater Treatment Facility in China. Significant reduction in the organic loads like biochemical oxygen demand (BOD) and chemical oxygen demand (COD) was found by the studies carried out by Gardner et al. (1997) on on-site effluent treatment by earthworms. Use of earthworms for the management of effluents containing heavy loads of BOD, total dissolved suspended solid (TDSS) and nutrients nitrogen were studied by Hartenstein and Bisesi (1989). The worms produced clean effluents and also nutrient-rich vermicompost. Studies were also carried out by Bajsa et al. (2003) on the vermifiltration of domestic wastewater using earthworms.

2.2.1 Wastewater Treatment Options

Reports by Lodge et al. (2000) stated that biological aerated filter (BAF) was used to treat grey water at the largest water recycling treatment plant in Europe, at the

Millennium Dome. This BAF technology was reported to remove suspended solids (SS) and carbonaceous organics with microorganisms. It was also reported that there was reduction of BOD from 50–240 mg/L BOD to 1–15 mg/L and reduction of SS, from 48–124 to 2–5 mg/L by BAF. Jefferson et al. (2000) reported that membrane bioreactors (MBR) were able to stabilize variations in influent water quality effectively. MBR had the highest efficacy of treatment for wastewater far beyond the performance of membrane aerated bioreactors (MABR) and BAFs. Traditional method of domestic grey water treatment like reed bed or pond systems has been utilizing many types of aquatic macrophytes. Mars et al. (1999) examined submergent macrophytes such as *Schoenoplectus validus* and *Triglochin huegelii* in Western Australia. From the test it was proved that efficient removal of nitrogen and phosphorous was done by *T. huegelii*. The authors suggested that for efficient nutrient stripping lagoons, wetlands and constructed basins filled with plants like this can be used. Although this has its positive points (eco-friendly and cost-effective), it has its limitation like requirement of large area and unaesthetic. Jasti et al. (2006) carried out a study on treatment of corn processing wastewater in an attached growth system of *Rhizopus oligosporus* fungi. A maximum chemical oxygen demand (COD) removal of 78% was achieved at a 5 h HRT, while COD removal reduced to 70% at a 2.5 h HRT. Dixon et al. (1999) studied the water-saving potential of a combination of wastewater reuse and rainwater harvesting on an urban housing environment. The data collected from a small-scale study of domestic water appliance usage was analysed, from which cumulative frequency distributions were derived for each hour of the day and for occupancy.

Effect of the application of a water extract of *M. oleifera* seed to a wastewater treatment sequence comprising coagulation–flocculation–sedimentation–sand filtration was studied by Bhuptawat et al. (2007). At the application of 50 and 100 mg/L *M. oleifera* doses, there was an overall COD removal of 50%. However, COD removal increased to 58 and 64%, respectively, when 50 and 100 mg/L seed doses were applied in combination with 10 mg/L of alum. It was observed that the majority of COD removal occurred during the filtration process. Ghangrekar et al. (2007) demonstrated an effective treatment of the sewage by using natural treatment systems and using of treated wastewater for aquaculture. Morari and Giardini (2009) studied the treatment effect of two pilot-scale vertical flow constructed wetlands (VFCWs) on municipal wastewaters and their suitability for irrigation reuse. Two mycrophytes *Typha latifolia* and *Phragmites australis* were grown on each VFCW. Good efficiency in wastewater pollutant removal was reported in the VFCWs, and the removal rates of nutrients were also reported to be particularly high due to the massive macrophyte growth. Suitability of a nine-chambered modified anaerobic baffled reactor (MABR) for the treatment of municipal wastewater was evaluated by Bodhkhe (2009). HRT of 0.25 day was reported to be the most suitable minimum HRT for economic operation of MABR. Suspended solids (SS), BOD and COD were reduced efficiently by 86%, 87% and 84%, respectively, resulting to conclude that MABR technology can be adopted as on-site wastewater treatment plant for individual houses or small cluster in Indian climatic conditions. Some

Table 2.1 Conventional wastewater treatment options in rural areas

Treatment system	Prerequisite	Benefits	Limitations
Cesspools	• Location of a cesspool should be downhill from a well; in any case, a distance of 15 m to prevent bacterial pollution of the well • The distance between a well and a cesspool placed directly uphill from it should be not less than 45 m, to prevent chemical pollution, too • Cesspools should be kept at least 6 m away from dwelling foundations as it may leach out	• No power requirement • No quality control is required • Simple operational mechanism that hardly goes wrong • Safe from the danger of intermittent usage • No immediate negative environmental impact	• In densely inhabited communities where wells are used as sources of drinking water supply their construction is not permitted • Regular emptying is required • Capital and operational cost are high • Expensive construction of underground storage
Septic tank	• Partial treatment • Requirement of seepage	• No power requirement • Poor effluent quality • Pipe blockage due to thick scum formation	• Possibility of groundwater contamination • Limited quality control • Suitable for individual houses and small communities • Most useful and satisfactory unit for rural/hilly areas
Seepage pit	• Location of seepage pits should preferably be downhill and, in any case, at least 15 m (50 ft.) away from drinking water sources and wells • Adequate contact area with the surrounding soil to absorb the effluent in to the soil. In clayey soil, larger pits will be needed	• No power requirement • No quality control is required • Almost nil maintenance	• The construction of seepage pits is not permitted where groundwater is used for domestic purposes • Possibility of groundwater pollution

conventional wastewater treatment options being practised till now in rural areas are listed in Table 2.1.

Adopting vermifiltration system has advantages when compared with other commonly used biological wastewater treatment systems. According to Yang et al. (2008), there is no sludge formation in the vermifiltration system which makes it more convenient to operate when compared to conventional activated sludge treatment methods. Moreover, total nitrogen and total phosphorus was reported to be tripped off 10% higher in vermifilters compared to activated sludge systems; NH^4-N was also removed three times more in vermifilters (Li et al. 2009).

2.3 Vermifiltration

The houses in rural areas are not built in a planned manner and may be in the middle of large farms; hence, they are not close to one another like in urban areas. Hence, in rural areas decentralized wastewater treatment systems may be adopted as they are more suitable. Vermifiltration is one such decentralized technology which is eco-friendly and most suited for rural areas. Vermifiltration consist of harnessing the vermiculture ecosystem for utilization of organic wastewaters so as to produce biosoil with beneficial soil bacteria and water for reuse (Fig. 2.1). This is an extension of soil filtration with earthworms to speed up, the soil processes to utilize organics by combining with the rock particles and to produce fresh biosoil that can be support healthy plant production. Soil filtration is a biological process and not a process of mechanical filtration alone. Dissolved and suspended impurities (organics and inorganics) are trapped in the filter media as they percolate and are stabilized through complex bioprocesses that take place in living soil (biosoil). Earthworms added to the soil filtration intensify the already existing soil processes enabling the soil filtration system to become smaller in size. In addition, there is resource generation instead of sludge generation. Thus, vermifiltration could be a zero-waste technique.

Utilization of liquid organics (i.e. wastewater) with vermiculture is similar to utilization of solid organics. The vermifilter does not get choked up because earthworms cause speedy utilization of absorbed organics. They also utilize surplus microbial biomass and immobilize surplus plant growth factors on the new biosoil that is produced. According to Sinha et al. (2008), each earthworm which works as a 'biofilter' was able to remove the BOD, COD, total dissolved solids and total suspended solids by 80–90% from wastewater. Changes were also found in the properties of a biofilm present in the vermifilter due to the earthworms' ability to burrow, ingest and produce mucus and castings. In addition, the earthworms can change the (Li et al. 2013). In addition to nearly odour-free process, it is also

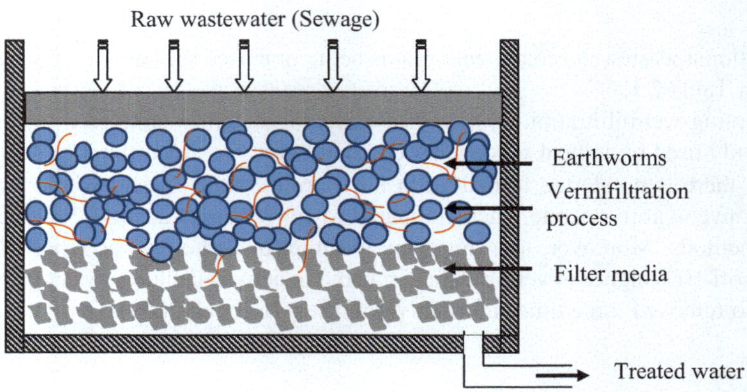

Fig. 2.1 Overview of vermifiltration

sludge-free, and the resulting vermifiltered water is clean enough to be reused for farm irrigation and in parks and gardens. They inhibit the action of anaerobic microorganisms creating an aerobic condition in the waste materials by their burrowing actions. Vermifiltration process satisfies multiple requirements, like zero-sludge generation, odour-free, simple design, simple equipment and comparatively lower operating and capital costs. Therefore, the application of a simple, efficient, low-cost and environment-friendly technology to treat wastewaters can solve the problems of sanitation issues in the rural areas in the long run.

2.3.1 Vermifiltration Mechanism

In the vermifiltration process, earthworms work in symbiosis with the microbes to treat the wastewater. In addition to the microbiological degradation, earthworms perform many beneficial activities by grinding of ingested soil along with the ingested organics, addition of gut microbes and excretion of soil plus organics as castings (Sinha et al. 2008; Xing et al. 2014). It acts as an aerator, grinder, crusher, chemical degrader and biological stimulator while promoting the growth of 'beneficial decomposer bacteria' in wastewater (Sinha et al. 2002). Earthworms improve aeration by the burrowing action and accelerate microbial activity by increasing the population of soil microorganisms. Earthworms host millions of decomposer (biodegrader) microbes in their gut and excrete them in soil along with nutrients nitrogen and phosphorus in their excreta called 'vermicast' (Singleton et al. 2003). The nutrients are further used by the microbes for multiplication and vigorous action. Singleton et al. (2003) also studied the bacterial flora associated with the intestine and vermicasts of the earthworms and found species like *Pseudomonas*, *Mucor*, *Paenibacillus*, *Azoarcus*, *Burkholderia*, *Spiroplasma*, *Alcaligenes* and *Acidobacterium* which has potential to degrade several categories of organics. Suspended solids are trapped on top of the vermifilter and processed by earthworms and fed to the soil microbes immobilized in the vermifilter (Komarowski 2001). Earthworms granulate the clay particles thereby increasing the 'hydraulic conductivity' of the system which further intensifies the organic loadings of wastewater in the vermifilter. The grinding activity of the earthworms leads to high total specific surface area to the filter media, enabling the absorption of the organic and inorganic from the wastewater. Such process is found ideal for diluted wastewater like sewage (Bhawalkar 1995). Earthworms devour the surplus harmful and ineffective microbes in the wastewater selectively, resulting to a pathogen-free vermifiltered effluent. The earthworms also prevent choking of the medium due to their burrowing nature and maintain a culture of effective biodegrader microbes to function. Individual households or a cluster of homes in cities can also treat their domestic wastewater with such decentralized system so as to reduce the burden on the sewage treatment plants down the sewer system.

2.3.2 *Earthworms*

Out of the 3000 species of earthworms in the world and about 500 species are available in India (Julka 1986). Earthworms are good decomposer and soil eater, and they have over 600 million years of experience in waste and environmental management. No wonder then, Charles Darwin called them as the 'unheralded soldiers of mankind', and the Greek philosopher Aristotle called them as the 'intestine of earth', meaning digesting a wide variety of organic materials including the waste organics, from earth (Darwin and Seward 1903; Martin 1976). Earthworms are long, cylindrical, narrow, bilaterally symmetrical, segmented animals without bones consisting of 100–200 almost cylindrical rings or segments, each covered with the minute bristles weighing around 1400–1500 mg after 8–10 weeks. Depending on the type of species and the ecological situation, the life span of an earthworm is practically 3–7 years. It is learned from the studies that their reproduction rate is high and within 60–70 days, they are able to double their number (Sinha et al. 2008). The gut of the earthworms is home to millions of good microbes like 'nitrogen-fixing' and 'decomposer microbes'. Availability of organic matter, soil moisture and pH of the soil plays an important role in the distribution of earthworms in soil. They breathe through their skin, and they prefer environment which is dark and moist. They can tolerate a temperature range between 5 and 29 °C. Earthworms can flourish comfortably in habitats with a temperature of 20–25 °C and moisture of 60–75% (Hand 1988). Their activity is reduced in low temperatures, but high temperature can dry them out killing them instantly. Earthworms are bisexual animals, and their multiplication rate is very high. Earthworms can multiply by 28, i.e. 256 worms every 6 months from a single individual if provided the optimal conditions of temperature, moisture and feeding materials. In their life cycle, they have the ability to produce 300–400 young ones (Hand 1988). Earthworms carry out the grinding of soil, debris, food, etc. with their muscular gizzard to a size of 2–4 microns. Earthworms play an important role in the vermifiltration system by consuming organic solids present in wastewater at a high rate. Therefore, their population density, maturity and health of the worms in the vermifiltration bed play an important role in the treatment process. For an efficient functioning of a vermifiltration system, maintaining an optimum worm density is very crucial (Li et al. 2008). Definitely the number of worms per unit area in the vermifilter bed affects the treatment efficiency of vermifiltration process. Therefore, initially enough worms should be added to vermicompost the incoming wastes and produce a suitable humus filter. From studies an estimate of at least 15,000–20,000 worms/m^3 of the vermifiltration system is required to start a vermifiltration system (Sinha et al. 2008).

2.3.3 *Classification of Earthworms*

Earthworms are broadly divided into three categories based on nature of feeding and defecation activity and their vertical distribution in soil strata.

2.3.3.1 Epigeic

Besides having a small body size, these earthworm species have high reproduction rate and are voracious eater of organic matter. Being surface dwellers they do not change the soil structure, and such earthworm species are predominantly used in vermicomposting. Examples include *Eisenia fetida* (brandling, red wiggler or manure worm), *Eisenia Andrei* (red tiger), *Lumbricus rubellus* (red worms), etc.

2.3.3.2 Endogeic

The endogeic species are fairly pigmented with varied body size and feed more on soil than organic matter, and they live in extensive horizontal burrows. These species are rarely used in vermicomposting although they efficiently decompose dead plant roots. Due to their burrowing nature, they take part actively in soil formation process such as mixing and aeration. Examples include *Allolobophora chlorotica* (green worms), *Aporrectodea caliginosa* (grey worm), etc.

2.3.3.3 Anecic

The anecic groups are burrowing earthworms forming complicated burrow system under soil surface helping in regular mixing of soil surface matter with the lower strata. They have large body size, relatively long-lived and have a longer growth and reproduction time than epigeic species. Anecic species may be used in vermicomposting but usually in combination with epigeic species. *Aporrectodea longa* (black-headed worms), *Lumbricus terrestris* (lob worms) are examples for anecic group.

The suitability of *Eisenia fetida*, *Eudrilus eugeniae* and *Perionyx excavatus* have been described by Reinecke et al. (1992) in Table 2.2.

2.3.4 *Studies on Vermifiltration*

Vermifiltration as discussed earlier is a decentralized wastewater treatment option which is recommended for adoption in small communities, colonies and rural areas owing to the fact that the system is eco-friendly, compact and cost-effective.

Table 2.2 Biology of earthworm species suitable for composting (Reinecke et al. 1992)

Parameters	*Eisenia fetida*	*Eudrilus eugeniae*	*Perionyx excavatus*
Duration of life cycle (days)	70	60	46
Growth rate (mg/worm/day)	7	12	3.5
Maximum body mass (mg/worm)	1500	4294	600
Maturation attained at age (days)	50	40	21
Start cocoon production (days)	55	46	24
Cocoon production (worm/day)	0.35	1.3	1.1
Incubation period (days)	23	16.6	18.7
Hatching success in water (%)	73	50	63.4
Mean number of hatching (cocoon)	2.7	2.7	1.1
Number of hatching from one cocoon	1-9	1-5	1-3

Vermifiltration technology ranging from small to pilot-scale levels were reported to efficiently treat sewage with different ranges of COD, BOD and SS, reducing N and P (Sinha et al. 2008). pH and ammonia which affect the earthworms' survival resulting to efficiency of wastewater treatment have also been studied. It was found that earthworms were able to neutralize pH in wastewater, and the process is nearly odour-free (Sinha et al. 2008; Hughes et al. 2009). Hughes et al. (2009) also reported that ammonia at high level was toxic to the earthworms in vermifilter. The effect of different hydraulic loading rates on vermifilter using riverbed material for the treatment of domestic wastewater has been investigated (Kumar et al. 2014). Taylor et al. (2003) found that filter bed depth affected the efficiency of small-scale vermifiltration system. The filter bed was found to be suitable environment for the earthworms and easily survived the surface loadings of solid organic waste and 130 L/day of raw domestic wastewater. Li et al. (2008) were able to reduce ammonia emissions by reusing the wastewater to flush out the manure in a swine facility. They were also able to treat diluted manure and produce earthworms and vermicompost. They reported that adopting decentralized vermifiltration system reduced transportation costs. Hughes et al. (2009) were able to conclude from their studies on toxicity of sodium during vermifiltration that juveniles were more susceptible and could be killed because of high NaCl in the wastewater. However, the predicted bed concentration (PEC) of NaCl was very low when compared to the average mass of sodium loaded into a vermifiltration system on a weekly basis. Treatment is done on various types of wastewater using vermifiltration process as listed in Table 2.3.

A study was also carried out to compare the performance of vermifilter containing the earthworm species (*Eisenia fetida*) and geofilter (without earthworms) for enhancing the treatment of the wastewater. Four different hydraulic loading rates of 1.5, 2, 2.5 and 3.0 $m^3 m^{-2}$ day^{-1} were applied to the vermifilter and geofilter using synthetic domestic wastewater. For the optimum hydraulic loading rate of 2.5 $m^3 m^{-2}$ day^{-1}, the removal efficiency values (%) of BOD, TSS and TDS with vermifilter were found to be 96, 90 and 82, respectively, while in geofilter the observed values (%) were 70, 79 and 56, respectively. In addition to this,

Table 2.3 Treatment of various wastewater using vermifiltration (Patel 2018)

Sr. no.	Types of wastewater	Earthworm species	Organics removal (%)	Nutrient removal (%)	Bed material and size	HLR (m^3/m^2d)	HRT (h)
1	Sewage	*Eisenia fetida*	BOD 98, COD 45, TSS 90	-	Pure soil, sand (10–12 mm), gravel (7.5, of 3.5–4.5)	-	1–2
2	Dairy industry effluent	*Earthworm*	BOD 98, COD 80–90, TSS 90–95, TDS 90–92	–	Pure soil, sand (10–12 mm), gravel (7.5, of 3.5–4.5)	–	6–10
3	Synthetic sewage	*Eisenia fetida*	COD 83.6	TN 63, TP 86.7, NH3-N 70.5	Cobblestones (6–10 cm), soil, sawdust	0.2	48, 72, 96
4	Urban wastewater	*Perionyx sansibaricus*	COD 80–90, TSS 88.6, TDS 99.8	No3 92.7, Po3-4 98.3	Surface vegetation, soil, dried leaves, sawdust, small stones (5–7 cm), large stones (10–15 cm)	–	1
5	Cheese whey	*Eisenia fetida*	BOD 76, COD 82, TSS 77	TN 60, TP 77	Ligneous mature compost, stones	0.04	–
6	Sewage	*Eisenia fetida*	BOD 98, COD 70, TDS 95	–	Garden soil, sand, aggregates (3–5, 7–8 cm)	–	2
7	Rural domestic sewage	*Eisenia fetida*	BOD 78, COD 67.6, TSS 89.8	NH+4 92.1	Ceramsite (3–5 mm)	4.2	1
8	Synthetic wastewater	*Eisenia fetida*	BOD 96, COD 90, TDS 82	–	Vermicompost, riverbed material (6–8 mm), sand (1– 2 mm), gravels (10–12.5 mm)	1.5, 2, 2.5, 3	
9	Urban wastewater	*Eisenia fetida*	BOD 98.5, COD 74.3, TSS 96.6	NH+4 99.1	Vermicompost, quartz sand, gravel (40 mm)	2.6, 1.3, 0.8, 0.6	2, 4, 6, 8

(continued)

Table 2.3 (continued)

Sr. no.	Types of wastewater	Earthworm species	Organics removal (%)	Nutrient removal (%)	Bed material and size	HLR (m^3/m^2d)	HRT (h)
10	Synthetic wastewater	*Eisenia fetida*	BOD 70–81, COD 59–72, TSS 55–75	–	Vermicompost, sand (2–4 mm), riverbed material, wood coal, glass balls, mud balls, gravel (10–12.5 mm)	1.5	–

characterization of the treated effluent and the vermicompost met the quality requirement for agricultural applications. Hartenstein and Bisesi (1989) carried out studies on the management of effluents using earthworms with wastewater having high organic contents resulting to clean effluents and nutrient-rich vermicompost.

2.4 Benefits and Limitations of Vermifiltration

2.4.1 Benefits

Vermifiltration treatment has various advantages over all the conventional biological wastewater treatment systems—the activated sludge process, trickling filters and rotating biological contactors in terms of energy usage, cost, ease of operation, etc.

- Vermifiltration process can be carried out with very less capital and operating costs and also can generate income with the nutrient-rich end product (vermicompost) and earthworm biomass. Xing et al. (2005) reported 1.16% nitrogen, 1.22% phosphorus and 1.34% potassium in the earthworm faeces.
- There is no sludge formation in the vermifiltration process which is not possible in any of the available other wastewater treatment technologies.
- Vermifiltered sewage is free of pathogens as earthworms devour on all the pathogens (bacteria, fungus, protozoa and nematodes) found in both the wastewater and the sludge as they are their loved food (Pierre et al. 1982).
- Vermifiltration process is odourless, and earthworm plays an important in doing so as they maintain aerobic conditions in the filter bed by their burrowing actions, inhibiting the action of anaerobic microorganisms.
- Vermifiltered sewage is free of toxic chemicals as earthworms have the ability to bioaccumulate high concentrations of toxic chemicals including heavy metals and also the 'endocrine disrupting chemicals' (EDCs) from sewage (Markman et al. 2007; Ireland 1983).
- Different varieties of wastewater can be treated with very less operational and maintenance cost.

2.4.2 Limitations

Vermifiltration process till now has been more successful in the lab-scale operation. Some limitations were observed during field-scale implementation which are listed as below:

- Most of the vermifiltration process are carried out in subsurface vertical/horizontal reactors as earthworms are cannot survive in submerged condition. In most of the reactors, the design is done in such a way that the influent enters in one point, and effluent is taken out from the other end in the horizontal reactors, and in the vertical reactors influent is from the top, and effluent is coming out from the bottom of the reactors.
- Earthworms do not like sunlight, and they breathe through their skin, so the movement of the earthworms are almost nil during the day time. Therefore, care should be taken while positioning the reactor by shunting it from sunlight either by covering it with gunny bags or keeping in shade.
- As most of the earthworms being employed are surface eaters, they concentrate mostly in the upper part of the reactors, and less activities are seen in middle and bottom layer of filter. This may lead to reduction in the efficiency of the filter.
- High moisture content in the filter might lead to anaerobic condition and may lead to the dead of the earthworms. So care has to be taken while deciding on the hydraulic loading rate of the filter.
- Wastewater with high NaCl cannot be treated in vermifilter as it is toxic for the earthworms (Hughes et al. 2007, 2008). The earthworms may die reducing the total biomass available to treat the wastewater, further affecting the efficiency of the system.
- Earthworms have the ability to accumulate heavy metals in their body. However, the amount of heavy metal uptake by the earthworm is limited (Sinha et al. 2008). Owing to which efficiency of treating heavy metals containing wastewater is less.
- The reproduction rate of earthworms is very high, and after certain time, there may be shortage of food and space as well, which may affect the efficiency of the treatment process.
- Cleaning of the vermifilter is required after the running period of the reactor is over which is really a cumbersome process. Proper care should be taken while handling the earthworms as they are sensitive and any injury may lead to their death.
- Vermifilters may get choked when wastewater with high organic loads are used. Another important limitation is the depth of the vermifilter as earthworms cannot survive in deeper depths.

2.5 Conclusions

Vermifiltration technology can be adopted and designed to treat wastewater from different sources or to suit a particular wastewater. Therefore, vermifiltration is found to be suitable technique with high efficiency. This is also a good alternative to treat wastewater in decentralized manner. The basic mechanism of vermifilter and the role of earthworms in treating the wastewater are explained in the paper. Although earthworms are sensitive in nature, they are highly productive organisms, and they play a significant role in breaking pollutants and aerating filter bed. Vermifiltration process is carried out in aerobic condition which is possible due to the presence of earthworms, also making it a favourable condition for aerobic decomposer microbes. The vermifiltered water had elevated value of nitrate and phosphate concentration which can be used for sewage farming or horticulture. The organic matter and solids present in the wastewater were consumed by earthworms transforming these into valuable vermicompost, and thus there is no sludge generation as seen in other treatment technologies. Hence, no sludge disposal is required instead the vermicompost obtained can be used as manure as it is having good content of nitrogen and phosphate. In addition, we also get earthworm biomass which can be used as feed for fishes and poultry. Although there are some limitations vermifilter has a huge potential to become a reliable treatment technology for wastewaters generated from different operations, especially for the countries facing severe challenges, such as lack of investment and technical labour, etc. Generally, this vermifiltration system has practical application value, although some technical details should be further researched. The results from various studies indicate that vermifiltration can be strongly recommended for rural settlements, small communities and small industries producing organic wastewater as a viable alternative to conventional treatment, provided that conditions are conducive for the growth of worms.

References

Arias CA, Brix H, Marti E (2005) Recycling of treated effluents enhances removal of total nitrogen in vertical flow constructed wetlands. J Environ Sci Health A Tox Hazard Subst Environ Eng 40:1431–1443

Bajsa O, Nair J, Mathew K, Ho GE (2003) Vermiculture as a tool for domestic wastewater management. Water Sci Technol 48(11–12):125–132

Bhawalkar U (1995) Vermiculture eco-technology. Bhawalkar Earthworm Research Institute (BERI), Pune

Bhuptawat H, Folkard GK, Chaudhari S (2007) Innovative physico-chemical treatment of wastewater incorporating *Moringa oleifera* seed coagulant. J Hazard Mater 142:477–482

Bodhkhe SY (2009) A modified anaerobic baffled reactor for municipal wastewater treatment. J Environ Manag 90:2488–2493

Carballeira T, Ruiz I, Soto M (2017) Aerobic and anaerobic biodegradability of accumulated solids in horizontal subsurface flow constructed wetlands. Int Biodeterior Biodegrad 119:396–404

Chyan JM, Senoro DB, Lin CJ, Chen PJ, Chen IM (2013) A novel biofilm carrier for pollutant removal in a constructed wetland based on waste rubber tire chips. Int Biodeterior Biodegrad 85:638–645

Darwin F, Seward AC (1903) More letters of Charles Darwin. In: A record of his work in series of hitherto unpublished letters. John Murray, London, p 508

Dixon A, Butler D, Fewkes A (1999) Water saving potential of domestic water reuse systems using greywater and rainwater in combination. Water Sci Technol 39(5):25–32

Gardner T, Geary P, Gordon I (1997) Ecological sustainability and onsite effluent treatment systems. Aust J Environ Manag 4(1):144–156

Ghangrekar MM, Kishor N, Mitra A (2007) Sewage reuse for aquaculture after treatment in oxidation and duckweed pond. Water Sci Technol 55:173–181

Hand P (1988) Earthworm biotechnology. In: Greenshields R (ed) Resources and application of biotechnology: the new wave. MacMillan Press Ltd, New York

Hartenstein R, Bisesi MS (1989) Use of earthworm biotechnology for the management of effluents from intensively housed livestock. Outlook Agric 18:72–76

Hughes R, Nair J, Mathew K, Ho G (2007) Toxicity of domestic wastewater pH to key species within an innovative decentralised vermifiltration system. Water Sci Technol 55:211–218

Hughes R, Nair J, Ho G (2008) The toxicity of ammonia/ammonium to the vermifiltration wastewater treatment process. Water Sci Technol 58:1215–1220

Hughes RJ, Nair J, Ho G (2009) The risk of sodium toxicity from bed accumulation to key species in the vermifiltration wastewater treatment process. Bioresour Technol 100(16):3815–3819

Ireland MP (1983) Heavy metals uptake in earthworms. In: Earthworm ecology. Chapman & Hall, London

Jasti N, Khanal SK, Pometto AL, Leeuwen JV (2006) Fungal treatment of corn processing wastewater in an attached growth system. Water Pract Technol 1(3):wpt2006069. https://doi.org/10.2166/WPT.2006069

Jefferson B, Laine AL, Judd SJ, Stephenson T (2000) Membrane bioreactors and their role in wastewater reuse. Water Sci Technol 41(1):197–204

Julka JM (1986) Earthworm resources in India. In: National seminar on organic waste utilization, pp 1–7

Komarowski S (2001) Vermiculture for sewage and water treatment sludges. Water Publication of Australian Water and Wastewater Association, pp 39–43

Kruzic AJ, Schroeder ED (1990) Nitrogen removal in the overland wastewater treatment process: removal mechanisms. J Water Pollut Control Fed 62(7):867–876

Kumar T, Rajpal A, Bhargava R, Prasad KSH (2014) Performance evaluation of vermifilter at different hydraulic loading rate using riverbed material. Ecol Eng 75:370–377

Kumar T, Rajpal A, Arora S, Bhargava R, Hari Prasad KS, Kazmi A (2016) A comparative study on vermifiltration using epigeic earthworm *Eisenia fetida* and *Eudrilus eugeniae*. Desalin Water Treat 57:6347–6354

Leach LE, Enfield CG (1983) Nitrogen control in domestic wastewater rapid infiltration systems. J Water Pollut Control Fed 55(9):1150–1157

Lens PN, Vochten PM, Speleers L, Verstraet WH (1994) Direct treatment of domestic wastewater by percolation over peat, bark and woodchips. Water Res 28:17–26

Li YS, Robin P, Cluzeau D, Bouché M, Qiu JP, Laplanche A, Hassouna M, Morand P, Dappelo C, Callarec J (2008) Vermifiltration as a stage in reuse of swine wastewater: monitoring methodology on an experimental farm. Ecol Eng 32(4):301–309

Li YS, Xiao YQ, Qiu JP, Dai YQ, Robin P (2009) Continuous village sewage treatment by vermifiltration and activated sludge process. Water Sci Technol 60:3001–3010

Li X, Xing M, Yang J, Lu Y (2013) Properties of biofilm in a vermifiltration system for domestic wastewater sludge stabilization. Chem Eng J 223:932–943

Lodge BN, Seager T, Stephenson T, English PH, Ford R (2000) A water recycling plant at the millennium dome. Civ Eng 138:58–64

Markman S, Guschinna IA, Barnsleya S, Buchanana KL, Pascoea D, Mullera CT (2007) Endocrine disrupting chemicals accumulate in earthworms exposed to sewage effluents. Chemosphere 70 (1):119–125

Mars R, Mathew K, Ho GE (1999) The role of submergent macrophyte *Triglochin huegelii* in domestic wastewater treatment. Ecol Eng 12(1/2):57–66
Martin JP (1976) Darwin on earthworms: the formation of vegetable moulds. Bookworm Publishing. ISBN 0-916302-06-7
Morari F, Giardini L (2009) Municipal wastewater treatment with vertical flow constructed wetlands for irrigation reuse. Ecol Eng 35:643–653
Patel JB (2018) Wastewater treatment by Vermifiltration: a review. Int J Latest Technol Eng Manag Appl Sci 6(1):2278–2540
Pierre V, Phillip R, Margnerite L, Pierrette C (1982) Anti-bacterial activity of the haemolytic system from the earthworms *Eisenia foetida andrei*. J Inverteb Pathol 40:21–27
Reinecke AJ, Viljoen SA, Saayman RJ (1992) The suitability of *Eudrilus eugeniae, Perionyx excavatus* and *Eisenia fetida* (Oligochaeta) for vermicomposting in southern Africa in terms of their temperature requirements. Soil Biol Biochem 24(12):1295–1307
Schudell P, Boller M (1989) On-site wastewater treatment with intermittent buried filters. International specialized conference on design and operation of small scale wastewater treatments plants, pp 95–104
Shao Y, Pei H, Hu W, Chanway CP, Meng P, Ji Y, Li Z (2014) Bioaugmentation in labscale constructed wetland microcosms for treating polluted river water and domestic wastewater in northern China. Int Biodeterior Biodegrad 95:151–159
Singleton DR, Hendrix PF, Coleman DC, Whitman WB (2003) Identification of uncultured bacteria tightly associated with the intestine of the earthworm *Lumbricus rubellus* (Lumbricidae; Oligochaeta). Soil Biol Biochem 35(12):1547–1555
Sinha RK, Herat S, Agarwal S, Asadi R, Carretero E (2002) Vermiculture and waste management: study of action of earthworms *Eisenia foetida*, *Eudrilus eugeniae* and *Perionyx excavatus* on biodegradation of some community wastes in India and Australia. Environmentalist 22:261–268
Sinha RK, Bharambe G, Chaudhari U (2008) Sewage treatment by vermifiltration with synchronous treatment of sludge by earthworms: a low-cost sustainable technology over conventional systems with potential for decentralization. Environmentalist 28:409–420
Taylor M, Clarke WP, Greenfield PF (2003) The treatment of domestic wastewater using small-scale vermicompost filter beds. Ecol Eng 21(2/3):197–203
Tomar P, Suthar S (2011) Urban wastewater treatment using vermi-biofiltration system. Desalination 282:95–103
Wang S, Yang J, Lou SJ, Yang J (2010) Wastewater treatment performance of a vermifilter enhancement by a converter slag-coal cinder filter. Ecol Eng 36:489–494
Wang Y, Xing MY, Yang J, Lu B (2016) Addressing the role of earthworms in treating domestic wastewater by analyzing biofilm modification through chemical and spectroscopic methods. Environ Sci Pollut Res 23:4768–4777
Xing M, Yang J, Lu Z (2005) Microorganism-earthworm integrated biological treatment process—a sewage treatment option for rural settlements. Five days ICID 21st European regional conference
Xing M, Li X, Yang J (2010) Treatment performance of small-scale vermifilter for domestic wastewater and its relationship to earthworm growth: reproduction and enzymatic activity. Afr J Biotechnol 9:7513–7520
Xing M, Zhao C, Yang J, Lv B (2014) Feeding behavior and trophic relationship of earthworms and other predators in vermifiltration system for liquid-state sludge stabilization using fatty acid profiles. Bioresour Technol 169C:149–154
Yang J, Xing M, Lu Z, Lu Y (2008) A decentralized and on-site option for rural settlements wastewater with the adoption of vermifiltration system. 2nd international conference on bioinformatics and biomedical engineering, pp. 3023–3026
Zhao L, Wang Y, Yang J, Xing M, Li X, Yi D, Deng D (2010) Earthworm-microorganism interactions: a strategy to stabilize domestic wastewater sludge. Water Res 44:2572–2582

Chapter 3
Treatment of Wastewater by Vermifiltration Integrated with Plants

Anu Bala Chowdhary, Jahangeer Quadar, Bhaskar Singh, and Jaswinder Singh

Abstract In the present scenario, due to industrialization and urbanization, wastewater management is a major issue. In order to overcome the issue of wastewater treatment, vermifiltration is one of the best remedies available. Vermifiltration is the procedure of wastewater remediation that includes the joint action of earthworms and microbes. Vermifiltration coupled with phytoremediation could prove to be an efficient and easy-to-adapt technology that could be adopted at a community level. The treated wastewater obtained after vermifiltration is reported to have its utility in gardening, irrigation, and other reuse applications. To make the process more efficient, there is a future scope of study on earthworm-microorganism-plant interaction in the vermifiltration process integrated with plants. Treatment by vermifiltration could also be adopted at a micro level by the local communities. The process could provide an efficient mode of treatment wherever there is a lack of proper and adequate conventional treatment facilities. Future perspectives of this technology will thus depend on a large extent on the cost-benefit analysis of the process. The process could be made more cost efficient by incorporating the plants in combination with vermifiltration.

Keywords Wastewater · Vermifiltration · Earthworm · Microorganisms · Plants

A. B. Chowdhary · J. Quadar
Department of Botanical and Environmental Sciences, Guru Nanak Dev University, Amritsar, Punjab, India

B. Singh
Department of Environmental Sciences, Central University of Jharkhand, Ranchi, Jharkhand, India

J. Singh (✉)
Department of Zoology, Khalsa College Amritsar, Amritsar, Punjab, India

S. A. Bhat et al. (eds.), *Earthworm Assisted Remediation of Effluents and Wastes*,
https://doi.org/10.1007/978-981-15-4522-1_3

3.1 Introduction

One of the most important resources on planet Earth is water. It comprises 71% of the Earth's surface and is important for existence of all known forms of life. However, only 2.5% of it is available as fresh water (Rajasulochana and Preethy 2016). The quality of natural water bodies is deteriorated by the wastewater discharged during industrial processing and urbanization (Marshall et al. 2007). Because of the poor treatment facilities, the wastewater generation and its treatment has become a major problem in the developing nations (Singh et al. 2015; Sinha et al. 2010a, b, c). Numerous technologies for the treatment of wastewater have been utilized, e.g., septic tanks, oxygen-consuming organic treatment units, fixed activated sludge treatments, constructed wetlands, soil infiltration trenches, vegetation-based wastewater treatment, and bioremediation through plants and others. Apart from these technologies, vermifiltration have been reported to be an appropriate technique for fairly efficient remedy for wastewater treatment (Choudhary and Medok 2017; Jatin 2018).

Vermifiltration also known as "lumbri-filtration" is a liquid-state vermi-change procedure through which domestic and industrial wastewater can be treated (Samal et al. 2017a, b). In other words, vermifiltration is the involvement of earthworms in filtration system with proper bedding materials to breakdown organic contaminants (Tomar and Suthar 2011; Arora et al. 2016). It was first recommended by Prof. Jose Toha at the University of Chile in 1992 and suggested to be an efficient alternative technology as it is a nearly odorless procedure producing a stable, purified, detoxified, and profoundly nutritive effluent (Wang et al. 2010a, b; Xing et al. 2010). Vermifiltration has a high effectiveness of removing contaminants from wastewater (Singh et al. 2017). It has been reported to be an efficient technology for diminishing biochemical oxygen demand (BOD), chemical oxygen demand (COD), as well as suspended solids. Bobade and Ansari (2016) reported 5 days BOD (BOD_5) reduction by over 90%, COD by 80–90%, total dissolved solids (TDS) by 90–92%, and the total suspended solids (TSS) by 90–95%. Table 3.1 depicts the potential of vermifiltration in treatment of various types of wastewater.

In this chapter, we present a systematic scientific literature and review the data with following objectives: (a) to study the prospects of vermifiltration technology in wastewater remediation and their design; (b) to review the treatment efficiency of different types of wastewater through vermifiltration; (c) to utilization of the treated wastewater for further applications; (d) to assess the effectiveness of plants in vermifiltration technology. The limitation and future perspectives have also been discussed.

Table 3.1 Potential of vermifiltration in the treatment of wastewater

Earthworm species	Plant species	Type of wastewater	Treatment efficiency	References
Eisenia fetida	Not used	Distillery wastewater	COD-89.4% BOD_5-91.1% TKN-94.9% TSS-92.4% TDS-91.9%	Manyuchi et al. (2018)
Eisenia fetida	Not used	Sewage waste water	BOD_5-98% COD-70%, TDSS-95% Turbidity-98%	Manyuchi et al. (2013a)
Perionyx excavatus	Not used	Domestic sewage	BOD_5-25.86% COD-24.38% TDS-25.26% TSS-35.47% Turbidity-17.06%	Bhise and Anaokar (2015)
Eisenia fetida	Not used	Dairy waste water	BOD_5-97.95% COD-91.64% TSS-76.39% TDS-84.27% Oil and grease-84.13%	Telang and Patel (2015)
Earthworm species	Wetland weed *Cyperus rotundus*	Urban wastewater	TSS-88.6% TDS-99.8% COD-90% NO_3^--92.7% $PO4^{3-}$-98.3%	Tomar and Suthar (2011)
Eisenia fetida	Not used	Rural domestic waste water	COD-78.0% BOD_5-98.4% Ammonia nitrogen ($NH4^+$-N)-90.3% Phosphorus-62.4%	Wang et al. (2010a, b)
Lumbricus rubellus	*Canna indica*	Liquid effluents from gelatin industry	BOD_5-89.24% COD-90.08% TDS-81.97% TS-77.56%	Ghatnekar et al. (2010)
Eisenia fetida	Not used	Ayurvedic industrial effluents	COD-98.03% BOD_5-98.43% TSS-95.8% TDS-78.66% Oil and grease-92.58%	Das et al. (2015)
Earthworm species	*Cyperus rotundus*	Domestic waste water	TSS-30.34 mg/L TDS-93.48 mg/L BOD_5-22.15 mg/L COD-88.35 mg/L Nitrate-25.85 mg/L Phosphate-9.14 mg/L	Reddy et al. (2018)

(continued)

Table 3.1 (continued)

Earthworm species	Plant species	Type of wastewater	Treatment efficiency	References
Eisenia fetida, Perionyx excavatus, and Eudrilus eugeniae	Not used	Sewage sludge	BOD_5-98% COD-45% TSS-90% Turbidity-98%	Sinha et al. (2008)
Eisenia fetida	Not used	Petroleum industry waste water	Hydrocarbons C10–C14-99.9% C15–C28-99.8% C29–C36-99.7%	Sinha et al. (2012)
Eisenia fetida	Not used	Community waste water	Total solids by 85–90% TSS-90–95% TDS-80–85% COD-90–95% BOD_5-93–95%	Misal and Mohite (2017)
Eudrilus eugeniae	Not used	Dairy effluent	BOD_5-95% COD-83% Total suspended solids-87% Oil and grease-97.8%	Jeevitha et al. (2016)

3.2 Vermifilter System

Vermifilter system provides the perfect natural system for earthworms to stimulate and accelerate microbial degradation of wastewater organics inside the vermibedding (Sinha et al. 2008). Filter materials are critical to separate pollutants from wastewater and to make ideal condition for earthworms and microbial networks (Xing et al. 2011). Vermifilter system is structured in a way that the wastewater is supplied on the top and the treated wastewater is collected in the base of the system. In a vermifilter system reported by Bhise and Anaokar (2015), the sprinkler system comprised of a simple 16 mm polypropylene pipe with holes for circulation of wastewater uniformly on filter bed.

3.2.1 Vermifilter Design

The design of a vermifilter system reported by Lakshmi et al. (2014) consisted of three layers—the bottom layer made up of gravel aggregates of size 16–20 mm and filled up to the depth of 10 cm. Above this, the second layer consisted of aggregates of 10 mm sizes filled up to another 10 cm. At the top of this, 10 cm layer of 5 mm aggregates was reported to be blended with sand. The third layer was the highest layer comprising of about 10 cm that comprised soil bed in which the earthworms are released. A layer of net of wire mesh lied underneath the layer of soil bed to enable water to stream down while holding the earthworms in the soil bed because it can slither down to filter materials. It is reported that around 1 week settling time in the soil bed is required for earthworms to acclimatize in the new environment as they play an important role in wastewater purification (Manyuchi et al. 2013a, b). After acclimatization, the burrows and castings are reported to form as the earthworm ingestion activities help in the removal of pollutants from the wastewater (Domínguez 2004). The wastewater is sprinkled on the upper top layer comprising of soil that is rich in microbes and earthworms. The passage of the wastewater through several layers of the vermifiltration system helps in the removal of pollutants. The treated wastewater can be collected at the bottom in the effluent collection tank (Ghatnekar et al. 2010; Singh et al. 2017). Vermifilters can be considered as an appropriate small-scale wastewater treatment, particularly in economical terms (Jiang et al. 2016). Figure 3.1 depicts a schematic representation of vermifiltration system.

3.2.2 *Vermifiltration Process*

The process of vermifiltration entails amalgamation of three steps, i.e., primary (removal of grit, silt, and so forth), secondary (biological degradation), and tertiary (removal of pathogens) treatment. Vermifiltration is the procedure of wastewater remediation that includes the joint action of earthworms and microbes. The treated wastewater might be reused for further applications (Dash 1978; Sinha et al. 2008). The ingestion and grinding of soil particles along with pollutants by earthworms results in the alteration of physical, chemical, and biological characteristics of wastewater (Edwards and Arancon 2004; Liu et al. 2012).

Vermibed and earthworms play an exquisite function in the removal of pollutants from wastewater. Apart from being a food supply to the earthworms, it helps in multiplication of microbes (Liu et al. 2013; Zhao et al. 2010). Burrowing of earthworms involves actions consisting of ingestion, grinding, digestion, and excretion (Aira et al. 2007; Suthar and Singh 2008). The joint effect of these actions results in numerous physical, chemical, and biological changes on the vermi-atmosphere. The fine particles are obtained through the grinding of the ingested substances within the excreted casting that offer higher filtration efficiency of solids (Komarowski 2001; Wang et al. 2011).

Burrowing and channeling activity of earthworms helps in expanding water-driven conductivity of bedding media to get an ideal time of interaction resulting in an efficient wastewater treatment (Aira and Domínguez 2009; Bhawalkar 1995). This is also one of the central points behind maintaining oxygen-consuming condition inside the vermibed. The addition of earthworms to the bedding consequently increases the microbial population inside vermifilter, as earthworms have massive variety of microbial flora in their gut and discharge them directly to the soil (Arora et al. 2014). The discharged nutrients in the form of vermin-castings are further

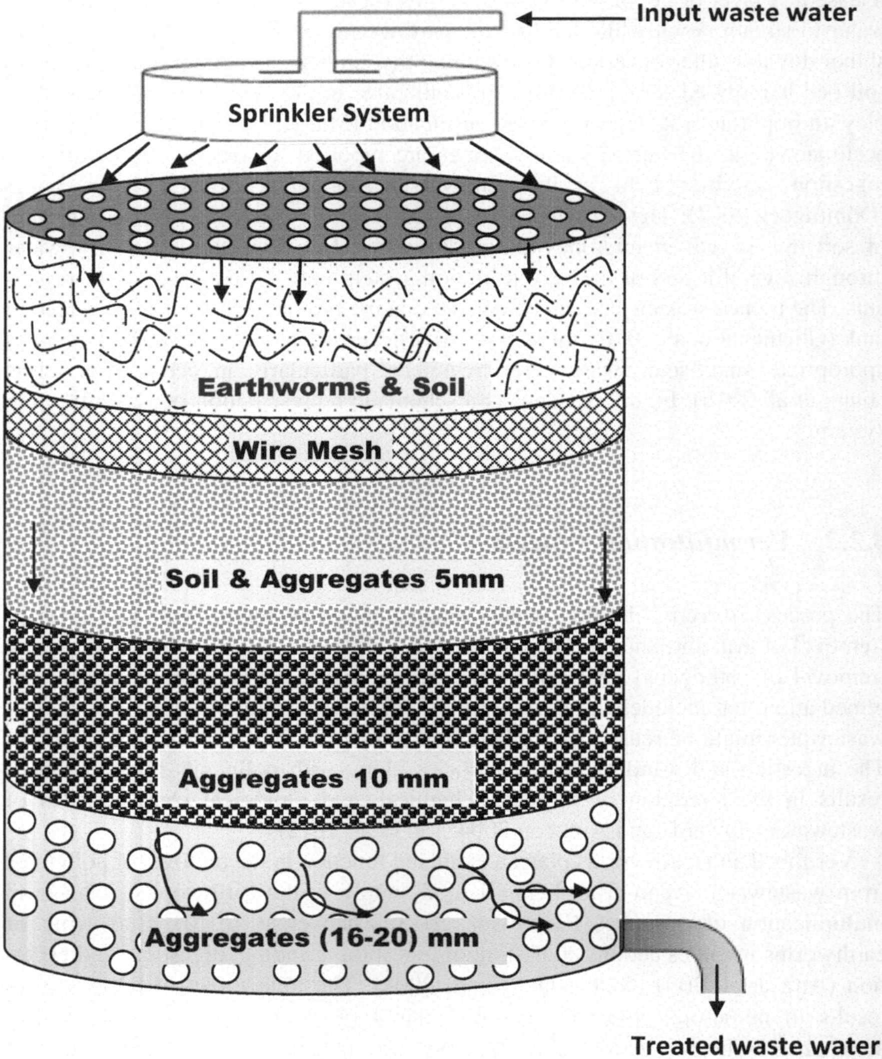

Fig. 3.1 Schematic representation of vermifiltration system

being absorbed by microorganisms for their populace improvement. This consequently increases the degradation potential of earthworms (Edwards and Fletcher 1988). The treated sludge with vermifiltration had a high rate of stabilization dependent on elemental analysis. This indicates that earthworms assist to make the proficiency and soundness of vermifiltration performance certain for the treatment of excess sludge that is an unavoidable by-product of the aerobic organic wastewater treatment technique (Liu et al. 2012).

3.2.3 Pathogen Removal and Role of Microbes and Earthworms in Vermifiltration

In vermifiltration, microorganisms are responsible for biochemical degradation of organic matter, and earthworms act as controllers (Arora and Kazmi 2015; Liu et al. 2012). The antibacterial action of the isolated microorganisms, viz., *Bacillus*, *Alcaligenes*, *E. coli*, *Enterobacter*, and *Klebsiella*, has been reported to oppose or avert the improvement of different pathogens found in vermifiltration due to release of some antibacterial substances. These findings further exhibit the purification capacity of vermifiltration to expel pathogens from wastewater. Subsequently, vermifiltration generation can be considered to endure minimum effort, feasible and plausible choice for wastewater remedy, that bring about a clean pathogen-free effluent (Arora et al. 2014).

Earthworms are also recognized to secrete mucus from their body which assist in mineralization of toxins found in wastewater (Arora et al. 2014; Ellis and McCalla 1978; Sahariah et al. 2015; Goswami et al. 2014). Mucus additionally maintains up an ideal condition for bacterial populace and higher growth. Besides, mucus made from amino acids, glucoproteins, and small glucosidic and proteic molecules allows in keeping up the satisfactory viable C/N proportion for progressed biochemical exercises of the vermifiltration system (Bajsa et al. 2004; Wang et al. 2011). These earthworms can possibly consume harmful and ineffective microbes of the wastewater and aid in pathogen elimination (Sinha et al. 2008). Earthworms have an intrinsic potential of changing organics and nutrients from insoluble structures to soluble structures. In this way, it makes the insoluble nutrients bioavailable for further debasement or union of new cells (Gogoi et al. 2015; Goswami et al. 2016; Sahariah et al. 2015; Singh and Kaur 2015).

In earlier days, vermifiltration has been utilized to treat household and city wastewater. Afterward, application of vermifiltration extended to other domains, e.g., rural wastewater, urban runoff (Tomar and Suthar 2011), sewage sludge (Xing et al. 2012), and to treat various industry effluents, viz., gelatinous effluents (Ghatnekar et al. 2010), dairy industry (Bharambe 2006), herbal pharmaceutical effluents (Dhadse et al. 2010), piggery effluents (Luth et al. 2011), and effluents from other processing industries (Sinha et al. 2007).

3.2.4 Role of Plants in Vermifiltration

There are several studies available on the use of either vermifiltration or only plants to extract pollutants from wastewater. However, there are very few reports available on the use of both systems potential to develop an efficient integrated system (Samal et al. 2017a, b). Tomar and Suthar (2011) used earthworm *Perionyx sansibaricus* and plant species *Cyperus rotundus* to treat urban runoff. Chen et al. 2016 developed a vermifilter supported by *Canna indica* to treat sludge and reported the removal of

COD and total phosphorus (TP) in the range 62–81% and 72–80% respectively. In a single reactor, Wang et al. 2010a, b used plant *Phragmites australis* and *Eisenia fetida* to treat domestic wastewater. Nuengjamnong et al. (2011) used earthworm *Pheretima peguana* in a two-stage vermifilter, macrophyte filter for effective treatment of swine wastewater, and reported more than 90% COD and Total Nitrogen (TN) removal efficiency. Xu et al. (2013a, b) explored the potential of earthworm for nitrification-denitrification and TN removal efficiency by adding *Eisenia fetida* to the vertical flow wetland, planted with various plants. Through planting macrophyte in vermifilter, some structural and functional problems are reported to be solved. Planting macrophytes in vermifilters or introducing earthworms in built wetlands have been reported in laboratory and pilot-scale experiments (Tomar and Suthar 2011; Wang et al. 2010a, b). Using plant-mediated system can benefit both the plants and earthworms as it is a well-known fact that both act in a symbiotic manner. The overall system performance also increases as does the filter's lifetime. Macrophytes also extract organics, nitrogen, and phosphorus from wastewater and make it easier for earthworms to work (Xu et al. 2013a). Oxygen leakage from underground tissue produces oxidized layer or 1–4 mm oxidizing protective film on the root surface primarily from root tips (Brix and Schierup 1990; Bezbaruah and Zhang 2005). Thus, the aerobic condition maintained in filter media is considered favorable for earthworm's survival (Ye et al. 2012). Earthworms create macro pores and loosen soil so that plant roots penetrate the media and form new tubular pores into deep soil. For this reason, the rate of infiltration in the top and bottom layers is maintained uniformly. A variety of processes can lead to clogging such as accumulation of suspended solids (SS), surplus production of sludge, chemical precipitation and deposition in substratum pores, growth of plant rhizomes and roots, gas generation, and clogging layer compaction (Caselles-Osorio et al. 2007; Sun et al. 1999). Earthworms digest the accumulated organic suspended solids for a blocked filter and their burrowing operation gradually loosens the substrates to restore the system smoothly (Xing et al. 2010). Plant root system provides a higher surface area for the production of various heterotrophs, autotrophs, nitrifiers, and ammonium-oxidizing bacteria (Vymazal 2005; Konnerup et al. 2009). Organic pollutants are reported to degrade very rapidly due to the enrichment of the diversified microbial population in the rhizospheric environment. For denitrification, plant litter and root exudates, i.e., citrate, malate, acetate, and oxalate, act as carbon sources (Lin et al. 2002). The addition of plants in the vermifilter also leads to pollutant removal and improves the efficiency of the process (Samal et al. 2017a). Samal et al. (2018a) also stated that the main benefit from the introduction of plants is the elimination of nutrients (nitrogen and phosphorus from the wastewater). Penetration of plant root in bedding helps to aerobicize the environment by producing cracks throughout the bedding (Bezbaruah and Zhang 2005). The formed cracks make it easy for earthworms to burrow deeper. Because of the root expansion and cracks, the aerobic depth of the bedding profile increases to three- to fourfold as compared to the plant-free filter (Samal et al. 2017b). The increase in aeration within the bedding also helps to increase the organics oxidation and nitrification. The plant roots often release oxygen that is useful in the removal of pollutants from wastewater. The plant root exudates contain several sources of carbon and nutrients that help to extract nitrogen by encouraging denitrification of nitrified nitrogen (Lin et al. 2002). The plant root exudates provides habitat to the microbes and thus allow for a higher rate of degradation. The root exudates also possess many antibacterial properties that help in killing harmful pathogens that are brought along with the wastewater (Nokes et al. 2003; Tunçsiper et al. 2012). The killing of the pathogens could also be due to the sticky and slimy nature of the exudates, which makes it impossible for the bacteria to escape or move to new places for food consumption.

3.3 Different Types of Wastewater Treated Through the Vermifiltration

The vermifiltration process is a combined action of the earthworms, soil, sand, and gravel particles. The vermifiltration technology was proposed as an alternative wastewater treatment method in developing countries that face different wastewater treatment challenges.

3.3.1 Urban/Domestic Wastewater Treatment

The vermifilter may be an efficient technology for stabilization of excess sludge from domestic, municipal wastewater treatment plants (Wei et al. 2003). Vermifiltration is an environment-friendly and economically feasible technology for domestic sewage treatment in rural areas and for herbal pharmaceutical wastewater treatment (Li et al. 2009; Dhadse et al. 2010). *Eudrilus eugeniae* earthworm could operate in the filter above 40 °C and increased the vermifilter conditions. It has been reported that the elimination of BOD_5, COD, TSS, and coliform organisms with the vermifilter could perform better than with the filter without earthworms (Adugna et al. 2014). The integrated vermi-biofiltration reactor has been reported to be more efficient than the traditional biofiltration system in removing key chemical pollutants from wastewater. The efficacy of the vermi-biofiltration system in wastewater treatment is well reported in the literature. However, further detailed studies are needed to address some of this system's key issues, such as hydraulic load, retention time impact, vermibed microbial ecology, earthworm-microbial interactions, etc. (Tomar and Suthar 2011). The proportion of more than 0.3 g of earthworm was significantly associated with COD, BOD_5, SS, and NH_4-N removal levels, which suggested that the larger earthworms (*Eisenia fetida*) could play a more important role in vermifilter wastewater treatment compared to the smaller one. The hydraulic loading has been reported to have little influence on the reproduction of earthworms and the increase of juveniles, the decrease in adults and larger earthworms contributed to the decrease in vermifilter treatment efficiency (Xing et al. 2010). Authors have reported good efficiency of the vermifiltration method in wastewater treatment. The characteristic of sewage water was greatly improved by applying sawdust to the soil, which could increase the soil porosity. The development of earthworms, growth, breed, and survival in the moist environment has been reported by Garkal et al. 2015. The high concentration of the chemicals can be tolerated by *Eudrilus eugeniae* and a high concentration of chemicals is not expected to affect the efficiency of vermifiltration. Other factors that fluctuate significantly in raw gray water such as pH, organic load, and temperature have not yet been studied and reported by researchers.

3.3.2 Dairy Wastewater Treatment

The dairy wastewater used as the influent and processed in the vermifiltration and was found to be excellent for removing COD, NH_4 and total nitrogen (Rodgers et al. 2006). Samal et al. (2017a, b) explored the efficiency and sustainability of an integrated system consisting of *Canna indica* and *Eisenia fetida*, found that macrophyte-assisted vermifiltration method has been able to reduce BOD and COD by 75–81%, and total nitrogen by 24–42% from dairy wastewater and it can be used as an effective and complementary technology for various decentralized wastewater treatment system. Sinha et al. (2007) studied the vermifiltration of brewery and milk dairy wastewaters having very high BOD_5 and TSS loadings. Earthworms reduced the high BOD_5 loads by 99% in both cases and TSS by over 98%. They reported that the hydraulic retention times (HRTs) of brewery wastewater and dairy wastewater was 3–4 h and 6–10 h respectively. The worms must act instantly in the process as the wastewater passes through their bodies (degrading the organics, ingesting the solids and heavy materials). That's why the wastewater has to be kept in the vermifilter bed for a reasonable time (HRT has to be within hours, not days) while the worms act on the wastewater (Sinha et al. 2007). Vermi-biofiltration is a combination of vermicomposting of traditional filtration methods. Earthworms body acts as a "biofilter" and the pollutant removal were reported to be BOD_5-97.95%, COD-91.64%, TSS-76.39%, and TDS-84.27%. It was reported that oil and grease content decreased by 84.13% (Telang and Patel 2015). Total dairy wastewater treatment using coagulation and decantation with the help of aluminum sulfate, ferrous chloride, and calcium hydroxide have shown the optimum dose concentrations (0.49–0.63 g/L), pH (11–12) and agitation duration (20 min) to obtain an effluent in accordance with the Moroccan standards for suspended matter, and total phosphorus with removal of 94% and 89% respectively (Hamdani et al. 2005). The vermifilter contains aerobic and anaerobic microhabitats that allow the enrichment of gene-containing microorganisms necessary for the complete transformation of organic nitrogen into N_2. The dual nature (aerobic/anaerobic) of vermifilter provides an ideal ecosystem to facilitate the microbial decomposition and removal of organic nitrogen from dairy wastewater reducing downstream N load on soil and groundwater without increasing emissions of gasses that accelerate climate change (Lai et al. 2018). Depending on the characteristics of the wastewater to be treated, vermifiltration can provide efficient treatment of dairy wastewater. Furthermore, for the treatment of voluminous wastewater by vermifiltration, economic optimization of these processes must be worked out (Natarajan et al. 2015). The combination of horizontal subsurface flow vermifilter system and vertical down flow vermifilter system planted with various macrophytes and the earthworm, *Eisenia fetida* was able to significantly reduce organics, nutrients, and solids from dairy wastewater (Samal et al. 2018a, b). Vermibed height is considered a significant variable in the process of vermifiltration.

3.3.3 Swine Wastewater Treatment

Vermifiltration can be used to treat diluted swine manure. The wastewater can be reused to flush the manure, inducing a reduction in ammonia emissions compared to rearing on a slatted floor with slurry accumulation (Li et al. 2008). Earthworms have a simple, although indirect, effect on gaseous emissions during pig fresh slurry vermifiltration. The abundance of earthworms can be used as a bioindicator to verify low-energy consumption and low greenhouse gas and ammonia production in fresh water recycled manure systems (Luth et al. 2011). The variable levels of pollutants in the influential hybrid constructed wetland (HCW) were measured. This may result from differences in the composition of pig waste, farm management, and storage conditions but may also be related to inefficiencies in the pretreatment used on the wastewater of raw piggery. This fluctuation may have had an impact on the HCW performance (Borin et al. 2013).

3.4 Limitations

In spite of being an effective technology in treatment of wastewater by vermifiltration, this technology still has some limitations. Among the species of earthworms, not all are reported to be equally efficient in the treatment of wastewater. Earthworms are categorized as epigeic (found on top surface of soil), endogeics (found in 10–30 cm of soil layer), and anecic (that burrow deep in the soil). Among the three, only the epigeic group of earthworms have been reported to be efficient in conversion of organic waste to vermicompost. Similarly, the process of vermifiltration is also reported to work efficiently when epigeic group of earthworms are used for treatment of wastewater. Among the epigeic group also, different species have varying capacity to remediate the pollutants from the wastewater. It has been reported that *Eisenia fetida* is more efficient than other worm in removal of pollutants from wastewater (Gupta 2015). At present, vermifiltration is not very efficient in removal of nutrients particularly nitrogen and phosphorus. A bio-vermifilter used by Nie et al. (2015) showed fluctuations in removal of nitrogen due to low ratio of C:N (carbon to nitrogen) in the biofilter. Thus, this requires an improvement in the design of the vermifiltration setup. Li et al. (2008) states that it is important to have an understanding of dynamics of earthworm population during the process of vermifiltration. Knowing the behavior of earthworms and their demographic behaviors will be useful for using them as bioindicators for managing the nutrient input in the system. The present chapter that reports the integration of plants with the process of vermifiltration could prove to be more effective in remediation of wastewater.

Another challenge for this process is to provide a reactor as the earthworms are said to be sensitive to light, touch, and moisture content. Though earthworms can adapt to low temperature conditions but high temperature becomes a deterrent in their survival (Patel 2018). Various studies state that earthworms may adversely be affected by some of the pollutants that are toxic in nature, viz., heavy metals and pesticides. Hence, it becomes important to ensure that no such toxic elements or substances are present in the wastewater that is detrimental to earthworm's health. However, if earthworms show any susceptibility toward any specific pollutants present in the wastewater, the process of vermifiltration may require the pretreatment of the wastewater through a conventional mode of treatment. This may escalate the cost of the overall treatment process.

3.5 Future Perspective

The concept of vermifiltration that combines the symbiotic relationship of earthworm and microorganisms to remove the pollutants from wastewater is a relatively new concept (Jiang et al. 2016). The concept of vermifiltration could be further extended by incorporating the plants as well to have a profound synergistic effect in the removal of pollutants from the wastewater (i.e., vermifiltration coupled with phytoremediation). Vermifiltration coupled with incorporation of plants can prove to be an effective mode of treatment of municipal and industrial wastewater. Apart from point source of pollutants, vermifiltration could also be utilized as a mode of treatment for diffuse source of wastewater (Wang et al. 2011). As the conventional mode of wastewater treatment (sewage treatment plant and effluent treatment plant) requires high investment, it becomes highly pertinent to employ an economical and manageable treatment plant (Tomar and Suthar 2011). Treatment by vermifiltration could also be adopted at a micro level by the local communities. The process could provide an efficient mode of treatment wherever there is a lack of proper and adequate treatment facilities. To make the process more efficient, there is a future scope of study on earthworm-microorganism interaction in the vermifiltration process (Liu et al. 2013). The treated wastewater obtained after vermifiltration is reported to have its utility in gardening, irrigation, and other reuse purposes (Singh et al. 2017). However, it is yet to be seen that how the treated wastewater arising from vermifiltration could affect the soil fertility. The acceptability of the vermifiltration technology coupled with phytoremediation thus depends on the extent of treatment achieved in the process and field studies on the agricultural crops. The future prospect of vermifiltration may also involve extension of the technology in treatment of wastewater to remove persistent organic pollutants, specific dyes, heavy metals, cations, anions, etc. as an alternative or in combination with other conventional modes of treatment. A low expenditure and minimal operation cost of the process is expected in the process. Future perspectives of this technology will thus depend to a large extent on the cost-benefit analysis of the process. The process could be made more cost efficient by incorporating the plants in combination with the vermifiltration.

3.6 Conclusion

The conventional mode of treatment for municipal wastewater and industrial wastewater requires construction of sewage and effluent treatment plants. These treatment plants require huge expenditure and availability of large land area. The operation cost of these treatment plants is high. Hence, it becomes pertinent to use ecologically safer and economically small-scale treatment plants. The process of vermifiltration coupled with phytoremediation could prove to be an efficient and easy-to-adapt technology that could be adopted at a community level. The best combination of selected earthworm and plant species could be explored for an efficient treatment for a given wastewater.

This manuscript describes the basic mechanism of vermifiltration, pathogen removal, treatment of urban/domestic wastewater, dairy wastewater, and swine wastewater. It is concluded that vermifilteration process is efficient in reduction of BOD, COD, as well as suspended solids from municipal and industrial wastewater. Based on the literature published, it is now realized that a vermifilter has enormous potential to become a reliable treatment technology for wastewater generated from various operations, especially for countries facing serious challenges such as lack of investment and technical labor.

References

Adugna AT, Andrianisa HA, Konate Y, Ndiaye A, Maiga AH (2014) Greywater treatment by vermifiltration for sub-Saharan urban poor. J Water Sanit Hyg De 4(4):625–632

Aira M, Domínguez J (2009) Microbial and nutrient stabilization of two animal manures after the transit through the gut of the earthworm *Eisenia fetida* (Savigny, 1826). J Hazard Mater 161 (2–3):1234–1238

Aira M, Monroy F, Domínguez J (2007) Earthworms strongly modify microbial biomass and activity triggering enzymatic activities during vermicomposting independently of the application rates of pig slurry. Sci Total Environ 385(1–3):252–261

Arora S, Kazmi AA (2015) The effect of seasonal temperature on pathogen removal efficacy of vermifilter for wastewater treatment. Water Res 74:88–99

Arora S, Rajpal A, Kumar T, Bhargava R, Kazmi AA (2014) Pathogen removal during wastewater treatment by vermifiltration. Environ Technol 35(19):2493–2499

Arora S, Rajpal A, Kazmi AA (2016) Antimicrobial activity of bacterial community for removal of pathogens during vermifiltration. J Environ Eng 142(5):04016012

Bajsa O, Nair J, Mathew K, Ho GE (2004) Vermiculture as a tool for domestic wastewater management. Water Sci Technol 48(11–12):125–132

Bezbaruah AN, Zhang TC (2005) Quantification of oxygen release by bulrush (*Scirpus validus*) roots in a constructed treatment wetland. Biotechnol Bioeng 89(3):308–318

Bharambe G (2006) Vermifiltration of wastewater from food processing industries (brewery and milk dairy) in Brisbane. In: 20 CP project submitted for the partial fulfillment of the degree of Master in Environmental Engineering. School of Environmental Engineering, Griffith University, Brisbane

Bhawalkar US (1995) Vermiculture ecotechnology. Bhawalkar Earthworm Research Institute, Pune, p 330

Bhise H, Anaokar G (2015) Design and suitability of modular vermifilter for domestic sewage treatment. Int J 44(4):44–51

Bobade AP, Ansari KS (2016) The use of Vermifiltration in wastewater treatment: a review. J Civil Eng Environ Technol 3(2):164–169

Borin M, Politeo M, De Stefani G (2013) Performance of a hybrid constructed wetland treating piggery wastewater. Ecol Eng 51:229–236

Brix H, Schierup HH (1990) Soil oxygenation in constructed reed beds: the role of macrophyte and soil-atmosphere interface oxygen transport. In: Constructed wetlands in water pollution control. Oxford, Pergamon, pp 53–66

Caselles-Osorio A, Puigagut J, Segú E, Vaello N, Granés F, García D, García J (2007) Solids accumulation in six full-scale subsurface flow constructed wetlands. Water Res 41 (6):1388–1398

Chen Z, Hu S, Hu C, Huang L, Liu H, Vymazal J (2016) Preliminary investigation on the effect of earthworm and vegetation for sludge treatment in sludge treatment reed beds system. Environ Sci Pollut Res 23(12):11957–11963

Choudhary MP, Medok P (2017) Vermifiltration: an efficient method of wastewater treatment. Inte J Modern Trends Enging Res 4(10):189–195

Das DC, Joseph M, Varghese D (2015) Integrated microbial-vermifiltration technique for Ayurvedic industrial effluents. Int J Eng Res Gen Sci 3(4):338–346

Dash MC (1978) Role of earthworms in the decomposer system. In: Glimpses of ecology. India International Scientific Publication, New Delhi, pp 399–406

Dhadse S, Satyanarayan S, Chaudhari PR, Wate SR (2010) Vermifilters: a tool for aerobic biological treatment of herbal pharmaceutical wastewater. Water Sci Technol 61(9):2375–2380

Domínguez J (2004) 20 state-of-the-art and new perspectives on vermicomposting research. In: Earthworm ecology. CRC Press, Boca Raton, pp 401–424

Edwards CA, Arancon NQ (2004) The use of earthworms in the breakdown of organic wastes to produce vermicomposts and animal feed protein. Earthworm Ecol 2:345–380

Edwards CA, Fletcher KE (1988) Interactions between earthworms and microorganisms in organic-matter breakdown. Agric Ecosyst Environ 24(1–3):235–247

Ellis JR, McCalla TM (1978) Fate of pathogens in soils receiving animal wastes—a review. Trans ASAE 21(2):309–0313

Garkal DJ, Mapara JV, Prabhune M (2015) Domestic waste water treatment by bio-filtration: a case StudY. Journal Geogr Geol 4(1):799–811

Ghatnekar SD, Kavian MF, Sharma SM, Ghatnekar SS, Ghatnekar GS, Ghatnekar AV (2010) Application of vermi-filter-based effluent treatment plant (pilot scale) for biomanagement of liquid effluents from the gelatine industry. Dynam Soil Dynam Plant 4(1):83–88

Gogoi A, Biswas S, Bora J, Bhattacharya SS, Kumar M (2015) Effect of vermicomposting on copper and zinc removal in activated sludge with special emphasis on temporal variation. Ecohydrol Hydrobiol 15(2):101–107

Goswami L, Sarkar S, Mukherjee S, Das S, Barman S, Raul P et al (2014) Vermicomposting of tea factory coal ash: metal accumulation and metallothionein response in *Eisenia fetida* (Savigny) and *Lampito mauritii* (Kinberg). Bioresour Technol 166:96–102
Goswami L, Pratihar S, Dasgupta S, Bhattacharyya P, Mudoi P, Bora J et al (2016) Exploring metal detoxification and accumulation potential during vermicomposting of tea factory coal ash: sequential extraction and fluorescence probe analysis. Sci Rep 6:30402
Gupta H (2015) A review on effectiveness of earthworms for treatment of wastewater. Int J Eng Dev 3(3):1–4
Hamdani A, Mountadar M, Assobhei O (2005) Comparative study of the efficacy of three coagulants in treating dairy factory waste water. Int J Dairy Technol 58(2):83–88
Jatin BP (2018) Wastewater treatment by vermifiltration: a review. Int J Latest Technol Eng Manag Appl Sci 7:2278–2540
Jeevitha P, Kiran BM, Manohara B, Sham Sundar KM, Nagarajappa DP (2016) Study on application and validation of vermifiltration in diary effluent treatment. Int J Innovat Res Sci Eng Technol 5(8):14901–14906
Jiang L, Liu Y, Hu X, Zeng G, Wang H, Zhou L et al (2016) The use of microbial-earthworm ecofilters for wastewater treatment with special attention to influencing factors in performance: a review. Bioresour Technol 200:999–1007
Komarowski S (2001) Vermiculture for sewage and water treatment sludges. Water-Melbourne Then Artarmon 28(5):39–43
Konnerup D, Koottatep T, Brix H (2009) Treatment of domestic wastewater in tropical, subsurface flow constructed wetlands planted with Canna and Heliconia. Ecol Eng 35(2):248–257
Lai E, Hess M, Mitloehner F (2018) Profiling of the microbiome associated with nitrogen removal during vermifiltration of wastewater from a commercial dairy. Front Microbiol 9:1964
Lakshmi C, Ranjitha J, Vijayalakshmi S (2014) Waste water treatment using vermifiltration technique at institutional level. Int J Adv Sci Tech Res 1(4):581–590
Li YS, Robin P, Cluzeau D, Bouché M, Qiu JP, Laplanche A et al (2008) Vermifiltration as a stage in reuse of swine wastewater: monitoring methodology on an experimental farm. Ecol Eng 32 (4):301–309
Li YS, Xiao YQ, Qiu JP, Dai YQ, Robin P (2009) Continuous village sewage treatment by vermifiltration and activated sludge process. Water Sci Technol 60(11):3001–3010
Lin YF, Jing SR, Wang TW, Lee DY (2002) Effects of macrophytes and external carbon sources on nitrate removal from groundwater in constructed wetlands. Environ Pollut 119(3):413–420
Liu J, Lu Z, Yang J, Xing M, Yu F, Guo M (2012) Effect of earthworms on the performance and microbial communities of excess sludge treatment process in vermifilter. Bioresour Technol 117:214–221
Liu J, Lu Z, Zhang J, Xing M, Yang J (2013) Phylogenetic characterization of microbial communities in a full-scale vermifilter treating rural domestic sewage. Ecol Eng 61:100–109
Luth PR, Germain P, Lecomte M, Landrain B, Li Y, Cluzeau D (2011) Earthworm effects on gaseous emissions during vermifiltration of pig fresh slurry. Bioresour Technol 102 (4):3679–3686
Manyuchi L, Kadzungura L, Boka S (2013a) Vermifiltration of sewage wastewater for potential use in irrigation purposes using *eisenia fetida* earthworms. World Acad Sci Eng Technol 78:538–542
Manyuchi MM, Kadzungura L, Phiri A, Muredzi P (2013b) Effect of vermicompost, vermiwash and application time on Zea mays growth. Int J Sci Eng Technol 2(7):638–641
Manyuchi MM, Mbohwa C, Muzenda E (2018) Biological treatment of distillery wastewater by application of the vermifiltration technology. S Afr J Chem Eng 25:74–78
Marshall FM, Holden J, Ghose C, Chisala B, Kapungwe E, Volk J et al (2007) Contaminated irrigation water and food safety for the urban and peri-urban poor: appropriate measures for monitoring and control from field research in India and Zambia. Incpetion Report DFID Enkar, 8160

Misal N, Mohite NA (2017) Community wastewater treatment by using vermifiltration technique. Int J Eng Res Technol 10(1):363–365

Natarajan N, Kannadasan N, Krishnamoorthy R (2015) Effect of earthworms on dairy wastewater treatment through vermifiltration. Int J Nano Corr Sci Eng 2(5):91–98

Nie E, Wang D, Yang M, Luo X, Fang C, Yang X et al (2015) Tower bio-vermifilter system for rural wastewater treatment: bench-scale, pilot-scale, and engineering applications. Int J Environ Sci Technol 12(3):1053–1064

Nokes RL, Gerba CP, Karpiscak MM (2003) Microbial water quality improvement by small scale on-site subsurface wetland treatment. J Environ Sci Health A 38(9):1849–1855

Nuengjamnong C, Chiarawatchai N, Polprasert C, Otterpohl R (2011) Treating swine wastewater by integrating earthworms into constructed wetlands. J Environ Sci Health A 46(7):800–804

Patel JB (2018) Wastewater treatment by vermifiltration: a review. Int J Latest Technol Eng Manag Appl Sci 7(1):186–190

Rajasulochana P, Preethy V (2016) Comparison on efficiency of various techniques in treatment of waste and sewage water–a comprehensive review. Res Effic Technol 2(4):175–184

Reddy P, Vinuprakash KC, Arun S (2018) Treatment of domestic wastewater using vermi-biofiltration system with and without wetland plants. Int J Civil Eng Technol 9:412–423

Rodgers M, Xiao LW, Mulqueen J (2006) Synthetic dairy wastewater treatment using a new horizontal-flow biofilm reactor. J Environ Sci Health A 41(5):751–761

Sahariah B, Goswami L, Kim KH, Bhattacharyya P, Bhattacharya SS (2015) Metal remediation and biodegradation potential of earthworm species on municipal solid waste: a parallel analysis between *Metaphire posthuma* and *Eisenia fetida*. Bioresour Technol 180:230–236

Samal K, Dash RR, Bhunia P (2017a) Performance assessment of a *Canna indica* assisted vermifilter for synthetic dairy wastewater treatment. Process Saf Environ Prot 111:363–374

Samal K, Dash RR, Bhunia P (2017b) Treatment of wastewater by vermifiltration integrated with macrophyte filter: a review. J Environ Chem Eng 5(3):2274–2289

Samal K, Dash RR, Bhunia P (2018a) A comparative study of macrophytes influence on performance of hybrid vermifilter for dairy wastewater treatment. Journal of Environmental Chemical Engineering g 6(4):4714–4726

Samal K, Dash RR, Bhunia P (2018b) Effect of hydraulic loading rate and pollutants degradation kinetics in two stage hybrid macrophyte assisted vermifiltration system. Biochem Eng J 132:47–59

Singh J, Kaur A (2015) Vermicompost as a strong buffer and natural adsorbent for reducing transition metals, BOD, COD from industrial effluent. Ecol Eng 74:13–19

Singh NK, Kazmi AA, Starkl M (2015) A review on full-scale decentralized wastewater treatment systems: techno-economical approach. Water Sci Technol 71(4):468–478

Singh R, Bhunia P, Dash RR (2017) A mechanistic review on vermifiltration of wastewater: design, operation and performance. J Environ Manag 197:656–672

Sinha RK, Bharambe G, Bapat P (2007) Removal of high BOD and COD loadings of primary liquid waste products from dairy industry by vermifiltration technology using earthworms. Indian Journal of Environmental Protection 27(6):486

Sinha RK, Bharambe G, Chaudhari U (2008) Sewage treatment by vermifiltration with synchronous treatment of sludge by earthworms: a low-cost sustainable technology over conventional systems with potential for decentralization. Environmentalist 28(4):409–420

Sinha RK, Agarwal S, Chauhan K, Chandran V, Soni BK (2010a) Vermiculture technology: reviving the dreams of sir Charles Darwin for scientific use of earthworms in sustainable development programs. Technol Invest 1(03):155

Sinha RK, Herat S, Bharambe G, Brahambhatt A (2010b) Vermistabilization of sewage sludge (biosolids) by earthworms: converting a potential biohazard destined for landfill disposal into a pathogen-free, nutritive and safe biofertilizer for farms. Waste Manag Res 28(10):872–881

Sinha RK, Herat S, Valani D, Chauhan K (2010c) Earthworms—the environmental engineers: review of vermiculture technologies for environmental management and resource development. Int J Glob Environ Issues 10(3–4):265–292

Sinha RK, Chandran V, Soni BK, Patel U, Ghosh A (2012) Earthworms: nature's chemical managers and detoxifying agents in the environment: an innovative study on treatment of toxic wastewaters from the petroleum industry by vermifiltration technology. Environmentalist 32(4):445–452

Sun G, Gray KR, Biddlestone AJ (1999) Treatment of agricultural wastewater in a pilot-scale tidal flow reed bed system. Environ Technol 20(2):233–223

Suthar S, Singh S (2008) Comparison of some novel polyculture and traditional monoculture vermicomposting reactors to decompose organic wastes. Ecol Eng 33(3–4):210–219

Telang S, Patel H (2015) Vermi-biofiltration-a low cost treatment for dairy wastewater. Int J Sci Res 4(7):595–599

Tomar P, Suthar S (2011) Urban wastewater treatment using vermi-biofiltration system. Desalination 282:95–103

Tunçsiper B, Ayaz SÇ, Akça L (2012) Coliform bacteria removal from septic wastewater in a pilot-scale combined constructed wetland system. Environ Eng Manag J 11(10):1873–1879

Vymazal J (2005) Horizontal sub-surface flow and hybrid constructed wetlands systems for wastewater treatment. Ecol Eng 25(5):478–490

Wang DB, Zhang ZY, Li XM, Zheng W, Yang Q, Ding Y et al (2010a) A full-scale treatment of freeway toll-gate domestic sewage using ecology filter integrated constructed rapid infiltration. Ecol Eng 36(6):827–831

Wang S, Yang J, Lou SJ (2010b) Wastewater treatment performance of a vermifilter enhancement by a converter slag–coal cinder filter. Ecol Eng 36(4):489–494

Wang L, Guo F, Zheng Z, Luo X, Zhang J (2011) Enhancement of rural domestic sewage treatment performance, and assessment of microbial community diversity and structure using tower vermifiltration. Bioresour Technol 102(20):9462–9470

Wei Y, Van Houten RT, Borger AR, Eikelboom DH, Fan Y (2003) Minimization of excess sludge production for biological wastewater treatment. Water Res 37(18):4453–4467

Xing M, Li X, Yang J (2010) Treatment performance of small-scale vermifilter for domestic wastewater and its relationship to earthworm growth, reproduction and enzymatic activity. Afr J Biotechnol 9(44):7513–7520

Xing M, Wang Y, Liu J, Yu F (2011) A comparative study of synchronous treatment of sewage and sludge by two vermifiltrations using an epigeic earthworm *Eisenia fetida*. J Hazard Mater 185 (2–3):881–888

Xing M, Li X, Yang J, Lv B, Lu Y (2012) Performance and mechanism of vermifiltration system for liquid-state sewage sludge treatment using molecular and stable isotopic techniques. Chem Eng J 197:143–150

Xu D, Li Y, Howard A (2013a) Influence of earthworm *Eisenia fetida* on removal efficiency of N and P in vertical flow constructed wetland. Environ Sci Pollut Res 20(9):5922–5929

Xu D, Li Y, Howard A, Guan Y (2013b) Effect of earthworm *Eisenia fetida* and wetland plants on nitrification and denitrification potentials in vertical flow constructed wetland. Chemosphere 92 (2):201–206

Ye J, Wang L, Li D, Han W, Ye C (2012) Vertical oxygen distribution trend and oxygen source analysis for vertical-flow constructed wetlands treating domestic wastewater. Ecol Eng 41:8–12

Zhao L, Wang Y, Yang J, Xing M, Li X, Yi D, Deng D (2010) Earthworm–microorganism interactions: a strategy to stabilize domestic wastewater sludge. Water Res 44(8):2572–2582

Part II
Wastewater Sludge Alone

Chapter 4
Recycling of Municipal Sludge by Vermicomposting

Kui Huang, Hui Xia, Fusheng Li, and Sartaj Ahmad Bhat

Abstract Rapid development of wastewater treatment technique has continuously changed the lifestyle of people and promoted the living standard in the world. Municipal sludge as a by-product of wastewater treatment plant is often dewatered by physicochemical methods to reduce the volume of sludge before its final treatment. The dewatered sludge not only contains high organic matter and water content but also has lots of chemical and biological pollutants that are difficult to be treated with environmental-friendly methods. Vermicomposting is a biochemical approach for converting the sludge into high-valued organic microbial fertilizer by the joint action of earthworms and microorganisms. It has been considered as a sustainable and competitive technology for recycling sludge, because it is low cost, easy to handle, and a perfect product. As a green technology, plenty of vermicomposting practices for treating sludge have sprung up in recent studies. Hence, in this chapter, the methods for municipal sludge treatment are compared, and the mechanisms and operation conditions of vermicomposting for sludge are presented.

Keywords Municipal sludge · Earthworms · Sludge recycling · Vermicomposting

4.1 Introduction

Large amounts of wastewater treatment plants in megacity purify the water environment and thus promote the living standard of people in the world. Excess sludge is an essential by-product from the primary tank and secondary sediment tank in wastewater treatment plant with activated sludge method. After concentrating, the excess

K. Huang (✉) · H. Xia
School of Environmental and Municipal Engineering, Lanzhou Jiaotong University, Lanzhou, China
e-mail: huangkui@mail.lzjtu.cn

F. Li · S. A. Bhat
River Basin Research Center, Gifu University, Gifu, Japan

S. A. Bhat et al. (eds.), *Earthworm Assisted Remediation of Effluents and Wastes*, https://doi.org/10.1007/978-981-15-4522-1_4

sludge is converted into concentrated sludge and then submitted to be dewatered. After sludge dewatering, the resulted sludge also needs the later disposal processes, such as incineration, thermal dry, landfill, composting, vermicomposting, and anaerobic digestion (Yang et al. 2015; Zhang et al. 2017). However, the treatment and disposal of dewatered sludge from wastewater treatment plant have always been a headache in the world, due to the large amounts and unstable organic component in sludge (Yang et al. 2015; Christensen et al. 2015; Qu et al. 2019). Given that lots of nutrients and elements exist in dewatered sludge, the recycling of dewatered sludge is considered as a sustainable approach for resolving this big problem.

Vermicomposting is a biochemical approach for converting the sludge into high-valued organic microbial fertilizer by the joint action of earthworms and microorganisms (Huang et al. 2018). Compared to composting, vermicomposting is low cost and easy to handle for recycling of sludge. More importantly, vermicomposting product displays a much more nutrient content and diverse microbial community beneficial for soil improvement or plant growth (Bhat et al. 2018; Huang et al. 2018). As a consequence, vermicompost is deemed as a gold microbial fertilizer, and vermicomposting is becoming rapidly popular all over the world. Until now, plenty of studies have reported that the municipal sludge could be vermicomposted by earthworms (Gupta and Garg 2008; Suthar and Singh 2008; Domínguez-Crespo et al. 2012; Ludibeth et al. 2012; Reddy et al. 2012; Lv et al. 2015; Das et al. 2015; Xie et al. 2016; Zhao et al. 2018; Lv et al. 2018a, b; Bhat et al. 2018; Huang et al. 2018). However, among these previous publications, several different crafts of vermicomposting for dewatered sludge were carried out, and different earthworms' species were selected, which causes many controversies in vermicomposting of sludge. Thus, it is necessary to summarize these vermicomposting methods to better improve this green technology. In addition, it has been well understood that earthworms are the leaders of microorganisms in vermicomposting, which can significantly change the number, activity, and population of microorganisms directly or indirectly and thus regulate the path and process of organic degradation (Domínguez et al. 2010). However, the detailed mechanism regarding earthworms and microorganisms in vermicomposting still needs to be elaborated.

In a word, the objectives of this chapter were to compare the current methods for municipal sludge treatment, to introduce the different crafts and operation conditions of vermicomposting for dewatered sludge, and to illustrate the mechanisms of vermicomposting for sludge.

4.2 Sewage Sludge Treatment and Disposal

4.2.1 Characteristics and Amounts of Sewage Sludge

Sewage sludge is a solid/semisolid sediment produced by primary setting tank and secondary setting tank during the process of sewage treatment. The composition of sludge is complex, which is generally regarded as a polymer composed of zoogloeal

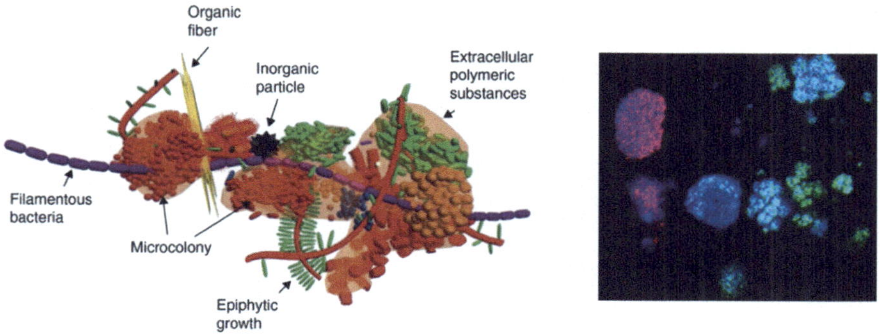

Fig. 4.1 Structural characteristics of sludge

formed by a variety of microorganisms and its adsorbed organic and inorganic substances, as given in Fig. 4.1 (Christensen et al. 2015). The sludge not only has high content of organic matter but also enriches protein, humus, nitrogen, phosphorus, and potassium and other nutrients and trace elements necessary for plant growth (Zhang et al. 2017). According to a survey (Yang et al. 2015), the average contents of organic matter, nitrogen, phosphorus, and potassium in sludge were 41.35%, 3.02%, 1.57%, and 0.69%, respectively, which exceeded the Chinese nutrient standard of compost. But at the same time, the sludge also contains parasitic eggs, pathogenic microorganisms, toxic and harmful heavy metals, and a large number of nondegradable substances (Yang et al. 2015; Zhang et al. 2017; Qu et al. 2019). These characteristics strongly affect the subsequent treatment and disposal of sludge.

With the development of the ecological civilization, the number of municipal sewage treatment plants in China has reached to 3781 in the end of 2017, and the treatment capacity of the sewage is over 186.1 million m^3/d (Statistical Yearbook of Urban and Rural Construction in China 2017). In general, the quality of the sludge accounts for 0.05–0.1% of the treated water (80% moisture content). It is estimated that the production of municipal sludge (including domestic and industrial sludge) in China is more than 40 million tons a year. Therefore, how to dispose the large quantity and complex sludge safely and effectively has become an important problem in the environmental protection industry in China.

4.2.2 Methods of Sludge Treatment and Disposal

The principle for sludge treatment and disposal refers to reduction, stabilization, and innocuity in the procession. Sludge treatment is a part of sewage treatment, including concentration, anaerobic digestion, dehydration, and other processes, in which concentration and dehydration are the core of sludge reduction. Sludge disposal is considered as the final process after sludge dewatering. The disposal methods are mainly comprised of sanitary landfill, aerobic composting, dry incineration, land use

Table 4.1 Municipal sludge disposal methods

Methods	Principle	Way out	Advantage	Shortcoming
Sanitary landfill	Transportation to landfill	Innocent treatment	Simple, low cost, adaptability	Foul smell, leachate
Aerobic composting	Oxidization and decomposition with microorganisms	Soil improvement or fertilizer	Simple operation, decentralized processing	Less reduction, large floor area, secondary pollution
Natural drying	Solar energy	Soil improvement or fertilizer	Low cost, simple and stable operation	Large floor area, high cost, biogas
Dry incineration	Direct incineration	Building materials, landfill	Complete reduction	High cost, pretreatment, complex operation and maintenance, large floor area, secondary pollution
Land use	Use in land	Soil improvement or fertilizer	Low cost, simple	Need of national agricultural standards, secondary pollution
Building materials use	Making building materials	Brick, concrete	Low cost and wide application	Low cost, poor quality

and building materials use, etc. (as given in Table 4.1). In recent years, some new sludge treatment and disposal technology and process have developed rapidly, and many of them have been applied in practice. However, due to limitation of management level and high-energy consumption problem, there is still no effective process route that can lead industries to have proper sludge treatment and disposal (Yang et al. 2015; Zhang et al. 2017; Qu et al. 2019).

According to statistics, the sludge disposal in China is dominated by sanitary landfill, accounting for 65%, followed by aerobic composting, natural drying, dry incineration, and other methods, accounting for 15%, 6%, 3%, and 11%, respectively (Yang et al. 2015). Agricultural use is considered to be a sustainable approach that is in line with the disposal of resources in China. However, a large number of pathogenic bacteria, toxic organic compounds, and heavy metals exist in sludge; thus, meeting the Chinese Standard of Agricultural Sludge Pollutant Control is difficult for its products (GB 4284-2018). Hence, the current sludge products can only be utilized in landscape or municipal greening, which lowers the agricultural value of sludge products. Therefore, how to detoxify sludge is a problem that hinders the wider agricultural use of its products.

4.3 Mechanism of Vermicomposting

Earthworm is an invertebrate living in soil, belonging to Oligochaeta of Annelida. As early as Darwin period, earthworm was found to have an important role in degrading forest dead leaves and improving soil (Graff 1983). After a hundred of years, it has been growingly known that earthworms contributed to the material circulation and energy transfer in the drilosphere through intestinal digestion, casting, mucus secretion, and burrowing (Brown 1995; Edwards and Bohlen 1996; Domínguez et al. 2010; Huang and Xia 2018). Vermicomposting is a kind of resource utilization technology for the degradation of biological organic waste, which refers to the ecological function of earthworms and microorganisms inhabited in the surround of earthworms (Huang et al. 2018). In this process, earthworms can convert soluble microbial products and aromatic compounds of organic matter into humus such as humic acid and fulvic acid. Meanwhile, the final product is rich in active nitrogen, phosphorus, potassium, and several beneficial microbial groups (Huang et al. 2017). Therefore, vermicompost is regarded as the gold of organic fertilizer.

Accordingly, earthworms are the leaders of microorganisms in the whole process of vermicomposting, which can significantly change the number, activity, and population of microorganisms directly or indirectly and thus regulate the path and process of organic degradation (Domínguez et al. 2010). As given in Fig. 4.2, the activities of earthworms in vermicomposting can be divided into three ways: feeding and digestion, mucus and casting secretion, and biological disturbance. In the feeding and digestion phase, the organic matrix is firstly preyed by earthworms

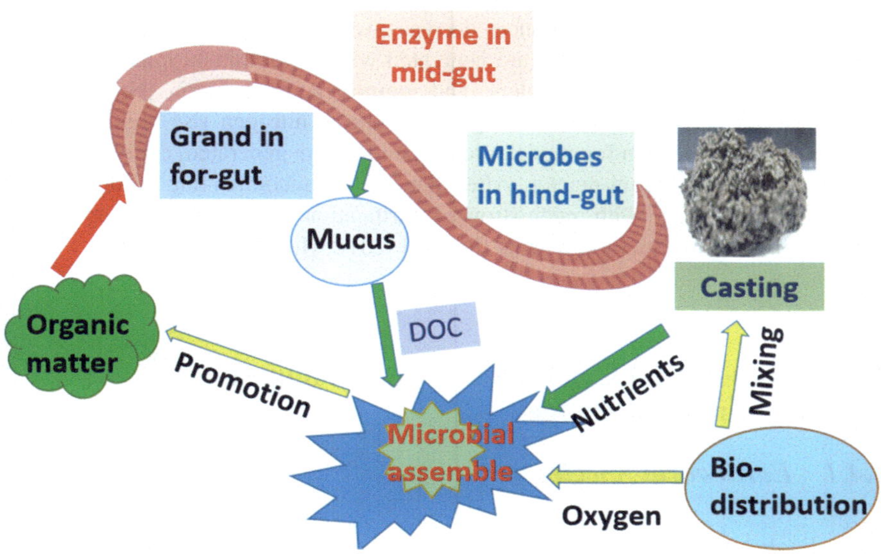

Fig. 4.2 Mechanism of vermicomposting

into their intestinal tract through passage of the grind by gizzard in the foregut, digestion by digestive enzymes in the midgut, and metabolism by microorganisms in the hindgut. Finally, it is crushed by the anus and then discharged in the form of granules (Curry and Schmidt 2007). This gut phase is also a critical factor affecting the performance of vermicomposting products. The resulting casting is a microbial agglomerate containing a granular surface with diverse functional microbial group and chemical mucus from earthworm intestines. Once this agglomerate adheres to the organic matter that has not been digested by the intestinal tract, it leads to the new degradation and transformation of organic matter (commonly known as fecal digestion phase) (Domínguez et al. 2010). Besides, the drilling of earthworms in vermicomposting can not only provide an aerobic environment but also disturb the microbial agglomerate, which continuously flips them into all spaces of vermicomposting, thus accelerating the degradation of organic matter (Meysman et al. 2006). Consequently, these three ways of earthworms may directly or indirectly affect the degradation and transformation of organic matter by regulating the involved microorganisms in vermicomposting system.

In addition, sludge vermicomposting can be divided into four stages: pretreatment stage, earthworm degradation stage, ammonification/nitrification stage, and decomposition stage (Fu et al. 2016). The methods of sludge pretreatment mainly include air drying, airing, adding assistant materials, creating pellet, and so on. These methods aim to remove excessive moisture content and ammonia from sludge and to increase the carbon/nitrogen ratio or oxygen of substrate to provide a better condition for earthworm's survival. In the degradation stage, earthworm and microorganism corporately degrade organic matter of sludge. In this process, the content of sludge organic matter is decreased gradually and the odor of sludge will disappear quickly (Fu et al. 2016). In the ammonification/nitrification stage, the proteinoid-like and microbial product-like substances decrease, while humic acid- and fulvic acid-like substances in organic matter are generated with the dominant microorganisms such as actinomycetes, azotobacters, and ammoxidation bacteria (Huang et al. 2018). Moreover, the ammonia nitrogen and nitrate nitrogen rise rapidly in this process. Therefore, when functional microorganisms are generated, earthworms can be taken out of the vermi-reactor. After that, the vermicomposting products enter into the maturity stage with much stronger nitrification and humification as well as widener diversity of microorganisms (Fu et al. 2016; Huang et al. 2018).

4.4 Technological Conditions of Vermicomposting for Municipal Sludge

4.4.1 Earthworm Species

Accordingly, more than 6000 kinds of earthworms are widely distributed in the world. Based on the ecological function, it can be divided into three types: epigeic earthworms, anecic earthworms, and endogeic earthworms (Domínguez 2004). Among them, epigeic earthworms like eating organic matter, showing a strong environmental adaptability and fecundity. Thus, they are considered to be suitable for sludge treatment. The main species used in vermicomposting of municipal sludge include *Eudrilus eugeniae*, *Eisenia fetida*, *Bimastus parvus*, *Lumbricus rubellus*, *Eisenia andrei*, and *Dendrobaena veneta*, as presented in Table 4.2. At present, *Eisenia fetida* is a common species used in vermicomposting of sludge. In general, the adult of *E. fetida* has the following features: is 60–160 mm long and 3–5 mm wide, has 80–110 segments, and is 0.5–1.2 g. However, it shows a longer length and higher growth and reproduction rate in the vermicomposting system, due to its rich nutrients of sludge. In our study, the maximum weight of earthworm in sludge can reach up to 2.3 g with a length of 155 mm. Given that earthworms are a valuable source of traditional Chinese medicine and aquaculture, vermiculture in sludge is also important income during vermicomposting.

4.4.2 Temperature

The environmental temperature can directly affect the metabolism, growth, reproduction, and activity of earthworm and microorganism, thus affecting their ability in degradation of organic matter. As shown in Table 4.2, the temperature set in vermicomposting showed a stronger inconsistency, and most studies selected room temperature (15–30 °C) as vermicomposting temperature, displaying a wide range of temperature for treating sludge (Fu et al. 2016; Huang et al. 2018; Domínguez-Crespo et al. 2012; Cesar et al. 2012; Molina et al. 2013; Lv et al. 2014; Zhang et al. 2015; Babić et al. 2015; Xie et al. 2016; Cui et al. 2018; Ozdemir et al. 2019). Only a few publications clearly reported that 20 °C was the optimal temperature condition for treating activated sludge (Hait and Tare 2012). Although the increase in temperature can accelerate the degradation and transformation rate of organic matter by promoting microbial activity, earthworms are prone to heatstroke and even to death if the temperature exceeds 30 °C. Hence, vermicomposting process is a biological and chemical decomposition of organic wastes under mesophilic condition to produce a stable bioorganic fertilizer.

4.4.3 Organic Matter Content Carbon/Nitrogen Ratio (C/N)

To provide proper nutrition for earthworms and microorganisms during vermicomposting, carbon and nitrogen in the substrates should be presented at the correct ratio (Ndegwa and Thompson 2000). The high ammonia nitrogen content can cause protein poisoning of earthworms, and even death. But lack of nitrogen may lead to a lower growth rate of earthworms in vermicomposting. In addition, vermicomposting substrate with low carbon/nitrogen ratio is better for older

Table 4.2 Vermicomposting operation condition for dewatered sludge

	Temperature (°C)	Pretreatment method	Earthworm species	Inoculation time (d)	Inoculation density	Reference
1	25	Sludge + cow dung	*Eisenia fetida*	56	5 worms/kg	Domínguez-Crespo et al. (2012)
2	10–30	Compost	*Eisenia fetida*	28	0.5–5.0 kg/m^2	Hait and Tare (2012)
3	20	Sludge + soil	*Eisenia andrei*	14	50 worms /kg	Cesar et al. (2012)
4	17	Sludge + cow dung	*Eisenia fetida*	50	62.5 worms /kg	Ludibeth et al. (2012)
5	25	Sludge + cow dung + dry	*Eisenia fetida*	60	30 worms/kg	Xing et al. (2012)
6	25	Sludge + rabbit dung	*Eisenia fetida*	56	30 worms/kg	Molina et al. (2013)
7	27	Dry	*Lumbricus rubellus*	105	10 worms/kg	Azizi et al. (2013)
8	25	Dry	*Eisenia fetida*	21	2.0 kg /m^2	Yang et al. (2014)
9	23	Dry + sawdust + cow dung	*Eisenia fetida*	120	1000 worms/kg	Lv et al. (2014)
10	23	Sludge + cow dung	*Eisenia fetida*	100	12.5 worms/kg	Lv et al. (2015)
11	23	Dry + sawdust + cow dung	*Eisenia fetida*	120	125 worms/kg	Zhang et al. (2015)
12	20	Pellet	*Bimastus parvus*	60	50 worms/kg	Fu et al. (2015a, b)
13	23	Dry + sawdust + dung	*Eisenia fetida*	56	125 worms/kg	Xing et al. (2015)
14	20	Sludge + vermicompost	*Eisenia fetida*	28	70 worms/3 L	Babić et al. (2015)
15	25	Pellet	*Eisenia fetida*	60	25 worms/kg	Fu et al. (2016)
16	20–28	Sludge + cow dung + swine manure	*Eisenia fetida*	60	100 worms/kg	Xie et al. (2016)
17	20–22	Dry	*Eisenia fetida*	31	125–375 worms/kg	Cui et al. (2018)
18	25–26	Sludge + cattle manure + sawdust	*Eisenia fetida*	84	375 worms/kg	Zhao et al. (2018)
19	21–25	Sludge + cattle dung	*Eisenia fetida*	80	13 worms/kg	Lv et al. (2018a, b)
20	18–26	Pellet	*Eisenia fetida*	80	25 worms/kg	Huang et al. (2018)
21	21–23	Sludge + rice straw	*Eisenia fetida*	84	167 worms/kg	Lv et al. (2018a, b)

22	21–25	Sludge + rice straw	*Eisenia fetida*	60	38 worms/kg	Lv et al. (2019)
23	23–27	Pellet	*Eisenia fetida*	60	3 kg/m^2	Ozdemir et al. (2019)
24	22–28	Precompost	*Eisenia fetida*	100	100 worms/kg	Ahadi et al. (2019)

earthworms, whereas high carbon/nitrogen ratio is suitable for the growth of small or young earthworms (Aira et al. 2006). Accordingly, the ratio of carbon to nitrogen of 25:1 is beneficial for the growth of earthworms and microorganisms, and their composting products had a high stability used as fertilizer (Ndegwa and Thompson 2000). Given that the carbon to nitrogen ratio of municipal sludge is often lower than 25:1, many scholars promoted carbon/nitrogen ratio using mixed materials such as straw and rice husk. However, vermicomposting for dewatered sludge without the addition of other carbon source was also completed by previous studies (Fu et al. 2015a, b; Huang et al. 2018). As a result, the optimal carbon to nitrogen rate in initial sludge should be taken into account for vermicomposting of dewatered sludge.

4.4.4 Density and Retention Time of Earthworms

The density and retention time of earthworms are important parameters for vermicomposting of sludge, which are associated with time, cost, and fertilizer performance. Undoubtedly, the optimal conditions for vermicomposting and vermiculture are different (Ndegwa et al. 2000). Based on the survey, the density of earthworms was 0.5–5.0 kg/m^2 or 10–375 worms/kg sludge for vermicomposting (Table 4.2). In addition, the retention time of earthworm ranged from 14 to 120 days in vermicomposting system (Table 4.2). As a consequence, there are different densities and retention times of earthworms in the previous vermicomposting, which may cause the treatment efficiency of sludge and properties of vermicompost are difficult to compare. For this, a uniform standard for optimal condition is highly expected for vermicomposting of sludge. However, due to the variable in the influents of wastewater treatment plant in every day, the optimal density and retention time of earthworm are strongly with the properties of dewatered sludge used in vermicomposting.

4.4.5 Pretreatment Methods

Fresh dewatered sludge easily generates poisonous gas and gives rise to an anaerobic environment where earthworms cannot survive (Gupta and Garg 2008; Suthar 2010; Hait and Tare 2012). Therefore, suitable pretreatment method of dewatered sludge is deemed to be of essence for earthworms. Pretreatment of sludge mainly includes mixing with other materials, dry, pelletized, and vermi-bedding used, as displayed in Table 4.2. The dry method such as sunshine and oven can reduce the water and toxicant gas content in sludge (Xing et al. 2012; Azizi et al. 2013; Cui et al. 2018). The blending bulking materials such as cow dung (Gupta and Garg 2008; Suthar and Singh 2008; Domínguez-Crespo et al. 2012; Ludibeth et al. 2012; Reddy et al. 2012; Lv et al. 2015; Das et al. 2015; Xie et al. 2016; Zhao et al. 2018; Lv et al. 2018a, b), sugarcane trash (Suthar 2010; Biruntha et al. 2019), and composting products (Hait and Tare 2012; Malińska et al. 2016; Babaei et al. 2016; Malińska et al. 2017) are employed for improving the live environment of earthworms and enhancing the activity of microbes by regulating the carbon to nitrogen rate. The pelletization method can increase the specific surface area of dewatered sludge and elevate its aerobic extent, which is used in recent vermicomposting (Fu et al. 2016; Huang et al. 2018). However, these methods are complicated; the effectively simplified pretreatment methods are desired to be explored in the future.

4.5 Conclusions

Vermicomposting is a biochemical degradation process of organic materials involving the interactions of earthworms and microbes, which is considered one of the environmentally friendly and sustainable approaches for recycling municipal sludge. In vermicomposting process, the feeding and digestion, mucus and casting secretion, and biological disturbance of earthworms directly affect the decomposition process of organic matter in sludge. Comparatively, the species, density, and retention time of earthworms along with temperature, carbon to nitrogen rate, and pretreatment method strongly affect the vermicomposting performance. However, the studies of optimal condition for vermicomposting of municipal sludge are still needed in the future.

References

Ahadi N, Sharifi Z, Hossaini SM, Rostami A, Renella G (2019) Remediation of heavy metals and enhancement of fertilizing potential of a sewage sludge by the synergistic interaction of woodlice and earthworms. J Hazard Mater 385:121573

Aira M, Monroy F, Domínguez J (2006) C to N ratio strongly affects population structure of *Eisenia fetida* in vermicomposting systems. Eur J Soil Biol 42:S127–S131

Azizi AB, Lim MPM, Noor ZM, Abdullah N (2013) Vermiremoval of heavy metal in sewage sludge by utilising *Lumbricus rubellus*. Ecotoxicol Environ Saf 90:13–20

Babaei AA, Goudarzi G, Neisi A, Ebrahimi Z, Alavi N (2016) Vermicomposting of cow dung, kitchen waste and sewage sludge with bagasse using *Eisenia fetida*. J Adv Environ Health Res 4 (2):88–94

Babić S, Dragun Z, Klobučar RS, Ivanković D, Bačić N, Fiket Ž et al (2015) Indication of metal homeostasis disturbance in earthworm *Eisenia fetida* after exposure to semi-solid depot sludge. Sci Total Environ 526:127–135

Bhat SA, Singh S, Singh J, Kumar S, Vig AP (2018) Bioremediation and detoxification of industrial wastes by earthworms: vermicompost as powerful crop nutrient in sustainable agriculture. Bioresour Technol 252:172–179

Biruntha M, Karmegam N, Archana J, Selvi BK, Paul JAJ, Balamuralikrishnan B et al (2019) Vermiconversion of biowastes with low-to-high C/N ratio into value added vermicompost. Bioresour Technol 297:122398

Brown GG (1995) How do earthworms affect microfloral and faunal community diversity? Plant Soil 170(1):209–231

Cesar R, Silva M, Colonese J, Bidone E, Egler S, Castilhos Z, Polivanov H (2012) Influence of the properties of tropical soils in the toxicity and bioavailability of heavy metals in sewage sludge-amended lands. Environ Earth Sci 66(8):2281–2292

Christensen ML, Keiding K, Nielsen PH, Jørgensen MK (2015) Dewatering in biological wastewater treatment: a review. Water Res 82:14–24

Cui G, Li F, Li S, Bhat SA, Ishiguro Y, Wei Y et al (2018) Changes of quinolone resistance genes and their relations with microbial profiles during vermicomposting of municipal excess sludge. Sci Total Environ 644:494–502

Curry JP, Schmidt O (2007) The feeding ecology of earthworms–a review. Pedobiologia 50 (6):463–477

Das S, Bora J, Goswami L, Bhattacharyya P, Raul P, Kumar M, Bhattacharya SS (2015) Vermiremediation of water treatment plant sludge employing *Metaphire posthuma*: a soil quality and metal solubility prediction approach. Ecol Eng 81:200–206

Domínguez J (2004) State–of–the–Art and New Perspectives on Vermicomposting Research. Earthworm Ecology

Domínguez J, Aira M, Gómez-Brandón M (2010) Vermicomposting: earthworms enhance the work of microbes. In: Microbes at work. Springer, Berlin, Heidelberg, pp 93–114

Domínguez-Crespo MA, Sánchez-Hernández ZE, Torres-Huerta AM, Negrete MDLLX, Conde-Barajas E, Flores-Vela A (2012) Effect of the heavy metals Cu, Ni, Cd and Zn on the growth and reproduction of epigeic earthworms (*E. fetida*) during the vermistabilization of municipal sewage sludge. Water Air Soil Pollut 223(2):915–931

Edwards CA, Bohlen PJ (1996) Biology and ecology of earthworms, vol 3. Springer Science & Business Media, Berlin

Fu X, Huang K, Chen X, Li F, Cui G (2015a) Feasibility of vermistabilization for fresh pelletized dewatered sludge with earthworms *Bimastus parvus*. Bioresour Technol 175:646–650

Fu X, Huang K, Cui G, Chen X, Li F, Zhang X, Li F (2015b) Dynamics of bacterial and eukaryotic community associated with stability during vermicomposting of pelletized dewatered sludge. Int Biodeterior Biodegradation 104:452–459

Fu X, Cui G, Huang K, Chen X, Li F, Zhang X, Li F (2016) Earthworms facilitate the stabilization of pelletized dewatered sludge through shaping microbial biomass and activity and community. Environ Sci Pollut Res 23(5):4522–4530

Graff O (1983) Darwin on earthworms—the contemporary background and what the critics thought. In: Earthworm ecology. Springer, Dordrecht, pp 5–18

Gupta R, Garg V (2008) Stabilization of primary sewage sludge during vermicomposting. J Hazard Mater 153(3):1023–1030

Hait S, Tare V (2012) Transformation and availability of nutrients and heavy metals during integrated composting–vermicomposting of sewage sludges. Ecotoxicol Environ Saf 79:214–224

Huang K, Xia H (2018) Role of earthworms' mucus in vermicomposting system: biodegradation tests based on humification and microbial activity. Sci Total Environ 610:703–708

Huang K, Xia H, Cui G, Li F (2017) Effects of earthworms on nitrification and ammonia oxidizers in vermicomposting systems for recycling of fruit and vegetable wastes. Sci Total Environ 578:337–345

Huang K, Xia H, Wu Y, Chen J, Cui G, Li F et al (2018) Effects of earthworms on the fate of tetracycline and fluoroquinolone resistance genes of sewage sludge during vermicomposting. Bioresour Technol 259:32–39

Ludibeth SM, Marina IE, Vicenta EM (2012) Vermicomposting of sewage sludge: earthworm population and agronomic advantages. Compost Sci Util 20(1):11–17

Lv B, Xing M, Zhao C, Yang J, Xiang L (2014) Towards understanding the stabilization process in vermicomposting using PARAFAC analysis of fluorescence spectra. Chemosphere 117:216–222

Lv B, Xing M, Yang J, Zhang L (2015) Pyrosequencing reveals bacterial community differences in composting and vermicomposting on the stabilization of mixed sewage sludge and cattle dung. Appl Microbiol Biotechnol 99(24):10703–10712

Lv B, Xing M, Yang J (2018a) Exploring the effects of earthworms on bacterial profiles during vermicomposting process of sewage sludge and cattle dung with high-throughput sequencing. Environ Sci Pollut Res 25(13):12528–12537
Lv B, Zhang D, Cui Y, Yin F (2018b) Effects of C/N ratio and earthworms on greenhouse gas emissions during vermicomposting of sewage sludge. Bioresour Technol 268:408–414
Lv B, Zhang D, Chen Q, Cui Y (2019) Effects of earthworms on nitrogen transformation and the correspond genes (amoA and nirS) in vermicomposting of sewage sludge and rice straw. Bioresour Technol 287:121428
Malińska K, Zabochnicka-Świątek M, Cáceres R, Marfà O (2016) The effect of precomposted sewage sludge mixture amended with biochar on the growth and reproduction of *Eisenia fetida* during laboratory vermicomposting. Ecol Eng 90:35–41
Malińska K, Golańska M, Caceres R, Rorat A, Weisser P, Ślęzak E (2017) Biochar amendment for integrated composting and vermicomposting of sewage sludge–the effect of biochar on the activity of Eisenia fetida and the obtained vermicompost. Bioresour Technol 225:206–214
Meysman FJ, Middelburg JJ, Heip CH (2006) Bioturbation: a fresh look at Darwin's last idea. Trends Ecol Evol 21(12):688–695
Molina MJ, Soriano MD, Ingelmo F, Llinares J (2013) Stabilisation of sewage sludge and vinasse bio-wastes by vermicomposting with rabbit manure using *Eisenia fetida*. Bioresour Technol 137:88–97
Ndegwa PM, Thompson SA (2000) Effects of C–to–N ratio on vermicomposting of biosolids. Bioresour Technol 75(1):7–12
Ndegwa PM, Thompson SA, Das KC (2000) Effects of stocking density and feeding rate on vermicomposting of biosolids. Bioresour Technol 71(1):5–12
Ozdemir S, Dede G, Dede OH, Turp SM (2019) Composting of sewage sludge with mole cricket: stability, maturity and sanitation aspects. Int J Environ Sci Technol 16(5827):5834
Qu J, Wang H, Wang K, Yu G, Ke B, Yu HQ et al (2019) Municipal wastewater treatment in China: development history and future perspectives. Front Environ Sci Eng 13(6):88
Reddy SA, Akila S, Kale RD (2012) Management of secondary sewage sludge by vermicomposting for use as soil amendment. Glob J Biotechnol Biochem 7(1):13–18
Suthar S (2010) Pilot–scale vermireactors for sewage sludge stabilization and metal remediation process: Comparison with small–scale vermireactors. Ecol Eng 36(5):703–712
Suthar S, Singh S (2008) Comparison of some novel polyculture and traditional monoculture vermicomposting reactors to decompose organic wastes. Ecol Eng 33(3-4): 210–219
Xie D, Wu W, Hao X, Jiang D, Li X, Bai L (2016) Vermicomposting of sludge from animal wastewater treatment plant mixed with cow dung or swine manure using *Eisenia fetida*. Environ Sci Pollut Res 23(8):7767–7775
Xing M, Li X, Yang J, Huang Z, Lu Y (2012) Changes in the chemical characteristics of water-extracted organic matter from vermicomposting of sewage sludge and cow dung. J Hazard Mater 205:24–31
Xing M, Lv B, Zhao C, Yang J (2015) Towards understanding the effects of additives on the vermicomposting of sewage sludge. Environ Sci Pollut Res 22(6):4644–4653
Yang J, Lv B, Zhang J, Xing M (2014) Insight into the roles of earthworm in vermicomposting of sewage sludge by determining the water-extracts through chemical and spectroscopic methods. Bioresour Technol 154:94–100
Yang G, Zhang G, Wang H (2015) Current state of sludge production, management, treatment and disposal in China. Water Res 78:60–73
Zhang J, Lv B, Xing M, Yang J (2015) Tracking the composition and transformation of humic and fulvic acids during vermicomposting of sewage sludge by elemental analysis and fluorescence excitation–emission matrix. Waste Manag 39:111–118
Zhang Q, Hu J, Lee DJ, Chang Y, Lee YJ (2017) Sludge treatment: current research trends. Bioresour Technol 243:1159–1172
Zhao C, Wang Y, Wang Y, Wu F, Zhang J, Cui R et al (2018) Insights into the role of earthworms on the optimization of microbial community structure during vermicomposting of sewage sludge by PLFA analysis. Waste Manag 79:700–708

Chapter 5
Influence of Distillery Sludge-Based Vermicompost on the Nutritional Status of *Rapanus sativus* L. (Radish)

Susila Sugumar, Tamilselvi Duraisamy, Selvakumar Muniraj, Ramarajan Selvam, and Vasanthy Muthunarayanan

Abstract The efficiency of the distillery sludge-based vermicompost to enhance the growth of *Raphanus sativus* (radish) was tested with distillery sludge-based compost (DSC) and vermicompost (DSV), vegetable waste vermicompost (VV), and commercially available chemical fertilizer (CF). This study reveals the suitability of distillery sludge-based vermicompost to facilitate the growth of *Raphanus sativus*. An enhancement of the biomass, antioxidants (38.45–52.49%), total phenolics (22.56 mg GAE/g), and flavonoids (32.59–45.69 mg QE/g) was reported in the plants treated with the organic manures. However, reducing sugars (0.61 mg/50 g) was found to be in higher quantity in the plants treated with chemical fertilizer. Significant increase in the antioxidants including phenols and flavonoids along with enhanced DPPH and FRAP activity suggests the suitability of application of organic fertilizers to promote plant growth. Among the treatments offered, distillery sludge-based vermicompost was found to enhance the content of antioxidants (52.49 ± 0.002%), phenols (31.24 ± 0.002 mg GAE/g), flavonoids (45.69 ± 0.002 mg QE/g), mineral content (Ca, Mg, Na, K, P, Cu, Mn), and growth parameters too of *Raphanus sativus*, thus making the eco-friendly management of the distillery sludge possible. Further, there are no significant polymorphisms in the protein band of the DSV-grown radish.

Keywords DSV · CF · Antioxidant · SDS-PAGE

S. Sugumar · T. Duraisamy · S. Muniraj · R. Selvam · V. Muthunarayanan (✉)
Water and Solid Waste Processing Lab, Department of Environmental Biotechnology, Bharathidasan University, Tiruchirappalli, Tamil Nadu, India
e-mail: vasanthy@bdu.ac.in

S. A. Bhat et al. (eds.), *Earthworm Assisted Remediation of Effluents and Wastes*,
https://doi.org/10.1007/978-981-15-4522-1_5

5.1 Introduction

In developing countries like India, distillery industries are one of the major sources of pollution. About 3,50,000 L of effluent molasses, 20,000 L of yeast sludge/day, and 120,000 L of spent malt grain wash were disposed every day (Suthar 2008). Though the distillery effluent contains heavy metals and pathogens, the distillery sludge is reported to be a good source of nutrients too. Hence, the application of such distillery sludge after proper treatment such as composting or vermicomposting would improve soil fertility and facilitate plant growth. Vermicompost of varied wastes were found to improve the plant quality, plant yield, and soil fertility and thereby reduce the need for inorganic fertilizers. Researchers prove that the application of chemical fertilizers affects soil fertility and drastically reduces the antioxidant and micro- and macronutrient levels in plants. Further the application of organic composts were found to improve the antioxidant defense system ensuring sustainability (Carbonaro et al. 2002)

Among various crops, radish is found to be one of the easily cultivable plants, and the literature suggests its nature of indicating pollution (Mathe-Gaspar and Anton 2002). Radish (*Raphanus sativus* L.) is reported to hold high amount of vitamin C, carotene, and nutrient-rich compounds. Reports suggest *Raphanus sativus* L. as a good medicine for cold, flu, fever, cough, respiratory problems and digestive disorders, jaundice, piles, urinary disorder, weight loss and cancer, leukoderma, skin disorders, kidney disorders, respiratory disorders, bronchitis and asthma, acidity, obesity, sore throat, dyspepsia, liver disorder, and gallbladder stones and a good detoxifier.

Having known the toxic nature of various chemical fertilizers, usage of organic fertilizer has increased worldwide so as to conserve the environment. Consequently, many literatures cite that the application of vermicompost improves the imperative macronutrient (N, P_2O_5, K_2O, Ca, and Mg) and micronutrient (Fe, Mn, Zn, and Cu) content of the plants. Hence, the present investigation included the (a) comparison of the growth, yield, and mineral content, (b) antioxidant levels, and (c) protein expression in organic compost and chemical fertilizer-grown radish plants.

5.2 Materials and Methods

5.2.1 Organic Compost

Distillery sludge compost, vermicompost, and vegetable waste vermicompost were prepared using respective wastes collected from the industry situated near Trichy and the canteen of Bharathidasan University, Tiruchirappalli, Tamil Nadu, India.

5.2.2 Chemical Requirements

The following chemicals purchased from Sigma Chemical Co., 1,1-diphenyl-2-picrylhydrazyl (DPPH) radical, butylated hydroxytoluene (BHT), butylated hydroxyanisole (BHA) and methanol solvent (HPLC grade), Folin-Ciocalteu reagent, sodium carbonate (Na_2CO_3), gallic acid, methanol, 5% sodium nitrate, 10% aluminum chloride, 1 M NaOH, quercetin, methanol, DPPH, 1% potassium ferricyanide, 10% TCA, 0.1% ferric chloride, 1% ascorbic acid, 0.2 M dibasic sodium phosphate Na2HPO4 (35.61 g/1 L), 0.2 M monobasic sodium phosphate NaH_2PO_4 (31.21 g/1 L), aluminum chloride, Whatman Nylon Membrane Filter (0.45 μm and 47 mm diameter), 50 mM Tris-HCl (pH 7.5), formic acid, and 0.5 M NaCl, were used in this study.

5.2.3 Field Experiments

The field experiments were carried out at Virudhachalam, Cuddalore (Dt), Tamil Nadu, India. The Seeds of *Raphanus sativus* L. plants were sowed in 25 rows: 10 seeds per trial with an interval of 10 cm, 1.5 m long and 1 m wide. The recommended doses of composts, namely, distillery sludge compost (DSC), vegetable waste vermicompost (VV), distillery sludge vermicompost (DSV) (100 g/sq. ft), and inorganic fertilizers (urea, CF), were added before sowing the seed. A control with no fertilizer addition is also maintained (C). The experimental layout was a randomized complete grid, with three replicates of each setup. Results were subjected to suitable statistical analysis for the calculation of mean, variance, and standard error by employing Prism software.

5.2.4 Growth Parameters

The numbers of leaves were counted and the average mean values were calculated. The root length was also measured using scaling. Weight of fresh and dry leaves, root, and whole plant was also noted.

5.2.5 Sample Preparation

The edible parts of *Raphanus sativus* L. (50 g) were cut into small pieces (<0.5 cm), and it was grounded into fine powder after the addition of liquid nitrogen. The powder was further sieved followed by lyophilization and extraction of the sample with methanol (1:10) using Soxhlet apparatus. Finally, the air-dried sample was stored at −80 °C for further analysis.

5.2.6 Determination of Phenolic and Flavonoid Content

The phenolic compounds were analyzed using Folin-Ciocalteu method (FCM), and the total phenolic content was calculated as

$$\text{Gallic acid equivalent} = T = C \times V/M.$$

The total flavonoid was analyzed using aluminum chloride colorimetric method, and it was calculated as follows:

$$\text{Quercetin equivalent} = T = C \times V/M.$$

5.2.7 Determination of DPPH and FRAP Activity

The DPPH free radical scavenging assay was measured using DPPH free radical test. The FRAP scavenging assay was measured by the method of Yildirim et al. (2001).

5.2.8 High-Performance Liquid Chromatography (HPLC)-Pulsed Amperometric 109 Detectors

Preparation of standards: 1 mg of each standard compounds (ascorbic acid (***ASA***), gallic acid (***GA***), benzoic acid (***BA***), tannic acid (***TA***), quercetin (***QN***), resorcinol (***RL***), catechol (***CAT***)) was dissolved in 1 mL of HPLC grade methanol. It was sonicated and filtered by Whatman Nylon Membrane Filter (0.45 μm and 47 mm diameter) before injecting into the column.

Specification of the HPLC-PAD instrument: Analysis of all the standard samples was performed using 515 HPLC pumps and 2489 UV/VIS detectors of Waters company, USA, having reverse phase water guard Column: Symmetry C18 (5 μm, 4.6∗250 mm) and Hamilton microliter syringe using an injection volume of 20 μL. The data analysis was done using Empower software.

5.2.9 Methods

The prepared plant sample and each standard were injected in HPLC and the following running program was followed:

Wavelength	280 nm
Flow rate	1.0 mL/min
Total run time	25 min
Solvent A	Methanol
Solvent B	1% formic acid in water (Lin et al. 1996)

The gradient program was started with 60% of solvent A and obtaining peak was changed into 35, 10, and 60% of elute A at retention times 5, 10, 15, and 20 min, respectively (Boligon et al. 2012).

5.2.10 Protein Extraction

Total proteins were extracted by adding l ml of 50 mM Tris-HCl (pH 7.5) and 0.5 M NaCl at 4 °C to 30 mg of fresh radish for 60 min. Then it was frozen at −20 °C and thawed three times for 24 h to disrupt the tissue and to liberate the proteins (Miller 1972). It was centrifuged at 10000 rpm for 15 min.

5.2.11 SDS-PAGE Electrophoresis

Protein was extracted through SDS-PAGE (Laemmli 1970). 10 μL of extracted protein was mixed with 2.5 μL of 2-mercaptoethanol and 7.5 μL of 0.002% bromophenol blue in 0.0625 M Tris-HC1 (pH 6.8), containing 10% glycerol and 2% SDS, and was kept into SDS-PAGE gel apparatus (Medox). All protein staining was performed using Coomassie blue (Hames 1998).

5.3 Results and Discussion

5.3.1 Morphological Characteristics of Organically and Inorganically Grown Plant

Table 5.1 shows the physicochemical characteristics and heavy metal content of the soil. Growth parameters denote that the organically grown *Raphanus sativus*

Table 5.1 Characterization of soil

Parameters	Quantity
pH	8.87
EC (μs/mos)	936.39
Alkalinity (mg/L)	70
Phosphorus (%)	0.012
Organic carbon (%)	1.81
Total nitrogen (%)	0.21
Potassium (mg/g)	0.000002
Sodium (mg/g)	0.000003
Total magnesium (mg/g)	0.0013
Calcium (mg/g)	0.000002
Total organic matter	3.120
C/N ratio	8.61
Copper (μg/kg)	8
Cadmium (μg/kg)	0.1
Fe (μg/kg)	8.6
Manganese (μg/kg)	10
Zinc (μg/kg)	150

L. showed higher number of leaves and fruits when compared with the plant grown in inorganic fertilizer. However, differences are not statistically significant and it is shown in Table 5.2. The highest total length of *Raphanus sativus* L. is 13 cm in DSV and DSC. The lowest length of DSV and DSC is 10.1 cm in CF. Plants grown with DSV produced maximum number of leaves (21), and the minimum leaves is 8 in C. Also, the highest leaf length (15 cm) was found in DSV, and the lowest leaf length was noticed in CF (10.75 cm). Similar results were already reported (Lingaiah et al. 1992). Consequently, poultry manure at 1.25 t/ha (50%) + vermicompost at 1.75 t/ha (50%) treated plant has resulted in greater number of leaves. The authors have stated that the increased number of leaves may be due to the macro- and micronutrient availability in the vermicompost (Giraddi 1993). Thanunathan et al.'s (1997) studies state that the application of vermicompost appears to be a very effective amendment in case of onion too. The descending number of leaves and leaf area in Bhumyamalaki (*Phyllanthus amarus* Schum) with the application of poultry manures along with biofertilizers was reported earlier (Chezhiyan et al. 2003).

5.4 Yield Parameters

In this investigation, the root length of organically and inorganically grown *Raphanus sativus* L. was shown in Table 5.2. From the results maximum fresh weight of the whole plant and root was observed to be 182.33 and 87.39 g in DSV-grown plants, and minimum fresh weight noticed was 153.32 and 43.13 g in C, respectively. Treated DSV plant produced maximum dry weight of the root and

Table 5.2 Effect of organic and inorganic fertilizer on the growth parameters of radish plant

Sample name	Whole plant height (cm)	Leaf length (cm)	Leaf width (cm)	Root length (cm)	No. of leaf	Fresh weight of whole plant (g)	Fresh weight of leaf (g)	Fresh weight of root (g)	Total dry weight of whole plant (g)	Dry weight of leaf (g)	Dry weight of root (g)
WOF	27.66	11.33	8.58	10.19	8.16	153.32	85.19	43.13	18	10	8
VV	39.08	11.66	10.08	11	11.86	160.01	105.78	45.25	34	25	9
DSC	34.16	12.25	10.41	13	11.66	160.15	110.08	57.58	34	25	9
DSV	38.58	15	10.91	13	21.33	182.33	115.03	87.39	60	50	10
CF	28.66	10.75	8.83	10.1	10.83	160.5	105.17	54.66	28	20	8

it was 60 g. The minimum dry weight of the root was 18 and 8 g was found in C. Similar results were reported earlier (Pujari et al. 1977; Rojagopal et al. 1979; Parthasarathi 1998. Babalad's (2005) report highlights that the highest dry weight and total dry matter were yielded by applying poultry manure to the chili plant (Yadav and Vijayakumari 2003). Further, the authors have mentioned that the increase in fresh weight of leaves, roots, and whole plant may be due to the higher level of nitrogen from inorganic and biofertilizers which could stimulate amino acid synthesis. This might have resulted in building of complex proteins and promoting the growth of crop (Muthuswamy and Muthukrishnan 1971). Further reports prove that the root volume or root diameter of chili has increased due to the application of vermicompost along with other organic manures (Yadav and Vijayakumari 2003). The increase in bulb diameter and length may be attributed to solubilization of plant nutrients by the addition of vermicompost and FYM, leading to increased uptake of NPK (Subbaih et al. 1982). The results were in agreement with the findings reported by Renuka and Sankar (2001) and Giraddi (1993) in onion.

Finally, the root growth is reported to be dependent upon the water, the nutrient availability, and the external applications of hormones (Lopez-Velez et al. 2003). Root growth and lateral root formation were found to be good in nutrient-rich environments, and the presence of hormone such as auxins was found to stimulate plant growth, root development, butt formation, etc., (Forde and Lorenzo 2001). Therefore, evidence proved that the development of morphology in the *Raphanus sativus* L. was favored by the nutrient-rich and biologically active substrates like compost and vermicompost when compared with the CF and C. The results obtained in this study clearly indicate that *Raphanus sativus* L. respond to the application of organic manures. In general, the plant treated with DSC, DSV, and VV consequently resulted in greater values of root diameter, yield, and dry matter production of roots than with the chemical fertilizer.

5.5 Effect of Heavy Metals in Organically Grown Radish

Organically cultivated DSC-, DSV-, VV-, C-, and CF-grown *Raphanus sativus* L. was reported to accumulate heavy metals like Cu, Cd, Zn, Fe, and Mn. But the obtained results were found to be within the limit of Indian standards (Table 5.3). Hence, the distillery sludge-based vermicompost could be used for promoting plant growth when compared with the chemical fertilizer.

5.6 Minerals

In this study, the highest content of minerals was found to be present in DSV than in others (DSC, VV, C, and CF) (Table 5.3). The mineral content in the plant was as follows: phosphorus, 0.215 mg/g; potassium, 2.65 mg/g; sodium, 0.58 mg/g;

Table 5.3 Effect of different organic and inorganic fertilizers on the heavy metal and mineral content of radish plants

Parameters (μg/g)	WOF	VV	DSC	DSV	CF	Indian standard
Fe	4.6	4.9	5.1	5.6	5.8	–
Zn	2.2	2.3	2.1	2.5	2.0	50
Copper	0.69	0.7	0.7	0.9	1.0	30
Manganese	1.00	1.00	1.5	1.00	1.00	–
Cadmium	0.2	0.3	0.3	0.4	0.3	1.5
Phosphorus	0.178	0.196	0.192	0.215	0.185	–
Potassium	2.3	2.53	2.5	2.65	2.45	–
Sodium	0.39	0.54	0.49	0.58	0.49	–
Calcium	0.020	0.028	0.026	0.030	0.024	–
Magnesium	0.105	0.106	0.108	0.113	0.110	–

calcium, 0.030 mg/g; and magnesium, 0.113 mg/g. The minimum values recorded in C were phosphorus, 0.178 mg/g; potassium, 2.3 mg/g; sodium, 0.39 mg/g; calcium, 0.020 mg/g; and magnesium, 0.105 mg/g.

5.7 Total Phenolics and Flavonoids

FRAP and DPPH assays for total phenolics and flavonoids were shown in Fig. 5.1. DSV-grown vegetables were found to possess higher antioxidant properties than C-, VV-, DSC-, and CF-grown radish. Hence, the total phenolic (standard curve equation $= Y = 0.0301 \times -0.0112$, R2205 $= 0.9903$) observed for C was 21.33 ± 0.002 mg GAE/g; for VV, it was 22.67 ± 0.002 mg GAE/g; for DSC, it was 22.56 ± 0.002 mg GAE/g; and for DSV, it was 31.24 ± 0.002 mg GAE/g and 22.57 ± 0.002 mg GAE/g. The total flavonoid (standard curve equation $= Y = 0.0018 \times +0.0379$) observed for C was 16.77 ± 0.002 mg QE/g; for VV, it was 39.81 ± 0.002 mg QE/g; for DSC, it was 43.62 ± 0.002 mg QE/g; for DSV, it was 45.69 ± 0.002 mg QE/g; and for CF, it was 32.59 ± 0.002 mg QE/g. Consistent with this report, organically grown plant has the higher amount of antioxidant properties than the inorganically grown plant (Dumas et al. 2003). In addition, organic fertilizers have enhanced the content of ascorbic acid and total phenolics in tomato (Toor et al. 2006). Broccoli plant has showed high yield and good quality with organic fertilizer (Abou El-Magd et al. 2006). Organic fertilizer-grown tuber of cassava plants is found to have higher content of phenolics and flavonoids (Nur et al. 2013). In our study *Raphanus sativus* L. also confirms that organically grown plants showed high level of phenolics than with chemical fertilizer. Organically grown *kacip Fatimah* (*Labisia pumila* Benth) plant had a higher percentage of phenolic and flavonoid content (Ibrahim et al. 2013).

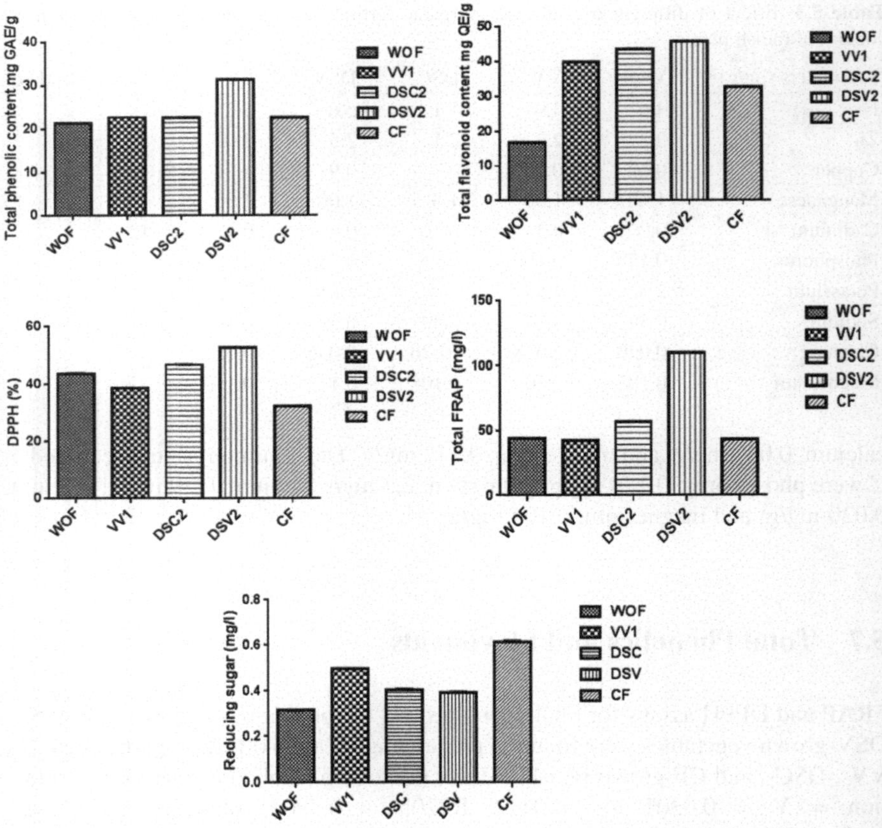

Fig. 5.1 Antioxidant status of organically and inorganically grown radish. Antioxidants such as DPPH, FRAP, and total phenolic and flavonoid content in without fertilizer (WOF)-, vegetable waste vermicompost (VV)-, distillery sludge compost (DSC)-, distillery sludge vermicompost (DSV)-, and chemical fertilizer (CF)-grown radish

5.8 Ferric Reducing Antioxidant Power (FRAP)

From the results, the total FRAP observed significantly increased in DSV plants than in WOF-, VV-, DSC-, and CF (chemical fertilizer)-grown plants. The total FRAP observed in C was 44.21 ± 0.007 mg/g; VV, 42.224 ± 0.0003 mg/g; DSC, 56.882 ± 0.0007 mg/g; DSV, 109.472 ± 0.0003 mg/g; and CF, 42.804 ± 0.0003 mg/g. The organically grown *kacip Fatimah* (*Labisia pumila* Benth) plant were found to possess higher percentage of FRAP scavenging activities than the NPK fertilization (Ibrahim et al. 2013). Similarly, FRAP assay for organically produced apples showed a 7–27% of higher antioxidant capacity than the conventionally produced apples in three comparable farm pairs (FP) in Switzerland (Stracke et al. 2009).

5.9 DPPH Scavenging Activities

The total DPPH scavenging activities significantly increased in DSV. The activities of the different type of organic- and chemical fertilizer-grown (radish) *Raphanus sativus* L. plants were shown in Fig. 5.1. Comparatively, DSV had higher level of antioxidant properties than C, VV, DSC, and CF (chemical fertilizer). Moreover, the total DPPH for C was observed to be 43.48 ± 0.11, for VV is 38.45 ± 0.0002, for DSC is 46.64 ± 0.002, for DSV is $52.49 \pm 0.002\%$, and for CF is $32.19 \pm 0.0002\%$. The obtained results reveal the highest percent of DPPH inhibition in vermicompost-treated Medan variety (47.58%) and vermicompost-treated cassava is reported to possess strong antioxidant activity, and hence it could be used as a good source of natural antioxidants (Omar et al. 2012). The same author reported that the vermicompost-treated cassava plant had higher scavenging activities than the plant grown with inorganic fertilizer. Further, organically grown *kacip Fatimah* (*Labisia pumila* Benth) plant had higher percentage of DPPH scavenging activities than the NPK fertilization (Ibrahim et al. 2013).

5.10 Reducing Sugar

A significant positive correlation is shown in Fig. 5.1 between the reducing sugar content and the type of fertilizer used. Hence, comparatively, CF shows higher level of reducing sugar than with C, VV, DSC, and DSV. And the total reducing sugar observed for C was 0.314 ± 0.0003 mg/50 g; VV, 0.496 ± 0.0003 mg/50 g; DSC, 0.406 ± 0.0007 mg/50 g; DSV, 0.394 ± 0.0003 mg/50 g; and CF, 0.61 ± 0.007 mg/50 g. The high sugar content in plants (young and mature galls) was interpreted, which may be due to increased metabolic activity under stress. It is reported to be responsible for additional synthesis of sugar (Marmit and Sharma 2008).

5.11 HPLC

The HPLC results were shown in Fig. 5.2. The total phenolic and flavonoid contents were examined, and the following compounds, tannic acids, gallic acid, ascorbic acid, quercetin, catechol, resorcinol, and benzoic acid, were identified. Hence, the present study confirms that the higher phenolic content was identified with DSV than with the other applications. Similar results were reported by Mradu et al. (2012).

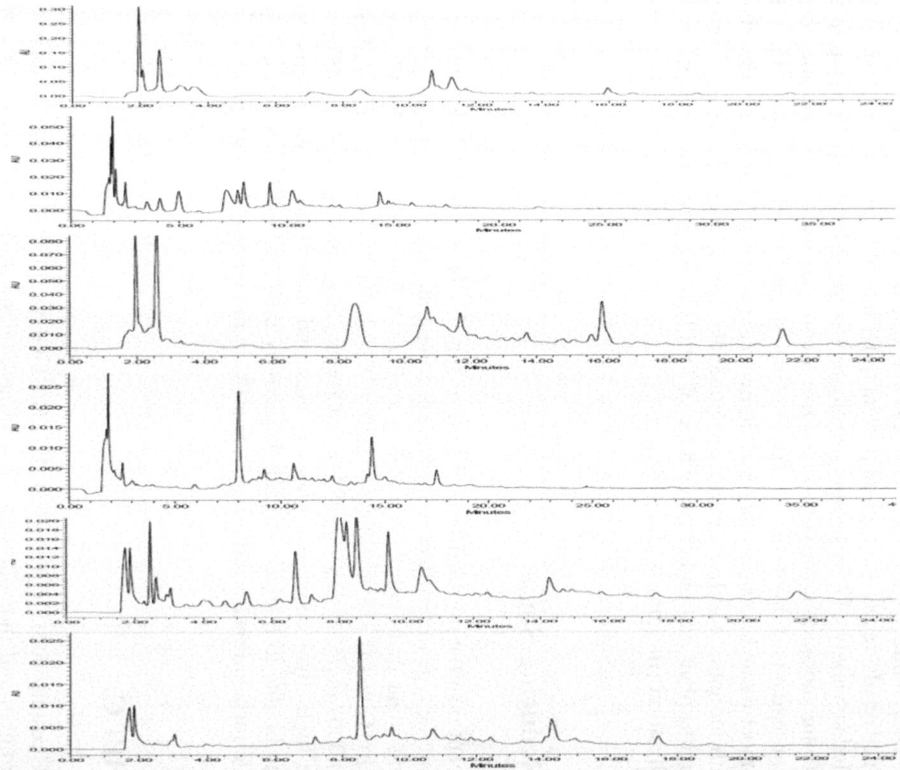

Fig. 5.2 HPLC chromatogram of different fertilizer-treated *Raphanus sativus L.* (white radish). Phenolic and flavonoid content in without fertilizer (WOF)-, vegetable waste vermicompost (VV)-, distillery sludge compost (DSC)-, distillery sludge vermicompost (DSV)-, and chemical fertilizer (CF)-grown radish

5.12 Protein Profile

The protein content of five different types of radish grown was estimated. The SDS-PAGE profile of C-, CF-, VV-, DSC-, and DSV-treated radishes was shown in Fig. 5.3. There are no significant polymorphisms. Similar results were observed and reported by Abedi et al. (2010). But very low differences to the total number of all quantified protein spots in vegetables were reported. The authors after studying the number of samples have reported that the protein expression either decreases or increases. The authors have noticed significant changes in most of the cases and recommend for further investigation (Brandt and Mølgaard 2001). Also, a higher number of metabolic differences, caused by organic and conventional cropping regimes, were detected during developmental stages of grains rather than in mature grains (Zorb et al. 2009). Therefore, the authors have stated that difference in protein expression happens not due to the cropping systems but due to their stages of growth. In conclusion, protein changes differ from plant to plant and differ with growing stages. So, further work is needed to confirm the protein changes, and the need exists for the optimization of farming methods for a variety of agricultural products.

5.13 Conclusion

We conclude that the influence of organic and inorganic fertilizer exists. The plant growth, yield parameters, mineral status, and antioxidants like phenolic, flavonoid, DPPH, and FRAP content are increased in the plant when grown with DSV than with other groups of organic fertilizer and control too. No protein profile changes were observed in all treatments. Based on the above results, it could be concluded that the distillery industrial sludge could be used as a good nutrient for plants than other types of compost and vermicompost. Hence, we could suggest that the usage or

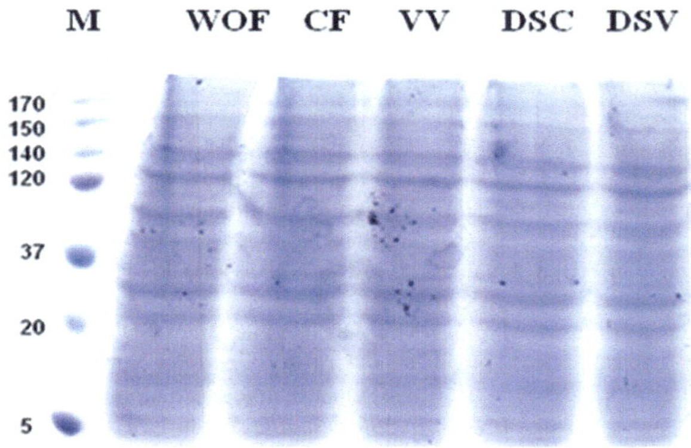

Fig. 5.3 SDS-PAGE protein profile of organically and inorganically grown radish. Protein profile in without fertilizer (WOF)-, vegetable waste vermicompost (VV)-, distillery sludge compost (DSC)-, distillery sludge vermicompost (DSV)-, and chemical fertilizer (CF)-grown radish

application of distillery sludge-based manure could reduce the chemical fertilizer usage and facilitate industrial sludge disposal issues too.

Acknowledgments The authors are grateful to the Ministry of Environmental, Forest and Climate Change (MoEF&CC), Department of Science and Technology (DST), UGC-SAP, DST-FIST, DST-PURSE, and MHRD-RUSA, Government of India, New Delhi, for their financial support and instrument facility for our department.

References

Abedi T, Alemzadeh A, Kazemeini SA (2010) Effect of organic and inorganic fertilizers on grain yield and protein banding pattern of wheat. Aust J Crop Sci 4(6):384

Abou El-Magd MM, El-Bassiony AM, Fawzy ZF (2006) Effect of organic manure with or without chemical fertilizers on growth, yield and quality of some varieties of broccoli plants. J Appl Sci Res 2(10):791–798

Babalad HB (2005) Network project on organic farming—annual report 2004–05, Adhoc Project, PDCSR

Boligon AA, De Brum TF, Frolhich JK, Froeder ALF, Athayde ML (2012) HPLC/DAD profile and determination of total phenolics, flavonoids, tannins and alkaloids contents of Scutia buxifolia Reissek stem bark. Res J Phytochem 6(3):84–91

Brandt K, Mølgaard JP (2001) Organic agriculture: does it enhance or reduce the nutritional value of plant foods. J Sci Food Agric 81(9):924–931

Carbonaro M, Mattera M, Nicoli S, Bergamo P, Cappelloni M (2002) Modulation of antioxidant compounds in organic vs conventional fruit (*peach, Prunus persica L.,* and *pear, Pyrus communis L.*). J Agric Food Chem 50(19):5458–5462

Chezhiyan N, Saraswathy S, Vasumathi R (2003) Studies on organic manures, biofertilizers and plant density on growth, yield and alkaloid content of bhumyamalaki (Phyllanthus amarus Schum. And Thonn.). South Indian Hort 51:96–101

Dumas Y, Dadomo M, Di Lucca G, Grolier P (2003) Effects of environmental factors and agricultural techniques on antioxidant content of tomatoes. J Sci Food Agric 83(5):369–382

Forde B, Lorenzo H (2001) The nutritional control of root development. Plant Soil 232(1–2):51–68

Giraddi RA (1993) Vermiculture and role in agriculture. In: Proc course on the officers of the state department of agriculture, Karnataka, 18–20 October 1993 by the Department of Agriculture microbiology University Agricultural Sciences Dharwad (M.S.), India, pp 50–54

Hames BD (ed) (1998) Gel electrophoresis of proteins: a practical approach, vol 197. OUP, Oxford

Ibrahim M, Jaafar H, Karimi E, Ghasemzadeh A (2013) Impact of organic and inorganic fertilizers application on the phytochemical and antioxidant activity of Kacip Fatimah (*Labisia pumila Benth*). Molecules 18(9):10973–10988

Laemmli UK (1970) Cleavage of structural proteins during the assembly of the head of bacteriophage T4. Nature 227(5259):680

Lin YL, Juan IM, Chen YL, Liang YC, Lin JK (1996) Composition of polyphenols in fresh tea leaves and associations of their oxygen-radical-absorbing capacity with ant proliferative actions in fibroblast cells. J Agric Food Chem 44(6):1387–1394

Lingaiah HB, Uthaiah BC, Gowda NAJ, Herle PS (1992) Evaluation of radish cultivars in the coastal region of Karnatakla. J Agric Sci 5(2):132–134

Lopez-Velez M, Martinez-Martinez F, Valle-Ribes CD (2003) The study of phenolic compounds as natural antioxidants in wine. Crit Rev Food Sci Nutr 43(3):233–244

Marmit KS, Sharma SL (2008) Quantitative estimation of some metabolites and enzymes in insect induced leaf galls of *Mangifera indica*. Asian J Experiment Sci 22(3):343–346

Mathe-Gaspar G, Anton A (2002) Heavy metal uptake by two radish varieties. Acta Biol Szeged 46 (3–4):113–114

Miller JH (1972) Experiments in molecular genetics. Cold Spring Harbor Lab Press, Cold Spring Harbor

Mradu G, Saumyakanti S, Sohini M, Arup M (2012) HPLC profiles of standard phenolic compounds present in medicinal plants. Int J Pharmacogn Phytochem Res 4(3):162–167

Muthuswamy S, Muthukrishnan CR (1971) Some growth response of radish (*Raphanus sativus L.*) to different nutrients. South Indian Hort 19:9–16

Nur FO, Siti AH, Umi KY (2013) Comparative evaluation of organic and inorganic fertilizers on total phenolic, total flavonoid, antioxidant activity and cyanogenic glycosides in cassava (*Manihot esculenta*). Afr J Biotechnol 12(18):2414–2421

Omar NF, Hassan SA, Yusoff UK, Abdullah NAP, Wahab PEM, Sinniah UR (2012) Phenolics, flavonoids, antioxidant activity and cyanogenic glycosides of organic and mineral-base fertilized cassava tubers. Molecules 17(3):2378–2387

Parthasarathi (1998) Response of radish (Raphanus sativus L.) varieties to varying levels of fertility. Doctoral dissertation, University of Agricultural Sciences, GKVK

Pujari MM, Jain BP, Mishra GM (1977) Investigations on morphological features and yield potential of promising radish (*Raphanus sativus L.*) cultivars. Haryana J Hort Sci 6(1–2):85–89

Renuka B, Sankar CR (2001) Effect of organic manures on growth and yield of tomato. South Indian Hort 49:216–219

Rojagopal AP, Rengaswamy P, Ashok Mehta V (1979) Varietal evaluation in radish. South Indian Hort 27:75–78

Stracke BA, Rüfer CE, Weibel FP, Bub A, Watzl B (2009) Three-year comparison of the polyphenol contents and antioxidant capacities in organically and conventionally produced apples (*Malus domestica Bork. Cultivar Golden Delicious'*). J Agric Food Chem 57 (11):4598–4605

Subbaih K, Helkaih J, Kumar VR, Gopal R (1982) Effect of combined application of organic and inorganic fertilizers on the yield and nutrition uptake of MDU chilies. South Indian Hort 30:45–47

Suthar S (2008) Bioremediation of aerobically treated distillery sludge mixed with cow dung by using an epigeic earthworm Eisenia fetida. Environmentalist 28(2):76–84

Thanunathan K, Natarajan S, Senthilkumar R, Arutmurugan K (1997) Effect of different sources of organic amendments on growth and yield of onion in mine spoil. Madras Agric J 84:382–383

Toor RK, Savage GP, Heeb A (2006) Influence of different types of fertilisers on the major antioxidant components of tomatoes. J Food Compos Anal 19(1):20–27

Yadav H, Vijayakumari B (2003) Influence of vermicompost with organic and inorganic manures on biometric and yield parameters of chilli *[Capsicum annuum (L.) var. Plri]*. Crop Res Hisar 25(2):236–243

Yildirim A, Oktay M, Bilaloglu V (2001) The antioxidant activity of the leaves of Cydonia vulgaris. Turkish J Med Sci 31(1):23–27

Zorb C, Steinfurth D, Seling S, Langenkämper G, Koehler P, Wieser H, Mühling KH (2009) Quantitative protein composition and baking quality of winter wheat as affected by late sulfur fertilization. J Agric Food Chem 57(9):3877–3885

Part III
Wastewater and Sludge/Solid and Liquid Waste

Chapter 6
Vermitechnology: A Sustainable Approach in the Management of Solid and Liquid Waste

Soubam Indrakumar Singh, Deachen Angmo, and Rahil Dutta

Abstract Solid and liquid waste (SLW) has become one of the serious problems and leads to significant damage to the environment and public health. With the increasing and fast-approaching water crisis, it is extremely essential to effectively treat the SLW before it enters the water bodies. The contamination of soil and water is increasing due to various anthropogenic activities, which has alarmed the scientific community. Different methods like thermal methods, landfilling, etc. for solid waste and primary, secondary, and tertiary treatment for liquid waste are being practiced worldwide. But, they seem to produce harmful gases, leachate, thereby affecting the environment and human health, and they are of high costs in installation and maintenance. Therefore, there is an urgent need for cost-effective and eco-friendly techniques for addressing these concerns. Thus, this chapter focuses on vermitechnology-using earthworms, like vermicomposting and vermifiltration to manage various types of solid and liquid waste like industrial sludge, food waste, animal waste, etc. into vermicompost (nutrient-rich manure) which acts as a soil conditioner and purifies the wastewater.

Keywords Solid waste · Liquid waste · Earthworms · Vermicomposting · Vermifiltration

6.1 Introduction

Waste is generated in every city in the world and, if not properly handled, results in spreading disease among the human and animal populations, and the amount of waste generated differs from place to place. Waste is any item beyond use in its current form and discarded as unwanted. Most businesses define waste as "anything that does not create value" (BSR 2010). However scientifically speaking, there is no

S. I. Singh · D. Angmo · R. Dutta (✉)
Department of Botanical and Environmental Sciences, Guru Nanak Dev University, Amritsar, India

S. A. Bhat et al. (eds.), *Earthworm Assisted Remediation of Effluents and Wastes*,
https://doi.org/10.1007/978-981-15-4522-1_6

waste as such in the world. Almost every component of waste possesses some potential and can become substrate in the production cycle if it is treated scientifically. Generally, there are two types of waste, i.e., solid and liquid waste, and both wastes have different properties. Solid waste means and includes solid or semisolid domestic waste including sanitary waste, commercial waste, institutional waste, catering, and marketing waste and other nonresidential wastes, street sweepings, silt removed or collected from the surface drains, horticulture waste, construction and demolition waste and treated biomedical waste excluding industrial hazardous waste, biomedical waste, and e-waste generated in an area under local body (MoEF&CC 2015). Solid waste generation is directly proportional to the growth of the population. Rapid industrialization and population explosion in India has led to the migration of people from villages to cities, which generate thousands of tons of MSW daily. Earlier the amount of waste generated in rural areas was much less and self-sufficient and easily manageable. But lately, the spread of consumerism into rural areas has led to an increased generation of solid waste that has different characteristics than the traditional waste that was generated earlier. As per the 2011 census, the rural population in India has increased from 743 to 833 million, which accounts for 68.84% of the total population. It is estimated that this rural population produces 0.3–0.4 million tons of waste per day (SWM in Rural India: A guide to Gram Panchayats 2016). Unplanned urbanization coupled with rapid population growth and changes in the standard of living in urban centers of India have led to tremendous increase in the amounts of municipal solid waste (MSW) which includes a mix of dry and wet waste, dumping in drains and open spaces, disposal without treatment for energy or resource recovery. On the other hand, liquid waste or wastewater is generated from industrial facilities, intensive agricultural methods, and municipalities (domestic sewage) in urban and rural areas of India. Subsequently, high population density also results in increased volumes of wastewater being produced from industries. Besides this, the majority of urban cities in developing countries dispose of the urban runoff and sewerage water directly into urban water bodies without any treatments or with partial treatments due to its high setup and maintenance cost of a treatment plant (Tomar and Suthar 2011). But, finding a proper solution to treat this waste has become a serious issue nowadays. As per the Central Pollution Control Board (CPCB Report 2009), about 8–51% of liquid waste is collected and treated in treatment facilities, and the treated effluent (industrial and domestic) is disposed of into our water bodies. These wastes are generated by commercial, residential agricultural, and industrial operation and release to the environment without undergoing proper treatment. Due to which, it is now mandatory for municipal corporation/urban local bodies (ULB) to collect 100% of the waste and to segregate and transport the waste to treatment and management facilities. There are 279 compost plants, 172 biomethanation plants, 29 refuse-derived fuel (RDF) plants, and 8 wastes to energy plants in India (CPCB Report 2009). Despite having so many treatment facilities, India still ranks high in waste accumulation. With proper technology, all this waste could be turned into wealth, but sadly waste is finally dumped in low-lying area at the periphery of city limits without cleaning it; thus leachate formed in landfill percolate deep down and tends to pollute underground water source and has posed great adverse impacts on the environment and public health. Dumping of waste also contributes to global warming and environmental degradation. Open dumping releases methane from the decomposition of biodegradable waste under anaerobic conditions. Methane is considered one of the most potent greenhouse gases and is a major contributor to global warming. There are problem arises with odor and migration of leachate to the nearby water body, and the leachate (cadmium, lead, mercury, chromium) not only pollutes water but also threatens aquatic animals and plants. The odor is a critical issue, particularly during the summer season when the average temperature in India can exceed 45 °C. The workers who engage in handling MSW without protective equipment are more susceptible to various infections. Also, discarded tires at dumping site store water and serve as a breeding ground for mosquitoes and other insects thereby increasing the risk of diseases such as malaria, dengue, etc. Uncontrolled burning of waste at dumping sites releases fine particles and toxic gases which are a major cause of respiratory disease and cause smog. Open burning of MSW and tires emits 22,000 tons of pollutants into the atmosphere around Mumbai every year (Ramachandra et al. 2018). The city of Delhi recently witnessed such a situation where the smog arises due to stubble burning in neighboring states. The impacts of poor waste management on public health are well documented, with increased incidences of nose and throat infections, breathing difficulties, inflammation, anemia, asthma, and other infections. It is very important to address the waste by means of proper technology and with the technology that has a minimum impact and maximum benefit to the environment. There are many technologies that greatly reduce the volume of waste. Some of the commonly used methods or techniques to combat the solid waste are thermal process (incineration, pyrolysis), landfilling, and biological process (composting, vermicomposting) and for liquid waste are primary treatment, secondary treatment, tertiary treatment, and vermifiltration. Also, many techniques for remediation of soil pollution due to agrochemicals have been reported using physical and chemical methods but, in turn, release the toxic compounds to the environment as by-product or final product.

The aim of this chapter is to focus on the cost-effective, hygienic, potent, and eco-friendly bioremediation techniques using earthworms like vermitechnology (vermicomposting and vermifiltration) for the management of solid and liquid waste, which has become the need of the hour in the present scenario. As earthworms are safe and have a high rate of degradation of solid waste and in the purification of liquid waste (polluted water), etc.

6.2 Classifications of Waste

Waste is broadly classified into solid and liquid waste depending on the physical state/appearance and the source from which the waste is generated. The classification of waste is shown in Fig. 6.1.

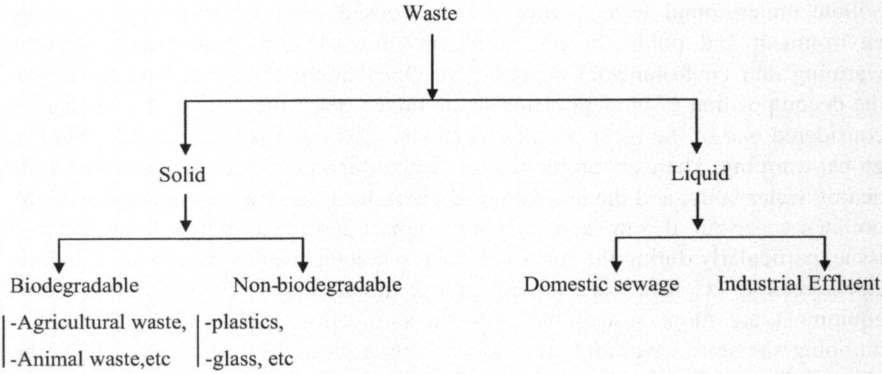

Fig. 6.1 Classification of waste

6.2.1 Types of Waste

6.2.1.1 Solid Waste

Any unwanted material, discarded by human activities, that can neither flow easily in the water streams and nor escapes into the atmosphere immediately in the gaseous form is called as solid waste.The quantity and characteristics of solid waste generation depends upon a place, season, climate, social behavior, income source, population size, etc.

They can be classified into two categories by its characteristics, namely: (1) Biodegradable and (2) nonbiodegradable waste

1. **Biodegradable waste:** Those solid wastes that can be easily degraded are called biodegradable waste. They are generally organic in nature which originated from living beings like plants and animals. Example of biodegradable solid waste includes agricultural waste, animal waste, food waste, paper waste, etc.
2. **Nonbiodegradable waste:** Those solid wastes that cannot be degraded are known as nonbiodegradable waste. They are inorganic in nature. They create environmental problems if they are not managed properly. They stay in the environment for a longer period of time. Example of nonbiodegradable solid waste includes plastics, metals, glass, thermocoa *etc.*

6.2.1.2 Liquid Waste

Wastes that are produced in the form of liquid are called liquid wastes. Wastewater generally originates from two sources: (1) domestic wastewater from gray water, toilets, and other domestic activities and (2) industrial wastewater, generated by industries during the normal course of activity, which often relies on the local sewerage systems for waste processing. Therefore, they mainly consist of wastewater that is generated from a home or community including toilet, bath, laundry, lavatory, and kitchen-sink wastes and surface runoff. Examples of liquid waste include industrial effluents, sewage, etc. They may be classified into:

1. Sanitary sewage
2. Industrial sewage

   1. **Sanitary sewage:** It is also called domestic sewage which contains human wastes and waste water from homes, public buildings, or commercial and industrial establishments. The untreated domestic sewage causes pollution to the receiving water body as it contains objectionable matter like organic matter, nutrients, and toxic elements. The organic matter leads to an increase in the biochemical oxygen demand that causes depletion in the dissolved oxygen of the receiving water body. Excessive nutrients (N&P) cause the problem of algal growth and eutrophication. Toxic elements include heavy metals, pesticides, and industrial xenobiotic pollutants that pose a serious threat to the living biota of receiving water bodies (CPCB 2009).
   2. **Industrial sewage:** These are also called industrial effluents that are produced or released as wastewater from manufacturing processes. Different industries produce different types of wastewater as the raw material used by the industrial units differs from one another. It usually carries a variety of chemical compounds. If this water is released into the soil and water sources, they pollute the soil and affect the aquatic life and also ultimately human health. A large amount of wastewater is released from many industrial units annually, but little of it is being treated. Some of the industries that produce industrial effluents include textile, leather, and beverage industry, pharmaceutical industry, etc.

6.3 Management of Solid and Liquid Waste

6.3.1 Solid Waste Treatment

1. **Thermal process:** It is the process in which waste is degraded by heating at high temperatures.

6.3.1.1 Incineration

It is referred to as the combustion of waste at high temperatures and ultimately ends up releasing toxic chemicals. Dioxins and furans are key toxic elements released as a by-product during incineration. Incineration not only releases toxic but consumes a lot of energy to run the process. The burning of solid, liquid, or gaseous wastes under "controlled" conditions is called incineration. The "controlled" conditions consist of a combustion chamber with enriched oxygen at high temperatures and the use of supplementary fuel and rapid agitation of incoming waste. The main purpose of incineration is the reduction in the volume of waste (>90%) that ultimately extends the lifetime of the land disposal facility. The second purpose is the generation of energy, i.e., "waste to energy." This energy can be used for heating and electricity generation purposes. The third benefit of this technique is that it leads to detoxification, i.e., killing of microbes and other pathogenic organisms within the waste. In modern incinerators, waste is burned at high temperatures under controlled conditions. This combustion of waste leads to the generation of gases like CO_2, SO_2, NO_X, and HCl and some toxic gases like dioxins and furans that are hazardous at very low concentrations. The process of incineration also produces a particular matter (fly ash, soot, and dust), incinerator residue (bottom ash), process water, and some heavy metals (Pichtel 2005). The modern incinerators are equipped with extensive emission control devices like fabric baghouse filters, acid gas scrubbers, and electrostatic precipitators to remove the fly ash and gaseous by-products. The ash is ultimately disposed into the landfill site, and if it is found to contain toxic metals, it should be treated as hazardous waste. The most critical factors that determine the combustion effectiveness in incineration are (1) the temperature in the combustion chamber, (2) the time duration for the combustion, (3) the rate of mixing or turbulence, and (4) the quantity of oxygen present for combustion process (Narayana 2009).

6.3.1.2 Pyrolysis

It is a thermal process, in which biomass is heated at a temperature ranging from 300 to 1000 °C in the absence of oxygen and air. It is used to treat the waste to obtain a solid residue, liquid oil, and gases. During this process, the biomass breaks into different phase state products, coexisting solids, liquids, and gases at the end of the process. The ratio is heavily dependent on the velocity of the process (Dhyani and Bhaskar 2019). It is also called as destrictive distillation as it is endothermic process.

6.3.1.3 Landfills

Landfills are the sites where waste is deposited onto or into the land that results in leachate leakage and release of landfill gas. Leachate leakage is dependent upon the relative permeability of the landfill liner. Landfill leachate is responsible for polluting groundwater supplies. As waste decomposes, a combination of chemical, thermal, and microbial reaction releases gases (US Environmental Protection Agency 1999). Landfill gas is a combination of methane and carbon dioxide in almost equal parts. The remaining 0.01–0.6% is composed of carcinogenic volatile organic compounds, such as benzene, toluene, xylenes, etc. A landfill site is an area of land where waste is deposited. It is intended to prevent or minimize the interaction of waste and its surroundings, especially the groundwater. It is the most common practice in India to tackle waste.

Landfills are generally classified into three categories as follows.

Open Dumps

These landfills are the most common means of waste disposal in developing countries where the waste is dumped into a low-lying area.

Semicontrolled landfills: In these types of landfills, the waste is covered with the covering layer to prevent the nuisance, but these are not engineered to collect and manage the toxic leachate and gases.

Sanitary or Secure Landfills

According to Narayana (2009), landfills having the provisions for control and proper management of gases and leachate from the waste decomposition are called sanitary or secure landfills. This type of landfill is considered to be the safest of all the three types of landfills, but its construction and management are costly. This type of landfill is mostly used in developed countries. The gases from the landfill are mainly composed of methane, carbon dioxide, and a small amount of some other non-methane organic compounds. Among them, methane is considered to be the most potent greenhouse gas, i.e., 28–36 times greater than carbon dioxide. This mixture of landfill gas is collected and processed which is ultimately used for energy generation.

6.3.1.4 Composting

It is considered to be one of the most preferred methods in solid waste management. Composting is one of the biological decompositions of organic waste under controlled conditions. It can be either aerobic (presence of oxygen) or anaerobic process (absence of oxygen). In aerobic composting, organic waste is consumed by aerobic thermophilic and mesophilic microorganisms which convert them into mineralized products such as carbon dioxide, water, ammonia, and relatively stable organic end product. The end product of composting is called compost which is a stable, humus-rich, and complex mixture that helps in improving the physical properties of soil and is a rich source of nutrients for plants (Watteau and Villemin 2011). Anaerobic composting is a low-temperature process as a result of which it leaves weed seeds and various pathogens intact. Anaerobic composting takes a longer time than aerobic composting. There are different factors that affect the process aerobic composting which include temperature, moisture, nutrients, aeration, and pH and C/N ratio. According to López-González et al. (2015), to provide an optimum environment for microorganisms to degrade the organic waste, the above factors are regulated and controlled. In this method, addition of organic material is provided to act as nutrient media for the growth of the microbial population. It can be used as an alternative to landfilling. The final compost obtained can be used as manure.

6.3.1.5 Vermicomposting

The term vermicomposting applies to the use of earthworms for the composting of organic residues. Vermicomposting is a joint interaction between earthworm and microorganisms found in its gut that helps in the breakage and degradation of the waste. The vermicompost so obtained is considered to be natural fertilizer and is capable of providing all the types of nutrients to the deficient soil. It not only offers the nutrient to the soil but also improves the soil structure by maintaining pH and EC. It is also helpful in detoxification of soil. In this chapter, a special emphasis will be given on the method that has the least limitations *i.e.,* vermicomposting. The approach of technology used for waste treatment varies from country to country and also on the type of waste generated. India is a country where there is a maximum generation of organic or wet waste and accordingly waste treatment technology should be used.

6.3.2 *Liquid Waste/Wastewater Treatment*

6.3.2.1 Preliminary Treatment

The main purpose of preliminary treatment is the elimination of coarse solids and other large materials frequently present in untreated wastewater. To increase the operation and maintenance of future treatment, removal of these materials is necessary. Typically coarse screening, grit removal, and, in some situations, large comminution are included in preliminary treatment procedures. In the grit tanks, the velocity of the water through the container is kept extremely high, or the air is used to avoid the sedimentation of most organic solids.

6.3.2.2 Primary Treatment

Primary treatment is aimed at removing settleable organic and inorganic solids by sedimentation and elimination of scum (floating material) by skimming. Primary treatment involves the use of physical processes like screening, skimming, and sedimentation, for the removal of unwanted materials that are settleable, floatable, or too large to pass through the screens. Primary treatment typically eliminates 25–50% of biochemical oxygen demand and 50–70% of total suspended solids, but it doesn't affect the colloidal and dissolved constituents. The effluent after the primary treatment can be used for irrigation of crops that are not consumed by humans; it can also be used to irrigate orchards and vineyards.

6.3.2.3 Secondary Treatment

Secondary treatment is also referred to as biological treatment as this involves the use of aerobic microorganisms for the removal of additional suspended solids and biochemical oxygen demand, beyond what can be accomplished through basic sedimentation. The microorganisms use these organic impurities as a food source and convert them into carbon dioxide, water, and energy for their own growth. The basic biological treatments used are trickling filters, activated sludge treatment, oxidation ponds, rotating biological contactors, and membrane bioreactors. Secondary treatment plants remove about 90 percent of the biochemical oxygen demand and 90 percent of the suspended solids (Masters and Ela 2004).

6.3.2.4 Tertiary Treatment

If the intended water supply is highly sensitive to the effects of pollution, many tertiary processes may further treat secondary water effluent. Advanced waste treatment methods in use or in implementation vary from microbial treatment capable of removing nitrogen and phosphorus to physicochemical extraction strategies such as filtration, carbon adsorption, distillation, and reverse osmosis. Such wastewater treatment systems, alone or in conjunction, may attain almost any degree of pollution control required; waste effluent treated by such application can be used for commercial, agricultural, or recreational purposes or even for drinking water.

6.3.2.5 Vermifiltration

The utilization of earthworms in wastewater or sludge treatment is called vermifiltration. According to Aguilera (2003), it was first advocated by Prof. Jose Toha at the University of Chile in 1992. Vermifiltration is a treatment method that incorporates traditional treatment and vermicomposting methods for the treatment of wastewater. It is an environmentally sustainable and economically viable wastewater treatment system (Singh et al. 2019a, b). The body of earthworms acts as biofilter as it contains different types of enzymes, decomposers, and nitrogen-fixing bacteria that helps in elimination of 5-day BOD (BOD5) by over 90%, COD by 80–90%, total dissolved solids (TDS) by 90–92%, and the total suspended solids (TSS) by 90–95% from wastewater by the process of ingestion, absorption, and biodegradation of organic wastes, heavy metals, and solids from wastewater (Sinha et al. 2008). Additionally, earthworms convert mineralized nitrogen to ammonium which is then transformed to nitrates by microbes and is then more readily available for plant uptake. The wastewater solids composted by earthworms can be useful soil additives that help in improving its fertility (Sinha et al. 2008).

6.4 Vermitechnology: Role of Earthworms in the Management of Solid and Liquid Waste

'Vermi' is a Latin term for "worm" and the technology using earthworm for the management and processing of waste is called vermitechnology. It has been classified into two types based on the management of waste:

(a) Vermicomposting—for the management of solid waste
(b) Vermifiltration—for the management of liquid waste or wastewater

6.4.1 Vermicomposting

Vermicomposting simply means composting using earthworms. Hand et al. (1988) reported that vermicomposting is a safe, hygienic, and cost-effective technique for the disposal or management of solid organic wastes. According to Hemalatha (2013), vermicomposting is the combination of biological processes, design, and techniques in which one or more species of earthworms are cultured in large quantities to speed up the stabilization of organic waste materials, when eaten, ground, and digested with the help of other microorganisms like bacteria, fungi, and other smaller animals. Earthworms are used to decompose or convert organic waste into a dark brown nutrient-rich by-product called vermicompost. According to Tomati et al. (1995), Lim et al. (2016) stated that earthworm during vermicomposting stabilized and decomposed the organic residue or organic matter and produces organic fertilizer. According to Sim and Wu (2010), vermicomposting process involves the associations between the earthworms and microorganisms in biodegradation of organic waste at a faster rate. According to Domínguez (2004), intensive interaction between earthworms with microorganisms and other fauna within the decomposer community, thereby strongly affecting decomposition processes, also accelerates the stabilization of organic matter and leads to great modification in its physical and biochemical properties. Vermicomposting is also called as the bio-oxidative process. Earthworms decomposed organic waste by fragmenting and modifying the substrate; as a result, they serve as the main drivers in this process. By doing so, the surface area of organic waste accessible to microorganisms is raised by earthworms. Thus, Lim et al. (2016) stated that the cycle of microbial development and decomposition of solid waste is improved during this vermicomposting. According to Lim et al. (2015), vermicompost or earthworm cast obtained after vermicomposting has a low C/N ratio, high porosity, and water-holding capacities and can be used as plant nutrients. Thiruneelakandan and Subbulakshmi (2014) reported that vermicompost obtained after vermicomposting contains a rich amount of nutrients like nitrogen, potassium, phosphorus, organic carbon, vitamins, enzymes, and antibiotics which are considered to improve the quality and yield of the plant. The performance of the vermicomposting process is also determined by various factors such as the initial C/N ratio, humidity, temperature, pH, and nature of the organic waste (Lim et al., 2016). Vermicomposting is considered one of the most sustainable methods to deal the organic waste with maximum benefits to the environment. Vermicomposting has been shown to be successful in processing sewage sludge and solids from wastewater, food industry waste, urban residues, and food and animal waste (Domínguez et al. 2003; Clarke et al. 2007; Pramanik et al. 2007, Suthar 2007, Nogales et al. 1999, 2005; Dominguez and Edwards 1997; Atiyeh et al. 2000; Garg et al. 2006; Aira et al. 2006a, b; Lazcano et al. 2008). There has been a report on the reduction of COD of about 87% through vermicomposting of human feces (Furlong et al. 2014).

6.4.1.1 Adaptations of Earthworms

Earthworms are adapted to survive in harsh environments and in moderately acidic to alkaline conditions with pH value ranging from 4.5 to 9. They can tolerate a temperature range of 5–29 °C. According to Hand (1988), the optimum condition for a worm to function is the temperature range of 20–25 °C and moisture level of 60–75%. Satchell (1983) found that *Eisenia fetida* can withstand extremely toxic chemicals like TCDD (2,3,7,8-tetrachlorodibenzo-*p*-dioxin).

According to Sinha et al. (2010), some of the earthworm species like *Eisenia fetida*, red tiger worm (*Eisenia andrei*), the Indian blue worm (*Perionyx excavatus*), the African night crawler (*Eudrilus eugeniae*), and the red worm (*Lumbricus rubellus*) are considered to be best suited for vermicomposting of sewage sludge at different climatic conditions. Recently, ICAR-Indian Veterinary Research Institute developed a new earthworm called "Jai Gopal" (*Perionyx ceylanesis*) through the selection and mating plan, which is better than exotic earthworms *Eisenia fetida* and *Eudrilus eugeniae*. Jai Gopal has great heat tolerance capacity of up to 43.0 °C ambient temperature and is very rich in protein (contain 67% protein and all functional amino acids). As compared to exotic earthworm species, "Jai Gopal" is considered to be more voracious feeder, produced super quality of vermicast and vermiwash, has a very good capability of inhabiting on animal and farm waste, produced high rate of cocoon per week, increases in number of hatching per cocoon, increases in number of earthworm population in 3 months, has long life span, etc.

6.4.2 Vermifiltration

Due to rapid industrialization and urbanization, water shortage is expected to rise in the years to come (Gan et al. 2018). Therefore, the management of unconventional water resources through wastewater treatment has become the essential need of the hour (García-Fernández et al. 2015). In wastewater management, reutilization of treated wastewater is one of the ways to be considered (Huang et al. 2016). More than 80% of the water supply reaches the sewers; then, treated wastewater can be considered as an unconventional water resource (Eriksson et al. 2002). Treated wastewater contains organic nutrients that have potential use in agriculture, but because of the high ionicity, the soil salinity is increased (Huang et al. 2018). The variety of techniques such as post-precipitation, adsorption, membrane filtration, electrochemical methods, nanocatalysis technology, and ion exchange have been applied to remove dissolved minerals and toxic compounds from the treated wastewater (Khan et al. 2012; Husain et al. 2014; Zhang et al. 2017). However, the application of some of these techniques is restricted by cost and economic factors, compatibility, efficiency, and existing regulations and standards (Khan et al. 2019a; Sinha and Chakma 2019). The increase in pollution along with the scarcity of water gives rise to the lookout for suitable treatment technologies for the removal of pollutants and reuse of treated effluents in operations requiring low-quality water. Among different options, physical unit operations are easy and economical, but it results in less success due to incomplete removal of contaminants (Sheehan and Greenfield 1980). The chemical treatment options have already been used in the remediation of wastewater (Huang et al. 2009). However, chemical treatments are mostly centered on the pH regulation for the removal of colloids. Despite being an effective solution for targeting pollutants of special concern, chemical treatments lack attention from the environmentalists as they require high capital investment in operation and maintenance (Rao et al. 2007). Lom (1977) stated that chemical treatments generate several by-products that require special attention and highly skilled manpower. On the other hand, compared to the other techniques mentioned above, biological remediation methods using earthworms (vermifiltration) seem to be a cost-effective, eco-friendly, hygienic, chemical-free method in the treatment of domestic and industrial wastewater. According to Kumar et al. (2015), vermifiltration process exploits the earthworms as a biofilter in combination with suitable bedding materials to treat domestic and industrial wastewater. Wastewater treatment by the vermifiltration system is carried out simultaneously by both germs and earthworms as earthworms host millions of microorganisms and their body walls are used to remove BOD, COD, and TDS through ingestion, biodegradation, and absorption (Singleton et al. 2003). Besides, these favorable conditions for the aerobic activity of microorganisms are being created by the movement of the earthworms in the filter bed, improve oxygen penetration (Khan et al. 2019b; Samal et al. 2019), and also prevent odors and sludge formation (Lourenço and Nunes 2017). Furthermore, the vermifiltration process required less area for the establishment and has the same efficiency as the activated sludge process (Samal et al. 2017).

6.4.2.1 Application of Vermifiltration

According to Sinha et al. (2008), a promising treatment method of vermifilter can now be used to treat the wastewater generated from domestic chores such as washing, cooking, toilet, etc. There has been a great result in the treatment efficiencies of BOD, COD, TDS, and TSS of the order greater than 90%, 80–90%, 90–92%, and 90–95%, respectively, with the application of vermifilters in treating effluents generated from domestic premises (Sinha et al. 2008). Arora et al. (2014) reported that COD and BOD are being removed off at the rate of 76% and 67%, respectively, from domestic premises when being treated using earthworm species *Eisenia fetida*. Xing et al. (2010) also observed the removal efficiencies of about 47.3–64.7%, 54.7–66%, 7.6–15%, and 2–62% against their influent COD, BOD, TN, and ammonium nitrogen (NH4+-N), respectively. Wang et al. (2016) stated that the use of vermifilter in the treatment of domestic wastewater has resulted in the BOD, COD, and SS removal efficiencies up to 84%, 81%, and 94%, respectively. Liu et al. (2013) also stated that using ceramsite as filter media during the treatment of rural domestic sewage has removal efficiencies of COD, BOD, and NH4+-N up to 67.6%, 78%, and 92.1%, respectively. According to Furlong et al. (2014, 2015), human feces can also be treated through vermifiltration.

Initially, vermifilters were applied to the treatment of the effluents generated from the domestic premises, but recently the use of vermifiltration has now been extended to the industrial context. However, the application of vermifiltration has been very limited to industrial wastewaters as earthworms have sensitivity toward high salinity, pH, heavy metals, and pesticides. A high concentration of these pollutants may turn out to reduce the efficiency and productivity of earthworms. In case of extreme concentrations, it may even be lethal to the earthworms. For example, according to Oste et al. (2001), the combination of heavy metals or a high concentration of heavy metals has a detrimental impact on earthworm's life cycle and productivity. In addition, according to Hughes et al. (2008), a high concentration of sodium salts causes an imbalance in the osmotic regulation of earthworms. Despite its high sensitivity, the application of vermifilter has exhibited overall good performance, and the results are encouraging to extend its application for many other industries also. According to Sinha et al. (2012), the average BOD, COD, and TDS removals were 90%, 60–80%, and 90–95%, respectively, on petroleum wastewater through vermifiltration. Dhadse et al. (2010) reported that a high retention time of 2 days resulted in BOD and COD removal efficiencies of up to 89.77–96.26% and 85.44–94.48%, respectively, on pharmaceutical wastewater. The introduction of earthworms can reduce the emission of NH3, N2O, and CH4 and can also be attributed to the reduction in denitrification through the dampening of anaerobic activities (Luth et al. 2011). Li et al. (2008) supported this report in another study. The dairy wastewater generally consists of organic molecules like carbohydrates (lactose), proteins (casein), and fats. The presence of the above composition in wastewater and its decomposition generates an obnoxiously odorous black sludge that has been proved to be lethal to many aquatic living beings. With the help of vermifiltration, it has also been applied successfully in treating wastewaters from the dairy industries. Sinha et al. (2007) reported that vermifiltration of dairy wastewater can remove BOD, COD, and solids efficiently up to 98%, 80–90%, and 90–95%, respectively. According to Merlin and Cottin (2009), 80–88% of organics and 70–80% of nutrients were removed from a compost biofilter containing earthworms in cheese whey industry effluent. Palm oil mill effluent (POME) was treated using earthworm *Eudrilus eugeniae* by mixing it with the substrates of soil with rice straw in varying ratios (Lim et al. 2014). Singh et al. (2019a) applied horizontal flow

Table 6.1 Application of vermifiltration in treating different types of liquid or wastewater

Sl. no.	Types of liquid/ wastewater	Earthworm species used	Organic matter removed (%)	References
1.	Swine wastewater	*Eisenia fetida*	BOD = 99.4 COD = 99.2	Manyuchi et al. (2019)
2.	Brewery wastewater	*Eisenia fetida*	COD =95	Singh et al. (2019a)
3.	Municipal wastewater	*Eisenia fetida, Eudrilus eugeniae*	BOD = 88 BOD = 70	Kumar et al. (2016)
4.	Municipal wastewater	*Eisenia fetida*	BOD = 81.3 COD = 83.5	Wang et al. (2016)
5.	Human feces	*Eisenia fetida*	COD = 87–90	Furlong et al. (2014, 2015)
6.	Municipal wastewater	*Eisenia fetida*	BOD = 67–75.9 COD = 66.7–76	Arora et al. (2014)
7.	Palm oil mill wastewater	*Eudrilus eugeniae*	COD = 19.7–88.7 VSS = 52.7	Lim et al. (2014)
8.	Municipal sewage sludge	*Eisenia fetida*	BOD = 61.06 COD = 53.01	Li et al. (2013b)
9.	Rural domestic sewage	*Eisenia fetida*	BOD =78, COD = 67.6	Li et al. (2013a)
10.	Petroleum industry wastewater	*Eisenia fetida*	BOD >90 COD =60–80	Sinha et al. (2012)
11.	Herbal pharmaceutical wastewater	*Eudrilus eugeniae*	BOD =89.77–96.26 COD = 85.44–94.48	Dhadse et al. (2010)
12.	Gelatine industry effluent	*Lumbricus rubellus*	BOD = 89 COD = 90	Ghatnekar et al. (2010)
13.	Dairy industry wastewater	*Eisenia fetida*	BOD = 98 COD = 80–90	Sinha et al. (2007)

vermifilters in treating brewery wastewater. The COD removals were observed to be up to 95%. In contrast to vertical flow vermifilters, horizontal flow vermifilters exhibit a greater potential in removing total nitrogen (TN) from the influent (8–20%). Because of its removing efficiencies, it can be suggested that the use of vermifilter can be extended to other wastewaters also. The application of vermifiltration in treating different types of liquid or wastewater is given in Table 6.1. This application can further be extended to treating wastewaters from agricultural and aquacultural operations. Biological technologies like vermifiltration can help in harvesting other benefits and by-products obtained can be applied to the field or ponds as fertilizer and feeds. Therefore, it can be concluded that even with high contamination or pollution strength, wastewaters can be treated by using vermifilters if high retention is offered to the system.

6.5 Future Scope and Conclusion

Vermitechnology, i.e., vermicomposting and vermifiltration using earthworms, is the advance and emerging technology used in the management of waste as earthworms can degrade the organic matter present in the waste and thereby convert this into manure or useful by-products. From the discussion in the chapter, we can conclude that the application of this technology can help convert the solid waste into a nutrient-rich manure (vermicompost) from different types of solid waste and also helps in purifying liquid waste or wastewater by removing the BOD, COD, and total solids present in the polluted water in a cost-effective, hygienic, eco-friendly manner and can solve the problem of solid waste and water scarcity in developing countries. Many researchers have reported its importance in the management of solid waste and wastewater only on laboratory-based analysis. It is necessary to apply this state-of-the-art technology on a large-scale basis especially by the urban local bodies (ULB), NGOs, and industrial units to solve the menace of MSW, domestic sewage, and industrial effluents, rather than treating with current methods which are costly, unhygienic, and less eco-friendly like incineration, landfilling, open dumping etc.

6.6 Acknowledgments

The authors acknowledge the Department of Science and Technology, New Delhi, for providing (DST-PURSE Fellowship under grant No. SR/PURSE Phase 2/18 (G)) and Department of Botanical and Environmental Sciences, Guru Nanak Dev University, Amritsar, for necessary laboratory facilities.

References

Aguilera ML (2003) 2003. Purification of wastewater by vermifiltration. Doctoral Thesis, University of Montpellier 2, France, p 188

Aira M, Monroy F, Domínguez J (2006a) Changes in microbial biomass and microbial activity of pig slurry after the transit through the gut of the earthworm Eudrilus eugeniae (Kinberg, 1867). Biol Fertil Soils 42(4):371–376

Aira M, Monroy F, Domínguez J (2006b) Eisenia fetida (Oligochaeta, Lumbricidae) activates fungal growth, triggering cellulose decomposition during vermicomposting. Microb Ecol 52 (4):738

Arora S, Rajpal A, Kumar T, Bhargava R, Kazmi AA (2014) A comparative study for pathogen removal using different filter media during vermifiltration. Water Sci Technol 70(6):996–1003

Atiyeh RM, Domínguez J, Subler S, Edwards CA (2000) Changes in biochemical properties of cow manure during processing by earthworms (Eisenia andrei, Bouché) and the effects on seedling growth. Pedobiologia 44(6):709–724

Business for Social Responsibility (BSR) (2010) The new frontier in sustainability. In: The business opportunity in tacking sustainable consumption. https://www.bsr.org/reports/BSR_New_Frontier_Sustainability.pdf

Central pollution control board (CPCB) (2009). cpcp.nic.in

Clarke WP, Taylor M, Cossins R (2007) Evaluation by respirometry of the loading capacity of a high rate vermicompost bed for treating sewage sludge. Bioresour Technol 98(13):2611–2618

Dhadse S, Satyanarayan S, Chaudhari PR, Wate SR (2010) Vermifilters: a tool for aerobic biological treatment of herbal pharmaceutical wastewater. Water Sci Technol 61(9):2375–2380

Dhyani V, Bhaskar T (2019) Pyrolysis of biomass. In: Biofuels: alternative feedstocks and conversion processes for the production of liquid and gaseous biofuels. Academic, Cambridge, pp 217–244

Domínguez J (2004) 20 state-of-the-art and new perspectives on vermicomposting research. In: Earthworm ecology. CRC Press, Boca Raton, pp 401–424

Dominguez J, Edwards CA (1997) Effects of stocking rate and moisture content on the growth and maturation of Eisenia andrei (Oligochaeta) in pig manure. Soil Biol Biochem 29(3–4):743–746

Domínguez J, Parmelee RW, Edwards CA (2003) Interactions between Eisenia andrei (Oligochaeta) and nematode populations during vermicomposting. Pedobiologia 47(1):53–60

Eriksson E, Auffarth K, Henze M, Ledin A (2002) Characteristics of grey wastewater. Urban Water 4(1):85–104

Furlong C, Templeton MR, Gibson WT (2014) Processing of human faeces by wet vermifiltration for improved on-site sanitation. J Water Sanit Hyg Dev 4(2):231–239

Furlong C, Gibson WT, Templeton MR, Taillade M, Kassam F, Crabb G, Kodgire M (2015) The development of an onsite sanitation system based on vermifiltration: the 'Tiger toilet'. J Water Sanit Hyg Dev 5(4):608–613

Gan D, Liu M, Huang H, Chen J, Dou J, Wen Y et al (2018) Facile preparation of functionalized carbon nanotubes with tannins through mussel-inspired chemistry and their application in the removal of methylene blue. J Mol Liq 271:246–253

García-Fernández I, Fernández-Calderero I, Polo-López MI, Fernández-Ibáñez P (2015) Disinfection of urban effluents using solar TiO2 photocatalysis: a study of significance of dissolved oxygen, temperature, type of microorganism and water matrix. Catal Today 240:30–38

Garg P, Gupta A, Satya S (2006) Vermicomposting of different types of waste using Eisenia fetida: a comparative study. Bioresour Technol 97(3):391–395

Ghatnekar SD, Kavian MF, Sharma SM, Ghatnekar SS, Ghatnekar GS, Ghatnekar AV (2010) Application of vermi-filter-based effluent treatment plant (pilot scale) for biomanagement of liquid effluents from the gelatine industry. Dynamic soil, Dynamic plant 4(1):83–88

Hand P, Hayes WA, Satchell JE, Frankland JC (1988) Vermicomposting of cow slurry. In: Edwards CA, Neuhauser EF (eds) Earthworms in waste and environmental management. SPB Academic Publishing, The Hague

Hemalatha B (2013) Application of vermicomposting for the biodegradation of MSW and crop improvement. Int J Adv Eng Technol 4(1):2–30

Huang H, Schwab K, Jacangelo JG (2009) Pretreatment for low pressure membranes in water treatment: a review. Environ Sci Technol 43(9):3011–3019

Huang J, Lou I, Li Y (2016) Integrated water resources management for emergency situations: a case study of Macau. J Environ Sci 50:72–78

Huang Q, Liu M, Zhao J, Chen J, Zeng G, Huang H et al (2018) Facile preparation of polyethylenimine-tannins coated SiO_2 hybrid materials for Cu^{2+} removal. Appl Surf Sci 427:535–544

Hughes RJ, Nair J, Ho G (2008) The toxicity of ammonia/ammonium to the vermifiltration wastewater treatment process. Water Sci Technol 58(6):1215–1220

Husain A, Javed I, Khan NA (2014) Characterization and treatment of electroplating industry wastewater using Fenton's reagent. J Chem Pharm Res 6(1):622–627

Khan NA, Khan KA, Islam M (2012) Water and wastewater treatment using nanotechnology. In: Chemistry of phytopotentials: health, energy and environmental perspectives. Springer, Berlin, pp 315–318

Khan, N. A., Khan, S. U., Ahmed, S., Farooqi, I. H., Hussain, A., Vambol, S., & Vambol, V. (2019a). Smart ways of hospital wastewater management, regulatory standards and

conventional treatment techniques. Smart and Sustainable Built Environment 10.1108/SASBE-06-2019-0079.
Khan NA, Khan SU, Islam DT, Ahmed S, Farooqi IH, Isa MH et al (2019b) Performance evaluation of column-SBR in paper and pulp wastewater treatment: optimization and bio-kinetics. Desalin Water Treat 156:204–219
Kumar T, Bhargava R, Prasad KH, Pruthi V (2015) Evaluation of vermifiltration process using natural ingredients for effective wastewater treatment. Ecol Eng 75:370–377
Kumar T, Rajpal A, Arora S, Bhargava R, Hari Prasad KS, Kazmi AA (2016) A comparative study on vermifiltration using epigeic earthworm Eisenia fetida and Eudrilus eugeniae. Desalin Water Treat 57(14):6347–6354
Lazcano C, Gómez-Brandón M, Domínguez J (2008) Comparison of the effectiveness of composting and vermicomposting for the biological stabilization of cattle manure. Chemosphere 72(7):1013–1019
Li YS, Robin P, Cluzeau D, Bouché M, Qiu JP, Laplanche A et al (2008) Vermifiltration as a stage in the reuse of swine wastewater: monitoring methodology on an experimental farm. Ecol Eng 32(4):301–309
Li X, Xing M, Yang J, Lu Y (2013a) Properties of biofilm in a vermifiltration system for domestic wastewater sludge stabilization. Chem Eng J 223:932–943
Li X, Xing M, Yang J, Zhao L, Dai X (2013b) Organic matter humification in vermifiltration process for domestic sewage sludge treatment by excitation-emission matrix fluorescence and Fourier transform infrared spectroscopy. J Hazard Mater 261:491–499
Lim SL, Wu TY, Clarke C (2014) Treatment and biotransformation of highly polluted agro-industrial wastewater from a palm oil mill into vermicompost using earthworms. J Agric Food Chem 62(3):691–698
Lim SL, Wu TY, Lim PN, Shak KPY (2015) The use of vermicompost in organic farming: overview, effects on soil and economics. J Sci Food Agric 95(6):1143–1156
Lim SL, Lee LH, Wu TY (2016) Sustainability of using composting and vermicomposting technologies for organic solid waste biotransformation: recent overview, greenhouse gas emissions, and economic analysis. J Clean Prod 111:262–278
Liu XQ, Yan BX, Liu SY, Zhu H (2013) Influence of outside environmental variations on ammonia nitrogen adsorption characteristics of HVMT/PC/EPDM composite. Environ Earth Sci 69 (8):2541–2548
Lom T (1977) A new trend in the treatment of alkaline brewery effluents. Tech Quart Master Brewers Assoc Am 14:50–58
López-González JA, del Carmen Vargas-García M, López MJ, Suárez-Estrella F, del Mar Jurado M, Moreno J (2015) Biodiversity and succession of mycobiota associated to agricultural lignocellulosic waste-based composting. Bioresour Technol 187:305–313
Luth L, Robin P, Germain P, Lecomte M, Landrain B, Li Y, Cluzeau D (2011) Earthworm effects on gaseous emissions during vermifiltration of pig fresh slurry. Bior Tech 102(4):3679–3686
Lourenço N, Nunes LM (2017) Optimization of a vermifiltration process for treating urban wastewater. Ecol Eng 100:138–146
Manyuchi MM, Mupoperi N, Mbohwa C, Muzenda E (2019) Treatment of wastewater using vermifiltration technology. In: Water conservation, recycling and reuse: issues and challenges. Springer, Singapore, pp 215–230
Masters GM, Ela WP (2004) Introduction to environmental engineering and science, 3rd edn. Pearson, London
Merlin G, Cottin N (2009) Performance of a compost biofilter containing earthworms to treat cheese whey. Environ Technol 30(10):995–1002
Ministry of Environment Forest and Climate Change (MoEF&CC) (2015). envis.nic.in
Narayana T (2009) Municipal solid waste management in India: from waste disposal to the recovery of resources. Waste Manag 29(3):1163–1166

Nogales R, Elvira C, Benitez E, Thompson R, Gomez M (1999) Feasibility of vermicomposting dairy biosolids using a modified system to avoid earthworm mortality. J Environ Sci Health B 34(1):151–169

Nogales R, Cifuentes C, Benitez E (2005) Vermicomposting of winery wastes: a laboratory study. J Environ Sci Health B 40(4):659–673

Oste LA, Dolfing J, Ma WC, Lexmond TM (2001) Cadmium uptake by earthworms as related to the availability in the soil and the intestine. Environ Toxicol Chem Int J 20(8):1785–1791

Pichtel J (2005) Waste management practices: municipal, hazardous, and industrial. CRC press, Boca Raton

Pramanik P, Ghosh GK, Ghosal PK, Banik P (2007) Changes in organic–C, N, P and K and enzyme activities in vermicompost of biodegradable organic wastes under liming and microbial inoculants. Bioresour Technol 98(13):2485–2494

Ramachandra TV, Bharath HA, Kulkarni G, Han SS (2018) Municipal solid waste: generation, composition and GHG emissions in Bangalore, India. Renew Sust Energ Rev 82:1122–1136

Rao AG, Reddy TSK, Prakash SS, Vanajakshi J, Joseph J, Sarma PN (2007) pH regulation of alkaline wastewater with carbon dioxide: a case study of treatment of brewery wastewater in a UASB reactor coupled with absorber. Bioresour Technol 98(11):2131–2136

Samal K, Dash RR, Bhunia P (2017) Treatment of wastewater by vermifiltration integrated with macrophyte filter: a review. J Environ Chem Eng 5(3):2274–2289

Samal K, Mohan AR, Chaudhary N, Moulick S (2019) Application of vermitechnology in waste management: a review on mechanism and performance. J Environ Chem Eng 7:103392. https://doi.org/10.1016/j.jece.2019.103392

Satchell JE (1983) Earthworm microbiology. In: Earthworm ecology. Springer, Dordrecht, pp 351–364

Sheehan GJ, Greenfield PF (1980) Utilisation, treatment, and disposal of distillery wastewater. Water Res 14(3):257–277

Sim EYS, Wu TY (2010) The potential reuse of biodegradable municipal solid wastes (MSW) as feedstocks in vermicomposting. J Sci Food Agric 90(13):2153–2162

Singh R, Bhunia P, Dash RR (2019a) Optimization of organics removal and understanding the impact of HRT on vermifiltration of brewery wastewater. Sci Total Environ 651:1283–1293

Singh R, Samal K, Dash RR, Bhunia P (2019b) Vermifiltration as a sustainable natural treatment technology for the treatment and reuse of wastewater: a review. J Environ Manag 247:140–151

Singleton DR, Hendrix PF, Coleman DC, Whitman WB (2003) Identification of uncultured bacteria tightly associated with the intestine of the earthworm Lumbricus rubellus (Lumbricidae; Oligochaeta). Soil Biology and Biochemistry 35(12):1547–1555

Sinha V, Chakma S (2019) Advances in the preparation of hydrogel for wastewater treatment: a concise review. J Environ Chem Eng 7(5):103295

Sinha RK, Bharambe G, Bapat P (2007) Removal of high BOD and COD loadings of primary liquid waste products from the dairy industry by vermifiltration technology using earthworms. Indian J Environ Prot 27(6):486

Sinha RK, Bharambe G, Chaudhari U (2008) Sewage treatment by vermifiltration with the synchronous treatment of sludge by earthworms: a low-cost sustainable technology over conventional systems with potential for decentralization. Environmentalist 28(4):409–420

Sinha RK, Chauhan K, Valani D, Chandran V, Soni BK, Patel V (2010) Earthworms: Charles Darwin's 'unheralded soldiers of mankind': protective & productive for man & environment. J Environ Prot 1(03):251

Sinha RK, Chandran V, Soni BK, Patel U, Ghosh A (2012) Earthworms: nature's chemical managers and detoxifying agents in the environment: an innovative study on the treatment of toxic wastewaters from the petroleum industry by vermifiltration technology. Environmentalist 32(4):445–452

Solid waste management in Rural India: a guide to gram panchayats (2016). nirdpr.org.in

Suthar S (2007) Vermicomposting potential of Perionyx sansibaricus (Perrier) in different waste materials. Bioresour Technol 98(6):1231–1237

Thiruneelakandan R, Subbulakshmi G (2014) Vermicomposting: a superlative for soil, plant, and environment. International Conference on Engineering Technology and Science-(ICETS 14). Int J Innov Res Sci Eng Technol 3(Special Issue 1), 930–938

Tomar P, Suthar S (2011) Urban wastewater treatment using vermi-biofiltration system. Desalination 282:95–103

Tomati U, Galli E, Pasetti L, Volterra E (1995) Bioremediation of olive-mill wastewaters by composting. Waste Manag Res 13(5):509–518

US Environmental Protection Agency (USEPA) (1999) National air toxics program: the integrated urban strategy. Fed Reg 64(137):38706

Wang Y, Xing MY, Yang J, Lu B (2016) Addressing the role of earthworms in treating domestic wastewater by analyzing biofilm modification through chemical and spectroscopic methods. Environ Sci Pollut Res 23(5):4768–4777

Watteau F, Villemin G (2011) Characterization of organic matter microstructure dynamics during co-composting of sewage sludge, barks, and green waste. Bioresour Technol 102 (19):9313–9317

Xing M, Li X, Yang J (2010) Treatment performance of small-scale vermifilter for domestic wastewater and its relationship to earthworm growth, reproduction and enzymatic activity. Afr J Biotechnol 9(44):7513–7520

Zhang X, Huang Q, Deng F, Huang H, Wan Q, Liu M, Wei Y (2017) Mussel-inspired fabrication of functional materials and their environmental applications: progress and prospects. Appl Mater Today 7:222–238

Chapter 7
Natural Biological Treatment of Effluent and Sludges to Combat the Burden of Waste

Deachen Angmo, Rahil Dutta, Soubam Indra Kumar, and Angelika Sharma

Abstract A huge amount of effluents and sludge (waste) are generated from different industrial, commercial, and domestic sectors, which are being discharged on land or into watercourse. These wastes are characterized by high concentrations of xenobiotic compounds, heavy metals, pesticides, high biochemical oxygen demand (BOD) and chemical oxygen demand (COD), which is responsible for degradation of environment and causes death of millions of people annually. It is very important to treat these wastes before being released into the environment. If the waste discharge is in liquid form, then it is called effluents, and if in solid form, it is called sludge. The process for the removal of impurities contained in waste can be done by conventional and natural treatment method. Majority of the conventional methods hold some limitations due to which natural treatment method is considered and proven to be one of the most prefer methods to deal with the waste generated due to its eco-friendly features. Natural treatment methods involve diverse microbes and plant to treat the waste. Degradation of waste is highly determined by the microbial and plant activity. The process that primarily relies on microbial population either to reduce or to transform the pollutant concentration to an innocuous state is called bioremediation. On the basis of site of treatment, there are two types of bioremediation: ex situ and in situ. Phytoremediation depends upon the plant to carry out the activity to treat the waste. There are several mechanisms involved in phytoremediation that depends on the type of pollutant (elemental or organic). It has been proven that sludge contains a good amount of nutrients, which makes it suitable for agricultural application.

Keywords Pollutants · Heavy metals · Effluents · Sludge · Bioremediation · Phytoremediation

---

D. Angmo · R. Dutta · S. I. Kumar
Department of Botanical and Environmental Science, Guru Nanak Dev University, Amritsar, India

A. Sharma (✉)
Department of Environmental Sciences, Central University of Jammu, Jammu, India

S. A. Bhat et al. (eds.), *Earthworm Assisted Remediation of Effluents and Wastes*,
https://doi.org/10.1007/978-981-15-4522-1_7

7.1 Introduction

Since the dawn of human civilization, man has been exorbitantly using natural resources, and this has given way to the enormous release of many wastewater in the environment. Man is behaving extensively as a polluter. And in this process, release of pollutants is high whether gaseous or liquid pollutants or solid pollutants. Liquid pollutants can be largely categorized as effluents, and solid pollutants can be categorized as sludges. Effluent is largely considered as liquid wastewater whether treated or untreated particularly from a sewage system, industrial unit, or a nuclear plant, whereas sewage is a type of wastewater produced by a community of people. They are also called as gray water and black water. Sewers carry it elsewhere or into an on-site sewage facility. The reality is that most wastewater produced globally remains untreated causing enormous problems of water pollution. The pollutants in the wastewater can be divided into two main groups:

- Potentially toxic elements—heavy metals (cadmium, chromium, copper, mercury, nickel, lead, zinc
- Organic pollutants—polyaromatic hydrocarbons (PAHs), polychlorinated biphenyls (PCB), dioxins, nutrients, and pathogens contained in human feces (bacteria, viruses, protozoans, parasites)

The sources of pollution can be categorized as:

Domestic wastewater—Wastewater from households, institutions
Industrial wastewater—Wastewater released from industries in the manufacture and production process of goods
Urban runoff—Wash-off water during rains from roads and drains

Wastewater carries various components and impurities that are discharged from diffuse and point sources. Pollutants generated by large point sources can be easily identified that result from particular activities in the area. It is much more difficult to identify and quantify the wastewater contribution from small point sources, such as households and small businesses, compared to point sources which are usually regulated. Sludges from different sources are found to be rich in nutrients (nitrogen and phosphorus), organic matter, and trace elements that are highly beneficial for the growth of plants. Due to high nutrient content in sludge, it acts as substitute for commercial fertilizers, and the requirement for fertilizer decreases. The application of sludge in agricultural practices depends on the amount of metal content in the sludge that decides the quality of sludge.

Objectives:

- To identify different toxic effluents and their sources
- To identify different biological and natural processes to treat toxic pollutants
- To find out the potential of selective organisms in remediation process

7.2 Sources of Toxic Effluents

- *Cadmium* primarily comes from launderettes, small electroplating and coating shops, and plastic manufacture industries, and it can be used in alloys, pigments, dental work, paints, photography, Ni-Cd batteries, glazes, artisanal shops, and car repair shops (ATSDR 1999).
- *Chromium* are found in alloys and is released from diffuse sources and products such as preservatives and dying and tanning activities. Chromium III is broadly used as a tanning agent in leather operation. Chromium VI being highly toxic, its uses are now restricted, and there are less commercial sources.
- *Copper* is used in electronics, plating, paper, textile, rubber, fungicides, printing, plastic, and brass and other alloy industries, and it can also be released from various small commercial activities and warehouses, as well as buildings with commercial heating systems (Chikwendu et al. 1989).
- *Lead* is often used as a fuel additive (now greatly reduced or banned in the EU) and in cigarettes, cosmetics, traditional medicine, and lead-glazed ceramics. Paint and pigments contain lead in the form of lead chromates (Romieu et al. 1994).
- *Mercury* is used in the manufacturing of electrical equipment, and it also acts as catalyst in chlor-alkali processes for chlorine and caustic soda production. The main sources in effluent are from clinical thermometers, dental practices, electrical equipment, glass mirrors, and traces in disinfectant products (bleach) and caustic soda solutions.
- *Nickel.* The primary sources of nickels are from corrosion of equipment from small electroplating shops, launderettes, and small and jewelry shops and from old pigments and paints. The mining and refining activities of nickel, burning of residual and fuel oils, and municipal waste incineration are the main anthropogenic sources of nickel emissions in the environment (WHO-1991). These sources account for about 90% of the total global emission, estimated to be 42.85×10^6 kg/year.
- *Zinc* is utilized for processes such as galvanization and brass and bronze alloy production; in products such as batteries, paints, plastics, rubber, fungicides, paper, textiles, taxidermy (zinc chloride) and embalming fluid (zinc chloride), building materials, and special cements (zinc oxide, zinc fluorosilicate); in dentistry (zinc oxide); and also in cosmetics and pharmaceuticals. The current trend towards electrolytic production of zinc which, in contrast to thermally produced zinc has virtually no cadmium contamination. This means that cadmium pollution due to the corrosion of galvanized steel will in time become negligible (Adams and Kramer 1999).
- *Platinum* and platinum group metals (PGMs) such as *palladium* and *osmium* often come from medical and clinical practices, mainly as anti-neoplastic drugs. The amount generated in hospital/clinical effluent has been estimated to be between 115 and 125 ng L^{-1} (Kümmerer and Helmers 1997; Kümmerer et al. 1999) giving a total emission of 84–99 kg per annum from hospitals. Other major

sources of platinum involve laboratories, glass manufacturing, jewelry shops, and petroleum/ammonia processing activities.
- *Silver* is generated mainly from small-scale household products such as polishes, domestic water treatment devices, etc. (Shafer et al. 1998; Adams and Kramer 1999). Studies in Spain showed the presence of elevated concentrations of Cd, Cu, Hg, Pb, and Zn in urban wastewater and in the coastal environment, with high concentrations of copper and zinc possibly due to the use of fungicides in glass houses.

7.3 Sewage Sludge

Sewage sludge is known to be rich in nutrients (nitrogen and phosphorus), organic matter, and trace elements that are beneficial for plant growth and better yield (Kauthale et al. 2005). Sewage sludge is also considered a suitable substitute for commercial fertilizers, and the use of sewage sludge as a fertilizer decreases the requirement for commercial fertilizers (Richards et al. 1998). Commercial fertilizers require large amount of phosphorus, whereas phosphorus is known to be a limited resource. Even though the nitrogen available in commercial fertilizers may not be a limited resource, the production of nitrogen fertilizers demands significant amount of energy. The organic carbon in sludge-amended soil can increase as far as threefold compared to inorganic fertilizer-amended soils (Seaker et al. 1988; Nyamangara and Mzezewa 2001). Inorganic fertilizers usually reduce soil pH and increase the rate of soil acidification as well as increase the percentage of aluminum saturation (Porta et al. 1999). The decrease in pH of soils is traced possibly to the ammonium in the fertilizer. Sludge needs to be treated before releasing it into environment because it contains potentially toxic elements such as zinc (Zn), copper (Cu), lead (Pb), chromium (Cr), and cadmium (Cd).

7.4 Environmental Consequences of Effluents and Sludge

Due to rapid establishment of industries during the last two decades, the disposal of industrial effluents has become a serious problem. The effluents and sludge are often discharged to nearby land or mainly in water bodies. The composition of effluents and sludge differs from industries to industries. Untreated water causes bad odor and foul smell near the point of disposal. It is known that organisms inhabiting areas influenced by effluent discharges can suffer deleterious somatic effects or genetic damage and that people using polluted water could be at higher risk of genotoxic effects and cancer development (Rajaguru et al. 2002). However, industrial wastewater and sludge could be used safely and effectively with proper precautions to increase soil productivity. The utilization of industrial wastes for agricultural purposes could also provide a solution to the disposal problems (Chhonkar et al. 2000).

The properties of water and soil change due to release of toxic chemicals during various operations carried out by industries. The effluents often contain high level of biological oxygen demand (BOD) and chemical oxygen demand (COD), low level of dissolved oxygen (DO), and high concentration of toxic metals such as cadmium, chromium, and lead, polyaromatic hydrocarbons (PAHs), petrochemicals, phenolic compounds, and dioxin. Sludges generated from various industrial sectors have also been utilized in various agricultural uses due to the fact that significant amount of nitrogen, phosphorus, and other organic matters are available in industrial as well as domestic sludge (Warman and Termeer 2005).

7.5 Biological and Natural Treatment of Effluents and Sludge

Wastewater from municipal sewage and industry effluent mainly consists of both organic and inorganic waste. It may also contain various forms of pathogenic organisms, heavy metals, carcinogens, and toxins. So proper treatment of this wastewater before discharging it into the land or to the water source is the need of the hour. The amount of contaminated wastewater can be treated before being discharged into the natural ecosystems (Escapa et al. 2015). In dry climates, a practical alternative to wastewater treatment process is recycling that can help to solve limited water resource problem (Almuktar et al. 2015). Over the past few decades, several methods like physical, chemical, and biological methods have been reported such as flotation, precipitation, oxidation, solvent extraction, evaporation, carbon adsorption, ion exchange, membrane filtration, electrochemistry, biodegradation, and phytoremediation (Berefield et al. 1982; Liu and Liptak 2000; Harvey et al. 2002; Chen 2004; Forgacs et al. 2004; Anjaneyulu et al. 2005; Crini and Badot 2010; Cox et al. 2007; Hai et al. 2007; Barakat 2011; Rathoure and Dhatwalia 2016; Morin-Crini and Crini 2017). But among the given methods, management of wastewater through biological treatment has become a trend in the recent year. It is basically depending upon the biological organisms to break down organic wastes into simpler forms using normal cellular processes. In contrast to physiochemical processes, biological treatment/processes of wastewater can remove organic contaminants like biodegradable organic material with the help of microbiological and plant metabolic activities. Biological treatment is also known as secondary treatment of wastewater in which microbial communities are mainly used, under varying growth conditions, to biochemically decompose organic compounds in the waste that have passed from primary treatment units. Biological treatment methods include aerobic treatment and anaerobic treatment sometimes through phytoremediation also is common. Depending on the availability and utilization of oxygen, the biological processes are classified as aerobic and anaerobic and as per the uses of plant and the organisms as bioremediation and phytoremediation.

(a) *Aerobic treatment*: This treatment is done with the help of aerobic microorganisms (aerobes) in the presence of air (oxygen) which break down the organic impurities present in the waste and convert them into carbon dioxide, water, and simpler biomass. Aerobic wastewater treatment processes include various treatments such as activated sludge process, trickling filters, lagoon-based treatments, aerobic digestion, etc. (Source: Mittal 2011).
(b) *Anaerobic treatment:* This treatment is done with the help of anaerobic microorganisms in the absence of oxygen, which can break down the organic waste matter and convert it into methane, carbon dioxide, and simple biomass (Source: Mittal 2011).

- The various biological and natural treatments of effluents and sludge are given as follows.

7.5.1 Activated Sludge

This process was first operated in the early 1900s with the word "activated" referring to solids that catalyze the degradation of the waste. Although there are many types of activated sludge process, but all have two things in common: they rely on pumping of surplus oxygen into the reactor, and they artificially accelerate the microbial mass in the reactor by isolating and returning back to the activated sludge tank a large portion of microbial cells in the reactor effluents. The activated sludge runs in two unit operations: the activated sludge bioreactor and the secondary clarifier. Since the production of microbial cells in the activated sludge reactor are particles; they can be separated from the water via sedimentation. This is done in the secondary clarifier. Much of the cell mass that is settled out in the clarifier is returned to the bioreactor via the return activated sludge (RAS) line. However, to maintain a steady-state concentration in the bioreactor, a portion must also be wasted through a waste activated sludge (WAS) line to compensate for the new cell mass being synthesized in the activated sludge tank (Master and Ela 1991).

7.5.2 Trickling Filters

Trickling filters were first used in 1893. This method is highly known for the contribution by diversity of life forms such as rotifers, annelid worms, snails, and many insects in treatment of wastewater. Wastewater is sprayed on a circular bed made of plastic packing or other coarse materials with the help of rotating distribution arm. The media used in trickling filter are of two types: rock media and synthetic media (Gonzalez 1996). "First size" rocks were used as packing media and are still in use due to low-cost operations and their filter depth being only 0.9–3 m (3–10 ft) and for synthetic media depth goes up to 12 m (40 ft) (Benefield and Randall 1980), but plastic media are highly preferred lately as they are easier to work with, provide taller structures, and can be manipulated to have much greater surface for microbial growth per volume of filter.

7.5.3 Membrane Bioreactor (MBR)

Membrane bioreactor has been in extensive usage for treatment of domestic waste, but for industrial wastewater treatment applications, its use has been limited or selective (Mittal 2011). MBR draw water from the mixed liquor into hollow fiber membranes submerged in the activated sludge aeration tank, thus avoiding the need for secondary clarifier in the conventional activated sludge process (Master and Ela 1991).

7.5.4 Bioremediation

Bioremediation is a branch of biotechnology that makes use of microorganism such as microbes and bacteria to clean up the contaminants, pollutants, and toxins from soil and water. In other words, it is a technology for removing pollutants from the environment, thus preventing the degradation of original natural surroundings and further reducing the pollution (Sasikumar and Papinazath 2003). Various microbes participate in the remedial process, and cleaning up of waste is highly depending on potential microbial action. Depending on the site where wastes are removed, it can be categorized as ex situ and in situ.

7.5.4.1 Ex Situ

It is a process of withdrawal or removal of polluted soil from ground.

Land Farming

Land farming is considered as one of the simplest techniques due to its low cost and lesser requirements of operational equipment. It allows effective removal of nearly all the constituents. Contaminated soil is removed and spread on the ground surface where the breakdown of constituents takes place by indigenous microbial activity.

Composting

Composting is one of the most widely accepted techniques in the world and is considered as a simple, rewarding way to convert the organic waste into a valuable soil amendments such as manure and organic fertilizer. The presence of organic material in waste promotes the growth of microbes for degradation process. Different microbes grow at different phases depending upon the temperature patterns. The phases involved are:

1. Psychrophilic phase—low-temperature phase
2. Mesophilic phase—medium-temperature phase
3. Thermophilic phase—high-temperature phase

Bio-piles

Bio-piles involve both the process of land farming and composting. Bio-piles are primarily used to treat the surface contaminated with petroleum hydrocarbons. Bio-piles assist a favorable environment for indigenous aerobic and anaerobic microorganisms.

7.5.4.2 In Situ

In Situ is considered as one of the most recommended technique due to its lower cost and less disruption as it allows on-site treatment.

Biosparging

Biosparging involves the pumping of air under pressure below the water table to stimulate the groundwater oxygen concentrations and intensify the rate of biological degradation of contaminants by naturally occurring bacteria (Harekrushna and Kumar 2012).

Bioventing

Bioventing is one of the most commonly used technologies among all in situ technologies. It encourages the growth of indigenous bacteria by supplying the air and nutrients through wells to contaminated soils. It is mostly applicable to treat the soil and groundwater. The technique is highly beneficial to remediate the hydrocarbon and can be used for the contamination in deep under the surface (Philp and Atlas 2005).

Bioaugmentation

Bioaugmentation frequently includes the insertion of microorganisms indigenous or exogenous to the contaminated sites. Two factors that limit the use of added microbial cultures in a land treatment unit: (1) non-indigenous rarely compete with an indigenous population to develop and sustain useful population levels and (2) most soils with long-term exposure to biodegradable waste have indigenous microorganisms that effectively degrade if the land treatment unit is well managed (Harekrushna and Kumar 2012).

7.5.5 Phytoremediation

Phytoremediation is defined as the reduction in the concentration of toxic elements with the help of plants and associated soil microbes in the environments (Greipsson 2011). This technique allows to improve the soil fertility with inputs of organic matter (Mench et al. 2009).

Techniques of phytoremediation consist of phytoextraction (or phytoaccumulation), phytofiltration, phytostabilization, phytovolatilization, and phytodegradation (Alkorta et al. 2004).

7.5.5.1 Phytoextraction

Phytoextraction (also known as phytoaccumulation, phytoabsorption, or phytosequestration) primarily involves the usage of metal-accumulating plants that absorb the contaminants from soil or water and further translocate and accumulate the contaminants in aboveground parts, i.e., shoots (Sekara et al. 2005; Yoon et al. 2006; Rafati et al. 2011). Biochemical process involves metal translocation to shoots which is crucial and desirable for an effective phytoextraction because it is not feasible to harvest the root part (Zacchini et al. 2009; Tangahu et al. 2011). Plants suitable for phytoextraction method are *Sedum alfredii* and *Rumex crispus* (Rani and Dhania 2014).

7.5.5.2 Phytofiltration

Phytofiltration is mainly applicable to treat contaminated surface water or wastewater; basically it involves the cleaning up of pollutants from contaminated sites by the utilization of certain plants (Mukhopadhyay and Maiti 2010). Phytofiltration may be rhizofiltration (use of plant roots) or blastofiltration (use of seedlings) or caulofiltration (use of excised plant shoots; Latin caulis = shoot) (Mesjasz-Przybylowicz et al. 2004).

7.5.5.3 Phytostabilization

Phytostabilization or phytoimmobilization is the use of metal-tolerant plants to inhibit the movement of metals, thus lowering the risk of mobility of leachate to groundwater (Singh 2012; Erakhrumen 2007). Plants used in this method are *Anthyllis vulneraria* and *Festuca arvernensis* (Rani and Dhania 2014). In this technique metals are chemically sequestered by complexation and sorption mechanism within the soils (Barceló and Poschenrieder 2003; Ghosh and Singh 2005; Yoon et al. 2006).

7.5.5.4 Phytovolatilization

Phytovolatilization is the process in which wate soluble contaminants are absorbed by plants root system and released into the atmosphere. The contaminants get modified along the way as the water travels from roots to leaves. This technique can be used for organic pollutants and some heavy metals like Hg and Se. Phytovolatilization is mainly favorable to remediate or treat groundwater, but it can be used to treat soil and sediments. Phytovolatilization is the most controversial of phytoremediation technologies (Padmavathiamma and Li 2007). In this technique *Arabidopsis thaliana* plants are primarily used for the treatment of toxic metals (Rani and Dhania 2014).

7.5.5.5 Rhizodegradation

Rhizodegradation refers to the degradation of organic pollutants by microbial action in the rhizosphere (Mukhopadhyay and Maiti 2010). The exudates released by plant's roots attract and enhance the growth of microbes in the rhizosphere. Rhizosphere extends about 1 mm around the root and is under the influence of the plant (Pilon-Smits 2005). The degradation of contaminates occurs in dynamic rhizosphere.

7.5.6 Constructed Treatment Wetlands

Constructed treatment wetlands are natural ways to treat wastewater by constructing filtration system often planted with reed, reed canary grass, and cattail. In order to prevent clogging and to maintain surface flow, it is important to have filter material with enough permeability. As the wastewater runs through the material, complex physical, chemical, and biological processes occur, thus treating the wastewater.

Vymazal and Kröpfelová (2009) conducted a study on the efficiency of constructed treatment wetlands.

7.5.7 *Water Stabilization Ponds*

Water stabilization ponds are an important phenomenon of natural wastewater treatment. Treatment occurs in aquatic environment under the influence of aquatic wetland bacteria, phytoplankton, and zooplankton and high vegetation. On the basis of availability of oxygen, there are two types of ponds (aerobic pond and anaerobic pond).

Organisms available in stabilization pond are as follows:

- *Producer*: includes mixotrophic green organisms and autotrophic organisms, for example, *Euglena*, *Chlamydomonas*, *Chlorella*, *Chlorogonium*, *Monoraphidium*, and *Cyanobacteria*
- *Consumer*: includes zooplankton such as rotifers, *Daphnia*, *Cladocera*, and copepods.
- *Decomposer*: includes both aerobic and anaerobic bacteria such as *Zoogloea ramigera*, *Chromatium*, coccus, and filamentous bacteria (Husák et al. 1989) (Table 7.1)

7.6 Conclusion

The presence of xenobiotic compounds and heavy metals in environment is natural and harmless if they are present below certain limit. However, human actions have accelerated the levels making these compounds more harmful to environment. Though there are many conventional treatment methods available, all ended up releasing certain chemicals as by-products. These by-products ultimately enter the food chain which lead to biomagnification, thus causing millions of death every year. It is very important to address these contaminants with proper treatment method in order to either remove or convert them into harmless compounds. Biological treatment method is the only technique which treat the waste without any detrimental impacts to the environment. Biological treatment involves diverse form of life to carry out the process. This method preferred to treat the waste because of its low operation cost and less labor requirements and it is an easy-to-operate process. This method is quite specific in nature; a particular bacterium or plant will act on particular contaminants.

Table 7.1 Potential of plants and organisms in remediation methods

Algae	Pollutant/contaminant	Reference
Chlorella vulgaris	Cadmium (II)	Aksu (2001)
Spirogyra	Copper	Lee and Chang (2011)
Ceramium virgatum	Chromium	Sari and Tuzen (2008)
Cladophora hutchinsiae	Selenium	Tuzen and Sari (2010)
Spirogyra condensate, Rhizoclonium hieroglyphicum	Chromium	Onyancha et al. (2008)
Chlorella vulgaris	Diazinon	Kurade et al. (2016)
Spirogyra neglecta	Lead(II), copper(II)	Singh et al. (2007)
Fungi	**Pollutant/contaminant**	**Reference**
Aspergillus versicolor	Chromium, nickel, copper	Taştan et al. (2010)
Cladonia rangiformis	Lead	Ekmekyapar et al. (2012)
Phlebia	Aldrin	Xiao et al. (2011a, b)
Streptomyces consortium	Lindane	Saez et al. (2014)
Lactarius scrobiculatus	Lead(II), copper(II)	Anayurt et al. (2009)
Saccharomyces cerevisiae	Manganese(II)	Fadel et al. (2017)
Plants	**Pollutant/contaminant**	**Reference**
Brassica juncea L.	Cadmium, copper, zinc, lead	Turan and Esringu (2007)
Eichhornia crassipes L.	Arsenic, chromium, zinc, cesium, cobalt	Alvarado et al. (2008), Mishra and Tripathi (2009), Saleh (2012)
Lantana camara L.	Lead	Alaribe and Agamuthu (2015)
Jatropha curcas L.	Iron, aluminum, copper, manganese, chromium, arsenic, zinc, mercury	Jamil et al. (2009), Yadav et al. (2009), Marrugo-Negrete et al. (2015)
Pisum sativum L.	Lead, copper, zinc, iron, cadmium, nickel	Sharma et al. (2010)
Spinacia oleracea L.	Cadmium, iron, nickel, zinc, chromium	Pathak et al. (2013)
Sorghum bicolor L	Cadmium, copper, zinc, iron	Pinto et al. (2004)
Zea mays L.	Cadmium, lead, zinc, copper	Tiecher et al. (2016), Mojiri (2011)
Organisms	**Pollutant/contaminant**	**Reference**
Bacillus subtilis	Chromium, zinc, selenium	Prasad and Freitas (2000)
Desulfovibrio desulfuricans	Chromium, zinc	Banik et al. (2014)
Pseudomonas aeruginosa	Uranium, copper, nickel	Prasad and Freitas (2000)
Pseudomonas veronii	Cadmium, zinc, copper	Vullo et al. (2008)
Sporosarcina ginsengisoli	Arsenic	Achal et al. (2012)
Shewanella oneidensis, Trichococcus pasteurii, and *Pseudomonas aeruginosa*	Chromium (VI)	Daulton et al. (2007), Tandukar et al. (2009)
Pseudomonas aeruginosa	Chlorpyrifos	Kharabsheh et al. (2017)

References

Achal V, Pan X, Fu Q, Zhang D (2012) Biomineralization based remediation of As (III) contaminated soil by Sporosarcina ginsengisoli. J Hazard Mater 201:178–184

Adams NW, Kramer JR (1999) Silver speciation in wastewater effluent, surface waters, and pore waters. Environ Toxicol Chem: Int J 18(12):2667–2673

Aksu Z (2001) Equilibrium and kinetic modelling of cadmium (II) biosorption by C. vulgaris in a batch system: effect of temperature. Sep Purif Technol 21(3):285–294

Alaribe FO, Agamuthu P (2015) Assessment of phytoremediation potentials of Lantana camara in Pb impacted soil with organic waste additives. Ecol Eng 83:513–520

Alkorta I, Hernández-Allica J, Becerril J, Amezaga I, Albizu I, Garbisu C (2004) Recent findings on the phytoremediation of soils contaminated with environmentally toxic heavy metals and metalloids such as zinc, cadmium, lead, and arsenic. Rev Environ Sci Biotechnol 3:71–90

Almuktar SAAAN, Scholz M, Al-Isawi RHK, Sani A (2015) Recycling of domestic wastewater treated by vertical-flow wetlands for irrigating chillies and sweet peppers. Agric Water Manag 149:1–22

Alvarado S, Guédez M, Lué-Merú MP, Nelson G, Alvaro A, Jesús AC, Gyula Z (2008) Arsenic removal from waters by bioremediation with the aquatic plants water hyacinth (Eichhornia crassipes) and lesser duckweed (Lemna minor). Bioresour Technol 99(17):8436–8440

Anayurt RA, Sari A, Tuzen M (2009) Equilibrium, thermodynamic and kinetic studies on biosorption of Pb (II) and Cd (II) from aqueous solution by macrofungus (Lactarius scrobiculatus) biomass. Chem Eng J 151(1–3):255–261

Anjaneyulu Y, Sreedhara Chary N, Samuel Suman Raj D (2005) Decolourization of industrial effluents—available methods and emerging technologies—a review. Arab J Chem 4:361–377

ATSDR U (1999). Toxicological profile for mercury. Agency for toxic substances and disease registry. Atlanta: US Department of Health and Human Services.

Banik S, Das K, Islam M, Salimullah M (2014) Recent advancements and challenges in microbial bioremediation of heavy metals contamination. JSM Biotechnol Biomed Eng 2(1):1035

Barakat MA (2011) New trends in removing heavy metals from industrial wastewater. Arab J Chem 4(4):361–377

Barceló J, Poschenrieder C (2003) Phytoremediation: principles and perspectives. Contrib Sci 2:333–344

Benefield LD, Randall CW (1980) Biological process design for wastewater treatment. Prentice hall

Berefield LD, Judkins JF, Weand BL (1982) Process chemistry for water and wastewater treatment. Prentice-Hall, Englewood Cliffs, 510p

Chen G (2004) Electrochemical technologies in wastewater treatment. Sep Purif Technol 38 (1):11–41

Chhonkar PK, Datta SP, Joshi HC, Pathak H (2000) Impact of industrial effluents on soil health and agriculture-Indian experience: part II-tannery and textile industrial effluents. J Sci Ind Res 59:446–454

Chikwendu VE, Craddock PT, Farquhar RM, Shaw T, Umeji AC (1989) Nigerian sources of copper, lead and tin for the Igbo-Ukwu bronzes. Sci Archaeol 30:27–36

Cox M, Négré P, Yurramendi L (2007) Industrial liquid effluents. INASMET Tecnalia, San Sebastian, p 283. Rev Environ Sci Biotechnol 4:245–273

Crini G, Badot PM (eds) (2010) Sorption processes and pollution. PUFC, Besançon, 489p

Daulton TL, Little BJ, Jones-Meehan J, Blom DA, Allard LF (2007) Microbial reduction of chromium from the hexavalent to divalent state. Geochim Cosmochim Acta 71(3):556–565

Ekmekyapar F, Aslan A, Bayhan YK, Cakici A (2012) Biosorption of Pb (II) by nonliving lichen biomass of Cladonia rangiformis Hoffm. Int J Environ Res 6(2):417–424

Erakhrumen AA (2007) Phytoremediation: an environmentally sound technology for pollution prevention, control and remediation in developing countries. Edu Res Rev 2:151–156

Escapa A, San-Martín MI, Mateos R, Morán A (2015) Scaling-up of membraneless microbial electrolysis cells (MECs) for domestic wastewater treatment: bottlenecks and limitations. Bioresour Technol 180:72–78
Fadel M, Hassanein NM, Elshafei MM, Mostafa AH, Ahmed MA, Khater HM (2017) Biosorption of manganese from groundwater by biomass of Saccharomyces cerevisiae. Hbrc J 13 (1):106–113
Forgacs E, Cserhati T, Oros G (2004) Removal of synthetic dyes from wastewaters: a review. Environ Int 30(7):953–971
Ghosh M, Singh SP (2005) A review on phytoremediation of heavy metals and utilization of it's by products. Appl Ecol Environ Res 3:1–18
Greipsson S (2011) Phytoremediation. Nat Edu Knowled 3(10):7
Gonzalez JF (1996) Wastewater treatment in the fishery industry. Food & Agriculture Org., p 355.
Hai FI, Yamamoto K, Fukushi K (2007) Hybrid treatment systems for dye wastewater. Crit Rev Environ Sci Technol 37(4):315–377
Harekrushna S, Kumar DC (2012) A review on: bioremediation. Int J Res Chem Environ 2(1):13–21
Harvey PJ, Campanella BF, Castro PM, Harms H, Lichtfouse E, Schäffner AR, Smrcek S, Werck-Reichhart D (2002) Phytoremediation of polyaromatic hydrocarbons, anilines and phenols. Environ Sci Pollut Res Int 9:29–47
Husák Š, Sládeček V, Sládečková A (1989) Freshwater macrophytes as indicators of organic pollution. Acta Hydrochim Hydrobiol 17(6):693–697
Jamil S, Abhilash PC, Singh N, Sharma PN (2009) Jatropha curcas: a potential crop for phytoremediation of coal fly ash. J Hazard Mater 172(1):269–275
Kauthale VK, Takawale PS, Kulkarni PK, Daniel LN (2005) Influence of flyash and sewage sludge application on growth and yield of annual crops. Int J Trop Agric 23(1–4):49–54
Kharabsheh HA, Han S, Allen S, Chao SL (2017) Metabolism of chlorpyrifos by Pseudomonas aeruginosa increases toxicity in adult zebrafish (Danio rerio). Int Biodeterior Biodegradation 121:114–121
Kümmerer K, Helmers E (1997) Hospital effluents as a source for platinum in the environment. Sci Total Environ 193(3):179–184
Kümmerer K, Helmers E, Hubner P, Mascart G, Milandri M, Reinthaler F, Zwakenberg M (1999) European hospitals as a source for platinum in the environment in comparison with other sources. Sci Total Environ 225(1–2):155–165
Kurade MB, Kim JR, Govindwar SP, Jeon BH (2016) Insights into microalgae mediated biodegradation of diazinon by Chlorella vulgaris: microalgal tolerance to xenobiotic pollutants and metabolism. Algal Res 20:126–134
Lee YC, Chang SP (2011) The biosorption of heavy metals from aqueous solution by Spirogyra and Cladophora filamentous macroalgae. Bioresour Technol 102(9):5297–5304
Liu DHF, Liptak BG (eds) (2000) Wastewater treatment. CRC Press, Boca Raton
Marrugo-Negrete J, Durango-Hernández J, Pinedo-Hernández J, Olivero-Verbel J, Díez S (2015) Phytoremediation of mercury-contaminated soils by Jatropha curcas. Chemosphere 127:58–63
Master GM, Ela WP (1991) Introduction to environmental engineering and science. Prentice Hall, Upper Saddle River
Mench M, Schwitzguebel J-P, Schroeder P, Bert V, Gawronski S, Gupta S (2009) Assessment of successful experiments and limitations of phytotechnologies: contaminant uptake, detoxification and sequestration, and consequences for food safety. Environ Sci Pollut Res 16:876–900
Mesjasz-Przybylowicz J, Nakonieczny M, Migula P, Augustyniak M, Tarnawska M, Reimold WU, Koeberl C, Przybylowicz W, Glowacka E (2004) Uptake of cadmium, lead, nickel and zinc from soil and water solutions by the nickel hyperaccumulator Berkheya coddii. Acta Biol Cracov Ser Bot 46:75–85
Mishra VK, Tripathi BD (2009) Accumulation of chromium and zinc from aqueous solutions using water hyacinth (Eichhornia crassipes). J Hazard Mater 164(2–3):1059–1063
Mittal A (2011) Biological wastewater treatment. Water Today 1:32–44

Mojiri A (2011) The potential of corn (Zea mays) for phytoremediation of soil contaminated with cadmium and lead. J Biol Environ Sci 5(13):17–22
Morin-Crini N, Crini G (eds) (2017) Eauxindustriellescontaminées. PUFC, Besançon, 513p (in French)
Mukhopadhyay S, Maiti SK (2010) Phytoremediation of metal enriched mine waste: a review. Glob J Environ Res 4:135–150
Nyamangara J, Mzezewa J (2001) Effect of long-term application of sewage sludge to a grazed grass pasture on organic carbon and nutrients of a clay soil in Zimbabwe. Nutr Cycl Agroecosyst 59(1):13–18
Onyancha D, Mavura W, Ngila JC, Ongoma P, Chacha J (2008) Studies of chromium removal from tannery wastewaters by algae biosorbents, Spirogyra condensata and Rhizoclonium hieroglyphicum. J Hazard Mater 158(2–3):605–614
Padmavathiamma PK, Li LY (2007) Phytoremediation technology: hyper- accumulation metals in plants. Water Air Soil Pollut 184:105–126
Pathak C, Chopra AK, Srivastava S (2013) Accumulation of heavy metals in Spinacia oleracea irrigated with paper mill effluent and sewage. Environ Monit Assess 185(9):7343–7352
Philp JC, Atlas RM (2005) Bioremediation of contaminated soils and aquifers: bioremediation. Am Soc Microbiol:139–236
Pilon-Smits E (2005) Phytoremediation. Annu Rev Plant Biol 56:15–39
Pinto AP, Mota AD, De Varennes A, Pinto FC (2004) Influence of organic matter on the uptake of cadmium, zinc, copper and iron by sorghum plants. Sci Total Environ 326(1–3):239–247
Porta J, López M, Roquero C (1999) Edafologıa para la agricultura y el medio ambiente (Edafophology for Agriculture and Environment). Mundi-Prensa, Madrid. (in Spanish)
Prasad MNV, Freitas H (2000) Removal of toxic metals from solution by leaf, stem and root phytomass of Quercus ilex L. (holly oak). Environ Pollut 110(2):277–283
Rafati M, Khorasani N, Moattar F, Shirvany A, Moraghebi F, Hosseinzadeh S (2011) Phytoremediation potential of Populus alba and Morus alba for cadmium, chromium and nickel absorption from polluted soil. Int J Environ Res 5:961–970
Rajaguru P, Vidya L, Baskarasethupathi B, Kumar PA, Palanivel M, Kalaiselvi K (2002) Genotoxicity evaluation of polluted ground water in human peripheral blood lymphocytes using the comet assay. Mutat Res 517(1–2):29–37
Rani K, Dhania G (2014) Bioremediation and biodegradation of pesticide from contaminated soil and water–a noval approach. Int J Curr Microbiol App Sci 3(10):23–33
Rathoure AK, Dhatwalia VK (2016) Toxicity and waste management using bioremediation. IGI Global, Hershey, 421p
Richards BK, Steenhuis TS, Peverly JH, McBride MB (1998) Metal mobility at an old, heavily loaded sludge application site. Environ Pollut 99(3):365–377
Romieu I, Palazuelos E, Hernandez Avila M, Rios C, Muñoz I, Jimenez C, Cahero G (1994) Sources of lead exposure in Mexico City. Environ Health Perspect 102(4):384–389
Saez JM, Álvarez A, Benimeli CS, Amoroso MJ (2014) Enhanced lindane removal from soil slurry by immobilized Streptomyces consortium. Int Biodeterior Biodegradation 93:63–69
Saleh HM (2012) Water hyacinth for phytoremediation of radioactive waste simulate contaminated with cesium and cobalt radionuclides. Nucl Eng Des 242:425–432
Sari A, Tuzen M (2008) Biosorption of total chromium from aqueous solution by red algae (Ceramium virgatum): equilibrium, kinetic and thermodynamic studies. J Hazard Mater 160 (2–3):349–355
Sasikumar CS, Papinazath T (2003) Environmental management: bioremediation of polluted environment. In: Proceedings of the third international conference on environment and health. Department of Geography, University of Madras, Chennai and Faculty of Environmental Studies, York University, New York, pp 465–469
Seaker EM, Sopper WE (1988) Municipal sludge for minespoil reclamation: II. Effects on organic matter. J Environ Qual 17(4):598–602

Sekara A, Poniedzialeek M, Ciura J, Jedrszczyk E (2005) Cadmium and lead accumulation and distribution in the organs of nine crops: implications for phytoremediation. Pol J Environ Stud 14:509–516

Shafer MM, Overdier JT, Armstong DE (1998) Removal, partitioning, and fate of silver and other metals in wastewater treatment plants and effluent-receiving streams. Environ Toxicol Chem: Int J 17(4):630–641

Sharma S, Sharma P, Mehrotra P (2010) Bioaccumulation of heavy metals in Pisum sativum L. growing in fly ash amended soil. J Am Sci 6:43–50

Singh S (2012) Phytoremediation: a sustainable alternative for environmental challenges. Int J Gr Herb Chem 1:133–139

Singh A, Kumar D, Gaur JP (2007) Copper (II) and lead (II) sorption from aqueous solution by non-living Spirogyra neglecta. Bioresour Technol 98(18):3622–3629

Tandukar M, Huber SJ, Onodera T, Pavlostathis SG (2009) Biological chromium (VI) reduction in the cathode of a microbial fuel cell. Environ Sci Technol 43(21):8159–8165

Tangahu BV, Abdullah S, Rozaimah S, Basri H, Idris M, Anuar N, Mukhlisin M (2011) A review on heavy metals (As, Pb, and Hg) uptake by plants through phytoremediation. International Journal of Chemical Engineering, 2011

Taştan BE, Ertuğrul S, Dönmez G (2010) Effective bioremoval of reactive dye and heavy metals by Aspergillus versicolor. Bioresour Technol 101(3):870–876

Tiecher TL, Ceretta CA, Ferreira PA, Lourenzi CR, Tiecher T, Girotto E et al (2016) The potential of Zea mays L. in remediating copper and zinc contaminated soils for grapevine production. Geoderma 262:52–61

Turan M, Esringu A (2007) Phytoremediation based on canola (Brassica napus L.) and Indian mustard (Brassica juncea L.) planted on spiked soil by aliquot amount of Cd, Cu, Pb, and Zn. Plant Soil Environ 53(1):7

Tuzen M, Sari A (2010) Biosorption of selenium from aqueous solution by green algae (Cladophora hutchinsiae) biomass: equilibrium, thermodynamic and kinetic studies. Chem Eng J 158 (2):200–206

Vullo DL, Ceretti HM, Daniel MA, Ramírez SA, Zalts A (2008) Cadmium, zinc and copper biosorption mediated by pseudomonas veronii 2E. Bioresour Technol 99(13):5574–5581

Vymazal J, Kröpfelová L (2009) Removal of organics in constructed wetlands with horizontal sub-surface flow: a review of the field experience. Sci Total Environ 407(13):3911–3922

Warman PR, Termeer WC (2005) Evaluation of sewage sludge, septic waste and sludge compost applications to corn and forage: yields and N, P and K content of crops and soils. Bioresour Technol 96(8):955–961

Xiao P, Mori T, Kamei I, Kiyota H, Takagi K, Kondo R (2011a) Novel metabolic pathways of organochlorine pesticides dieldrin and aldrin by the white rot fungi of the genus Phlebia. Chemosphere 85(2):218–224

Xiao P, Mori T, Kondo R (2011b) Biotransformation of the organochlorine pesticide trans-chlordane by wood-rot fungi. New Biotechnol 29(1):107–115

Yadav SK, Juwarkar AA, Kumar GP, Thawale PR, Singh SK, Chakrabarti T (2009) Bioaccumulation and phyto-translocation of arsenic, chromium and zinc by Jatropha curcas L.: impact of dairy sludge and biofertilizer. Bioresour Technol 100(20):4616–4622

Yoon J, Cao X, Zhou Q, Ma LQ (2006) Accumulation of Pb, Cu, and Zn in native plants growing on a contaminated Florida site. Sci Total Environ 368:456–464

Zacchini M, Pietrini F, Mugnozza GS, Iori V, Pietrosanti L, Massacci A (2009) Metal tolerance, accumulation and translocation in poplar and willow clones treated with cadmium in hydroponics. Water Air Soil Pollut 197:23–34

Part IV
General Organic/Inorganic and Chemical Waste

Chapter 8
Vermicomposts Are Biologically Different: Microbial and Functional Diversity of Green Vermicomposts

María Gómez-Brandón, Manuel Aira, and Jorge Domínguez

Abstract Vermicomposting, the process by which organic waste is broken down through the synergistic actions of earthworms and microbial communities, fulfils the dual purpose of environmental protection and fertilizer production. Vermicompost is considered a nutrient-rich and biologically active organic amendment with various beneficial effects when used as a soil conditioner and/or as a plant growth promoter. The underlying biological mechanisms involved in vermicomposting largely determine the dynamics of the process and, consequently, the properties of the end product. However, it is still necessary to further explore the plausible microbial-based mechanisms by which vermicompost may exert a positive influence on plant growth. Therefore, the purpose of this chapter is to compare and provide a detailed characterization of the microbiome of different green vermicomposts by applying 16S rRNA high-throughput sequencing. An in-depth characterization of the taxonomic and functional diversity of bacterial communities and their metabolic functions was assessed in the vermicomposts and the respective parent materials.

Keywords Sustainable agriculture · Plant material · Bacterial community composition · Organic amendment · Soil conditioner · Plant growth promoter

8.1 Introduction

The bio-conversion of biomass waste into organic amendments permits the recycling of large amounts of organic waste that could otherwise cause a serious threat to the environment (Chew et al. 2019). In the last two decades, vermicomposting technology has expanded to seek its potential role in sustainable agriculture (Lazcano and Domínguez 2011) and soil bioremediation (Sánchez-Hernández and Domínguez 2019). Vermicomposting, the process by which organic waste is broken down

M. Gómez-Brandón (✉) · M. Aira · J. Domínguez
Grupo de Ecoloxía Animal (GEA), Universidade de Vigo, Vigo, Spain
e-mail: mariagomez@uvigo.es

S. A. Bhat et al. (eds.), *Earthworm Assisted Remediation of Effluents and Wastes*, https://doi.org/10.1007/978-981-15-4522-1_8

through the synergistic actions of earthworms and microbial communities, yields a nutrient-rich and biological active organic amendment known as vermicompost (Domínguez et al. 2010). Vermicomposting therefore fulfils the purpose of disposing of and recycling increasing amounts of organic waste and at the same time offering an alternative to inorganic fertilizers (Lazcano and Domínguez 2011; Gómez-Brandón and Domínguez 2014).

Beneficial effects of vermicompost on plant growth have been documented in a wide variety of agronomic and horticultural crops (Lazcano and Domínguez 2011; Blouin et al. 2019). In addition, extracts and teas made from vermicompost also exhibited positive influence when used as plant growth promoters or soil conditioners (Gómez-Brandón et al. 2015), even though they do not impart the same physical properties as vermicompost amendment. These beneficial effects have been partially attributed to the physico-chemical properties of vermicomposts including their buffering capacity, their high porosity and water-holding capacity and low C to N ratio. The high content of nutrients both in their mineral and organic forms in vermicompost has also been proposed as a plausible, albeit not exclusive, mechanism by which such improvement is achieved (Blouin et al. 2019). Indeed, Lazcano et al. (2013) observed that the beneficial effects of vermicompost amendment on sweet corn yields persisted even when it represented a small portion (25%) of the total amount of nutrients supplied into the soil. This suggests that besides the amount of organic matter and nutrient content that the vermicompost provides, it is necessary to consider other concomitant mechanisms to get the whole picture about the potential role of vermicompost as a plant growth promoter and/or soil conditioner.

Organic amendments contain an endogenous active microbiome that may exert a long-term effect on the productivity and sustainability of agro-ecosystems (Mas-Carrió et al. 2018; Domínguez et al. 2019). Focusing on this aspect is therefore of particular interest in order to broaden our understanding about the role of vermicompost as a plant biostimulant and unravel how its addition into soil may enhance nutrition efficiency and/or crop quality traits, regardless of its nutrient content. In a recent meta-analysis, Medina-Sauza et al. (2019) emphasized the importance of how earthworm-induced changes in the soil microbiota largely impact soil processes, particularly those occurring in the rhizosphere and related to plant growth and health. However, it is still necessary to further evaluate the vermicompost microbiome to shed light onto the plausible microbial-based mechanisms by which the use of vermicompost as an organic amendment may exert a positive influence on plant growth, similar to that when earthworms are present.

The underlying biological mechanisms involved in vermicomposting largely determine the dynamics of the process and, consequently, the properties of the final vermicompost for its further use as a plant growth promoter and in plant disease suppressiveness (Gómez-Brandón and Domínguez 2014). Having this in mind, recent studies from our research group have deeply explored the compositional changes and functional capabilities of the bacterial communities over the course of the vermicomposting process of different plant materials including grape marc (Kolbe et al. 2019; Gómez-Brandón et al. 2019) and the leguminous shrub Scotch broom (Domínguez et al. 2019). Although these substrates have long been considered to have low value and have usually been discarded, if properly treated, they can be converted into high-nutrient biofertilizers and used as growing media for vegetables and ornamentals (Cai et al. 2018; Gong et al. 2018). Their transformation into green vermicomposts may therefore offer a dual purpose, i.e., environmental protection and fertilizer production.

The purpose of this chapter is therefore to compare and provide a detailed characterization of the microbiome of the green vermicomposts derived from raw and distilled grape marcs of the grape variety *Vitis vinifera* v. Albariño and those obtained from the leguminous shrubs Scotch broom and acacia. We evaluated the taxonomic and functional diversity of bacterial communities and their metabolic functions in the above-mentioned green vermicomposts and the respective raw materials by applying 16S rRNA high-throughput sequencing.

The four types of plant material were separately processed in a pilot-scale vermireactor (6 m^3) housed in a greenhouse with no temperature control located in our research facilities using the epigeic earthworm *Eisenia andrei*. This earthworm species has been widely used in vermicomposting facilities due to its high reproductive rate and its tolerance to a wide range of temperature and moisture conditions (Domínguez and Edwards 2011). The vermireactor set-up and sampling procedure were performed as described in Domínguez et al. (2019).

8.2 How Does Vermicomposting Influence Composition of Bacterial Communities from Dead Plant Material?

For the evaluation of the bacterial community composition of the fresh materials and the resulting vermicomposts, we amplified and sequenced a fragment ~250 bp long of the 16S rRNA gene covering the V4 region using a dual-index sequencing strategy as described by Kozich et al. (2013). DADA2 (version 1.9) was used to infer the amplicon sequence variants (ASVs) present in each sample (Callahan et al. 2016). Default settings were used for ASV inference and chimera detection following the DADA2 pipeline tutorial (https://benjjneb.github.io/dada2/tutorial.html). The taxonomic assignment was performed against the Silva v132 database using the *assignTaxonomy* function in dada2, which implements RDP naive Bayesian classifier (Wang et al. 2007; Quast et al. 2013). The minimum bootstrap confidence for assigning taxonomy was 80. By using an ASV-based approach, it provides a more accurate and reproducible description of amplicon-sequenced communities than an OTU-based approach (Callahan et al. 2017). This relies on the fact that an ASV is defined by a unique sequence that is different from sequences of other ASVs. Moreover, ASVs could differ by as little as one base pair allowing a direct and more straightforward comparison across studies whenever researchers use the same primer set and procedure to obtain the ASVs.

8.2.1 Compositional Changes of Bacterial Communities After Vermicomposting

Vermicomposting led to significant changes in bacterial community composition at phylum level (Fig. 8.1). The fresh plant materials and the respective vermicomposts were found to group separately in two different clades (Fig. 8.1), as shown in the phylogenetic tree inferred from the sequence data using FastTree 2.1 (Price et al. 2010). The raw plant materials were split into two additional clades (Fig. 8.1 top), one comprised by the fresh samples of Scotch broom and acacia substrates and another one by the initial samples of raw and distilled grape marcs. Within each of these clades, there was also a clear differentiation between the two leguminous shrubs and the two winemaking by-products (Fig. 8.1). The dissimilarity of bacterial communities among samples at ASV levels was based on the relative abundance of the dominant bacterial phyla in each sample (Fig. 8.1 barplot). *Proteobacteria* was the most prevalent phylum in the four types of plant material accounting for almost 100% of the sequences in acacia and Scotch broom initial samples and nearly 60–70% of the sequences in the fresh samples of raw and distilled grape marcs (Fig. 8.1). In the case of these two latter substrates, the bacterial community composition was also composed of *Firmicutes*, with minor contributions of *Bacteroidetes* and *Actinobacteria* (Fig. 8.1).

Proteobacteria continued to make up a significant proportion of the bacterial communities within the vermicomposts' microbial composition (Fig. 8.1). However, at ASV level, we observed that 90% of ASVs were exclusively found in the vermicomposts representing in average 90% of the sequences of each vermicompost. Additionally, we tested for differentially abundant bacterial taxa between the fresh plant materials and the respective vermicomposts by using the DESeq2 package (Love et al. 2014). In particular, there was a pronounced reduction in the differential abundance of *Firmicutes* in the vermicomposts derived from raw and distilled grape marcs when compared to the initial samples (*raw*: mean of −9.51 log2 fold change, $p < 0.0001$; *distilled*: mean of −8.56 log2 fold change, $p < 0.0001$, respectively). In contrast, the differential abundance of *Bacteroidetes* (*raw*: mean of 1.50 log2 fold change, $p = 0.029$; *distilled*: mean of 1.79 log2 fold change, $p = 0.007$); *Planctomycetes* (*raw*: mean of 7.81 log2 fold change, $p < 0.0001$; *distilled*: mean of 3.20 log2 fold change, $p < 0.0001$); and *Verrucomicrobia* (*raw*: mean of 4.86 log2 fold change, $p < 0.0001$; *distilled*: mean of 3.92 log2 fold change, $p < 0.0001$) significantly increased after vermicomposting of both raw and distilled grape marcs. Moreover, higher differential abundances of the phyla *Bacteroidetes* (mean: 1.69 log2 fold change, $p = 0.014$), *Verrucomicrobia* (mean: 4.99 log2 fold change, $p < 0.0001$) and *Planctomycetes* (mean: 5.17 log2 fold change, $p < 0.0001$) were also found in the Scotch broom-derived vermicompost when compared to the fresh plant material. Similarly, the acacia-derived vermicompost was characterized by higher abundances of the phyla *Verrucomicrobia* (mean: 3.15 log2 fold change, $p < 0.0001$) and *Bacteroidetes* (mean: 2.99 log2 fold change, $p < 0.0001$) and by

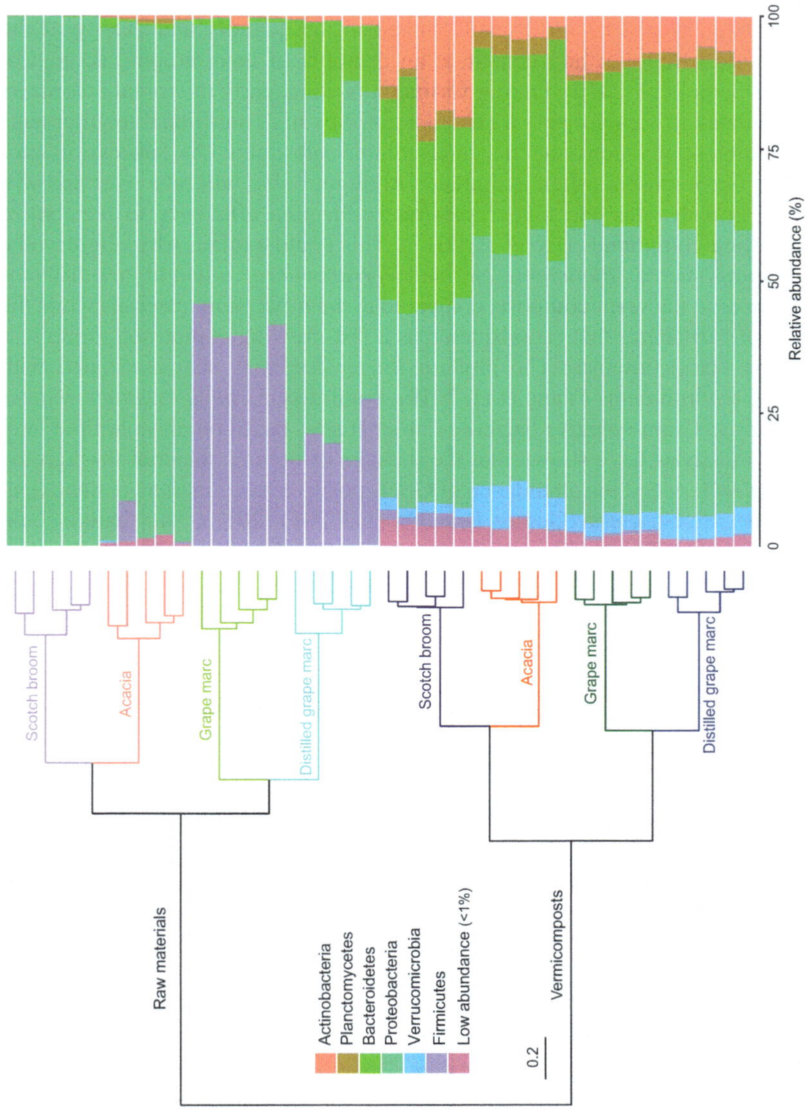

Fig. 8.1 Changes in the bacterial community composition (phylum level) of the fresh plant materials (Scotch broom, acacia, raw and distilled grape marcs) and the respective vermicomposts. The dendrogram represents the dissimilarity of bacterial communities between raw materials and vermicomposts at ASV levels (unweighted UniFrac distances, Ward method). Bars represent the relative abundance of dominant bacterial phyla. Bacterial phyla with relative abundances lower than 1% were grouped together

lower abundances of the phylum *Firmicutes* (mean: −5.14 log2 fold change, $p < 0.0001$).

These differences in microbial community composition between the vermicomposts and the respective raw materials might be a consequence of the successional changes that take place over the course of the vermicomposting process. Such changes will be inextricably linked to the quantity and quality of the available nutrient supplies and constitute an example of heterotrophic ecological succession (Fierer et al. 2010). During vermicomposting, early microbial colonizers (often assimilated to copiotrophs) are characterized by high nutritional requirements and will preferentially consume rich and soluble substrates. As succession progresses, late decomposers (often assimilated to oligotrophs) will take over and will exhibit low growth rates and higher substrate utilization efficiency to metabolize complex carbon compounds (Ho et al. 2017). This could explain why oligotrophic bacterial groups such as *Planctomycetes* and *Verrucomicrobia* had higher differential abundances in the vermicomposts compared to the fresh plant materials while the abundance of *Firmicutes* that is considered as a fast-growing copiotroph group was lower in the vermicomposts. The phylum *Bacteroidetes* is also associated with copiotrophic environments (Yang et al. 2019), but unlike *Firmicutes*, *Bacteroidetes* mostly comprise Gram-negative bacteria. Earlier studies based on phospholipid fatty acid (PLFA) analysis have shown that the passage of organic material through the gut of *E. andrei* reduced the abundance of Gram-positive bacteria to a greater extent than Gram-negative bacteria (Gómez-Brandón et al. 2011, 2012). This feature could have favoured that *Bacteroidetes*, in contrast to *Firmicutes*, appeared in higher abundances in the vermicomposts than in the initial substrates.

8.2.2 Compositional Changes of Bacterial Communities Among Vermicomposts

The vermicomposts could be differentiated from each other at phylum level in line with the type of plant material (Fig. 8.1 bottom). The percentage of shared ASVs and sequences between each pair of vermicomposts is shown in Fig. 8.2. Testing for differentially abundant bacterial taxa between pairs of vermicomposts was also achieved using the DESeq2 package. When comparing the bacterial community composition of the vermicompost samples derived from the raw grape marc to those obtained from distilled marc, we found that five bacterial phyla significantly differed in abundance. They comprised the phyla *Patescibacteria* (mean: −8.85 log2 fold change, $p = 0.003$), *Acidobacteria* (mean: 2.34 log2 fold change, $p = 0.006$), *Nitrospirae* (mean: 5.95 log2 fold change, $p = 0.011$), *Gemmatimonadetes* (mean: 3.71 log2 fold change, $p = 0.010$) and *Armatimonadetes* (mean: 5.09 log2 fold change, $p = 0.003$). The mean log2 fold change values were either positive or negative depending whether the above-mentioned phyla appeared in higher or lower differential abundances in the raw grape marc vermicompost samples.

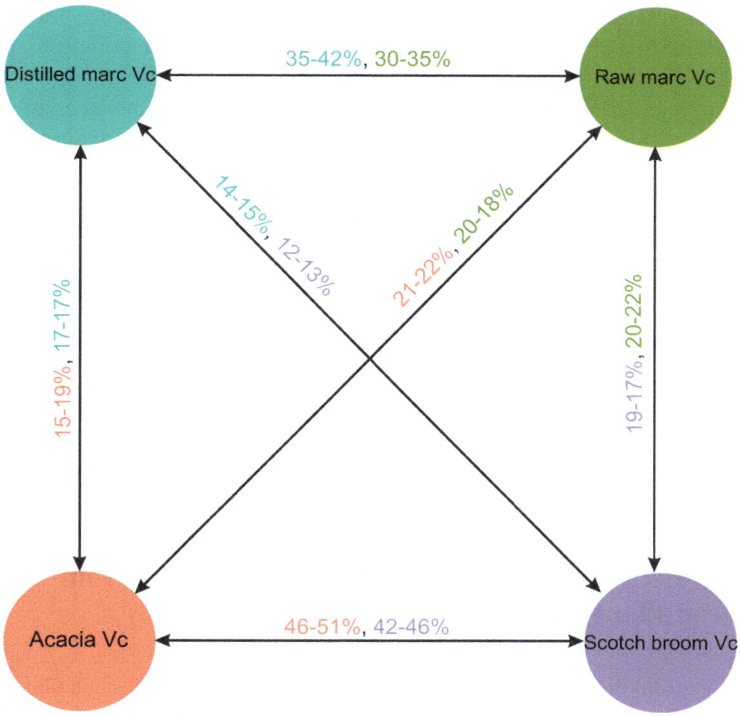

Fig. 8.2 Degree of sharing of native ASVs among vermicomposts (abbreviated as Vc). The shared ASVs and sequences between each pair of vermicomposts are shown as percentage. The colour given to the percentage values is in line with the type of plant material, and it represents the native ASVs (first value before the hyphen) and sequences (second value after the hyphen) shared by this type of vermicompost with respect to the other vermicomposts

Moreover, higher differential abundances of the phyla *Chloroflexi* (mean: 4.10 log2 fold change, $p = 0.0001$), *Proteobacteria* (mean: 1.02 log2 fold change, $p = 0.028$) and *Armatimonadetes* (mean: 4.89 log2 fold change, $p = 0.003$) were detected in the vermicompost samples obtained from the raw grape marc when compared to those derived from the Scotch broom plant material. The opposite trend was found for the phylum *Fibrobacteres* (mean: −7.82 log2 fold change, $p = 0.048$), with lower differential abundances in the raw grape marc-derived vermicompost. If we compared the raw grape marc- and acacia-derived vermicomposts, *Actinobacteria* (mean: 2.01 log2 fold change, $p = 0.004$), *Patescibacteria* (mean: −6.85 log2 fold change, $p = 0.026$), *Spirochaetes* (mean: −6.04 log2 fold change, $p = 0.002$) and *Armatimonadetes* (mean: 3.48 log2 fold change, $p = 0.049$) were found to significantly differ in terms of abundance between this pair of vermicomposts.

The comparison of the bacterial community composition of the distilled grape marc-derived vermicompost and that from Scotch broom- and acacia vermicomposts resulted in a total of ten and seven phyla that differed in abundance for the pairs distilled marc-Scotch broom and distilled marc-acacia vermicompost samples, respectively. Some of these phyla including *Acidobacteria*, *Spirochaetes*, *Nitrospirae* and *Gemmatimonadetes* had lower differential abundances in the distilled grape marc-derived vermicompost compared to acacia and Scotch broom vermicomposts. Others like *Chloroflexi*, *Verrucomicrobia*, *Patescibacteria* and *Actinobacteria* reached higher abundances in the vermicompost samples from distilled grape marc according to DESeq results.

While lower differential abundances of *Verrucomicrobia* (mean: −2.02 log2 fold change, $p = 0.002$) and *Patescibacteria* (mean: −6.92 log2 fold change, $p = 0.024$) were recorded in Scotch broom vermicompost samples, higher abundances of *Actinobacteria* (mean: 2.08 log2 fold change, $p = 0.003$) and *Firmicutes* (mean: 2.75 log2 fold change, $p = 0.003$) were found in these vermicompost samples compared to those obtained from the acacia plant material.

8.3 How Does Vermicomposting Influence Alpha- and Beta-Diversity of Bacterial Communities from Dead Plant Material?

In this chapter taxonomic α-diversity of bacterial communities was calculated as the number of observed ASVs, while phylogenetic diversity was assessed as Faith's phylogenetic diversity (Faith 1992). An increase in α-diversity was reported for all of the four vermicomposts in comparison with the respective fresh materials at both taxonomic and phylogenetic levels (Fig. 8.3a, b). The Scotch broom-derived vermicompost harboured the highest α-diversity at taxonomic level (689 ± 35), being 17 times greater than in the initial substrate (Fig. 8.3a), while no differences were observed among the other three vermicomposts (Fig. 8.3a). A similar pattern emerged for the Faith phylogenetic diversity (Fig. 8.3b), and in this case the highest values were reported for both the Scotch broom- and acacia-derived vermicomposts (Fig. 8.3b).

The increase in α-diversity was reflected in different patterns of taxonomic and phylogenetic β-diversity (Fig. 8.4a, b). Taxonomic β-diversity at the ASV level was estimated as the difference in the composition of the bacterial taxonomic community between the fresh plant materials and the respective vermicomposts. This was done by coupling principal coordinate analysis (PCoA) with distance matrices that take the abundance of ASVs into account (Bray–Curtis). Phylogenetic β-diversity was estimated by PCoA of weighted (considering abundance of ASVs) UniFrac matrix distances (Lozupone and Knight 2005) by using the phyloseq library (Love et al. 2014). Principal coordinate analysis showed that the fresh plant materials differed from each other in terms of β-diversity at both taxonomic and phylogenetic levels (Fig. 8.4a, b). In addition, they also grouped separately from the respective vermicomposts (Fig. 8.4a, b). This implies that the bacterial community present in

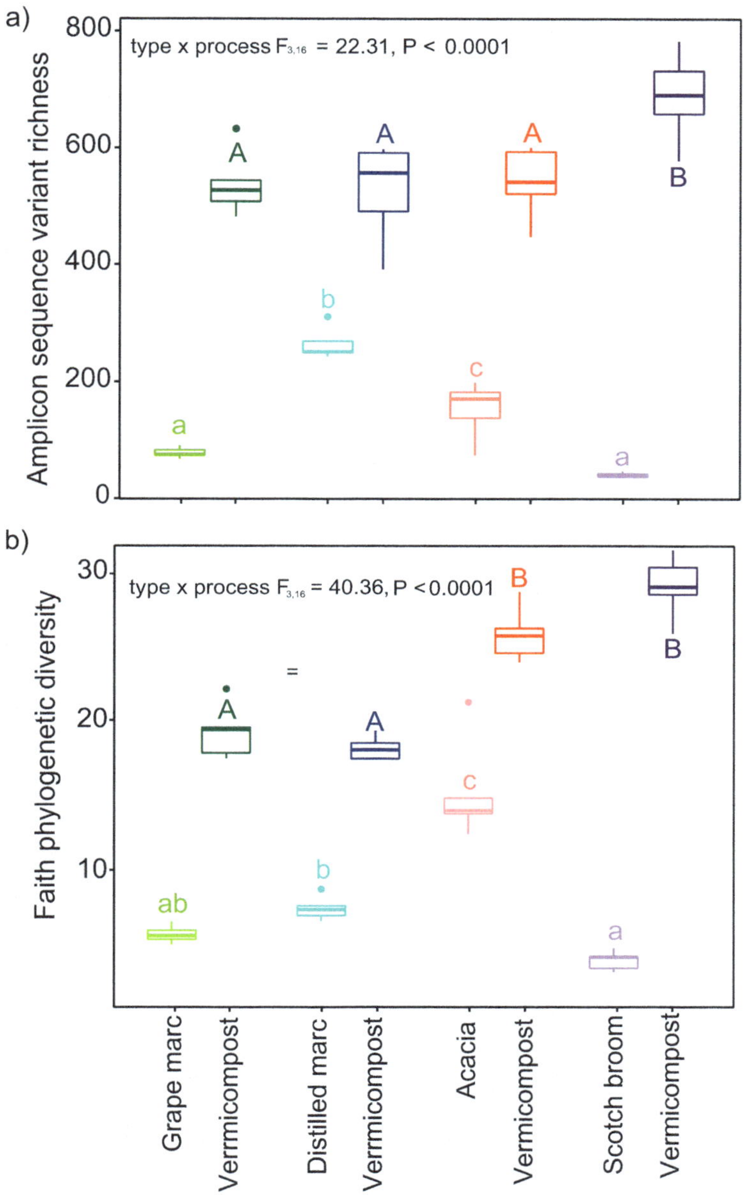

Fig. 8.3 Changes in bacterial α-diversity estimated as (**a**) observed ASV richness and (**b**) Faith phylogenetic diversity of the fresh plant materials (Scotch broom, acacia, raw and distilled grape marcs) and the respective vermicomposts. All of the four plant materials were significantly different from their respective vermicomposts ($P < 0.05$). Different lowercase and capital letters over the boxplots indicate significant differences among fresh plant materials and vermicomposts, respectively (Tukey HSD test, FDR corrected)

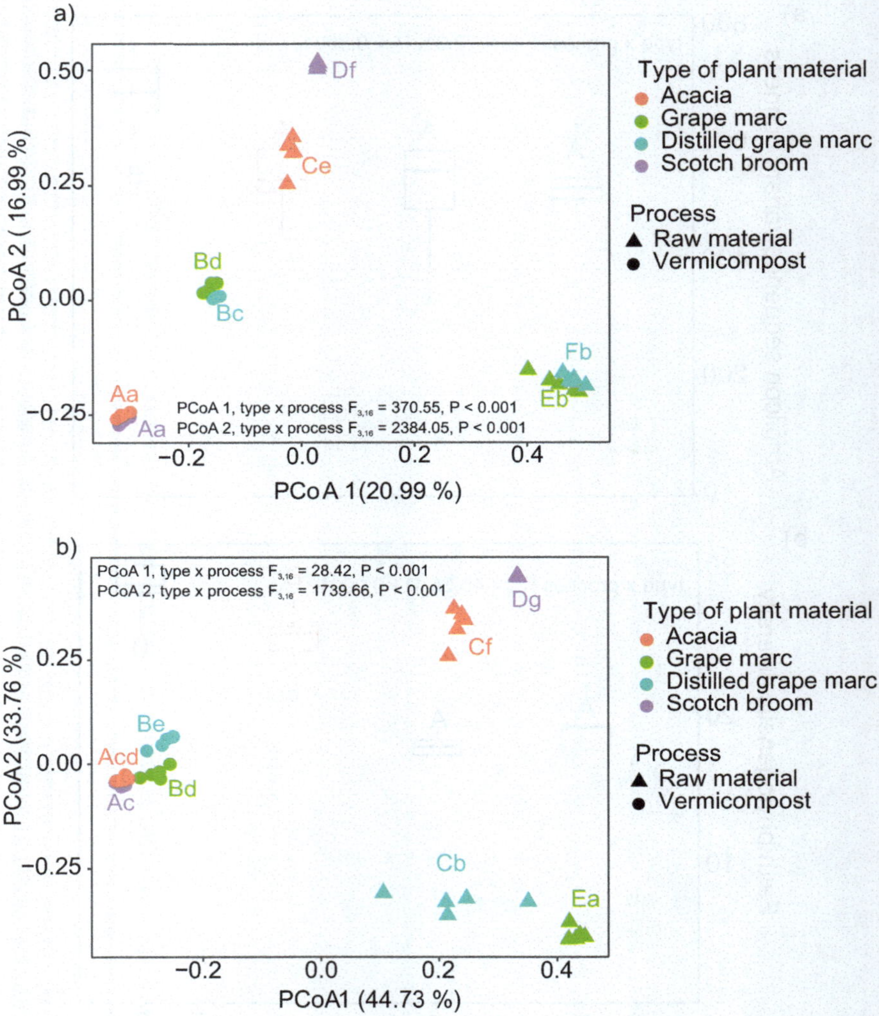

Fig. 8.4 Changes in bacterial β-diversity assessed as principal coordinate analysis of (**a**) Bray–Curtis and (**b**) weighted UniFrac distances of the fresh plant materials (Scotch broom, acacia, raw and distilled grape marcs) and the respective vermicomposts. Capital and lowercase letters indicate significant differences among fresh plant materials and vermicomposts in PCoA1 and PCoA2 scores, respectively (Tukey HSD test, FDR corrected)

the parent materials was strongly affected by the earthworms' activity over the course of the vermicomposting process. Earthworms can directly modify the microbial diversity of the substrate by selectively feeding on and/or stimulating specific taxa during transit through the earthworm gut (Aira et al. 2015). The changes occurred through the gut-associated processes may have further consequences on

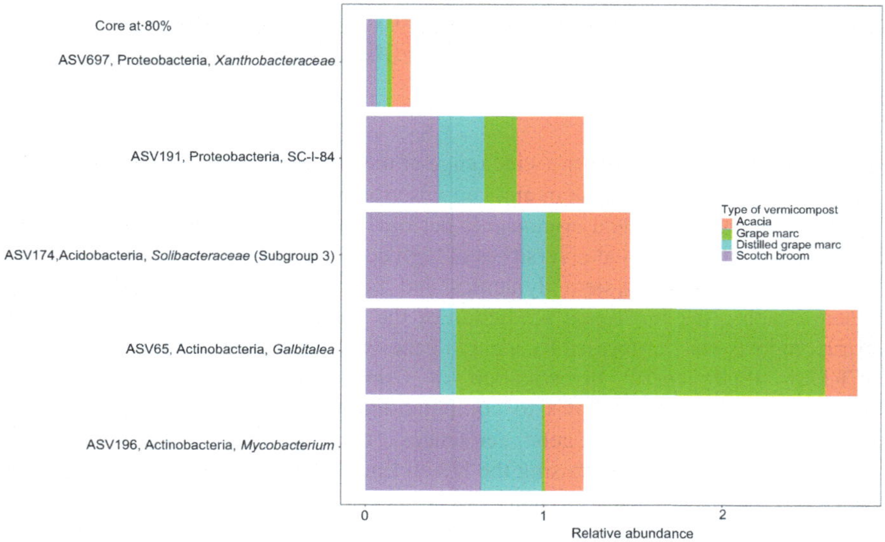

Fig. 8.5 Relative abundance (%) of ASVs (phylum and genus or most inclusive taxonomy found) from the core microbiome of vermicomposts obtained from Scotch broom and acacia plant materials and those derived from raw and distilled marcs of the grape variety *Vitis vinifera* v. Albariño

the decomposition process because microorganisms are released again into the vermicompost matrix as part of the earthworm casts (Aira and Domínguez 2011). Later on, natural ageing processes known as cast-associated processes take place during vermicomposting with implications on nutrient availability and, consequently, on the structure of the microbial community (Aira et al. 2010).

Furthermore, we found that the initial differences among bacterial communities in terms of β-diversity were markedly reduced, mainly at phylogenetic level, once transformed into vermicomposts (Fig. 8.4b). All of the four vermicomposts grouped closely together in the negative side of the first axis (Fig. 8.4b). These similarities at phylogenetic level also held true for the microbiome of fresh casts of *E. andrei* feeding on different types of animal manure as reported by Aira et al. (2015). The presence of a core of common bacteria was evidenced in the four vermicomposts (Fig. 8.5), even though this core only represented a small percentage (1.74%) of all sequences. Core microbiome can be defined arbitrarily at any threshold for presence and/or abundance (Shade and Handelsman 2012), but in our case, we chose the most restrictive option, that is, one ASV should be present in all of the four vermicomposts to be considered within the core microbiome regardless of its abundance. The fresh plant materials were not considered within the core microbiome because these samples were not subjected to the action of earthworms. After removing them, the core microbiome of the studied vermicomposts comprised only five ASVs that were affiliated to the phyla *Proteobacteria*, *Actinobacteria* and *Acidobacteria* (Fig. 8.5).

8.4 How Does Vermicomposting Influence Functional Diversity of Bacterial Communities from Dead Plant Material?

Functional diversity includes the wide range of metabolic activities carried out by the microorganisms in an ecosystem and can describe the way in which diverse microorganisms interact as a meta-organism to perform specific functions (Goswami et al. 2017). Assessing functional diversity is of high ecological importance because it can influence both the ecosystem dynamics and functioning (Tilman 1999). In the present chapter, the functional prediction of the fresh materials and the respective vermicomposts was performed by using a gene-based computational tool known as PICRUSt (Phylogenetic Investigation of Communities by Reconstruction of Unobserved States) that allows to computationally infer metagenome functional contents from 16S rRNA gene sequences (Langille et al. 2013). Predicted metagenomes were collapsed using the Kyoto Encyclopedia of Genes and Genomes (KEGG) Pathway metadata (Kanehisa et al. 2019). Putative functional genes involved in cellulose metabolism, hormone synthesis and antibiotic production, which can be considered as a proxy for plant growth and development, were also predicted using PICRUSt tool.

Overall, we detected significant predicted increases in genes classified only as "metabolism" in KEGG functional hierarchies in all four vermicomposts in comparison with the respective fresh plant materials (Fig. 8.6a). The same pattern emerged for specific genes related to cellulose metabolism (Fig. 8.6b). The degree of such effect varied with the type of plant material, and lower gene abundances were found in the distilled grape marc-derived vermicompost compared to the other three vermicomposts (Fig. 8.6a, b). This indicates that the bacterial communities initially contained in the parent materials had an influence on the metabolic functions in the resulting vermicomposts. Indeed, previous works have underlined the importance of the starting material for driving bacterial succession during vermicomposting (Fernández-Gómez et al. 2010; Yakushev et al. 2011). Other factors shaping the functional diversity of vermicompost microbiomes may include the earthworm species and/or the vermicomposting procedure (Domínguez et al. 2019).

With regard to the genes involved in the synthesis of plant hormones, the vermicomposts obtained from the raw grape marc and Scotch broom plant material had higher abundances than the respective initial substrates (Fig. 8.6c). An increase in the abundance of genes related to antibiotic production was also observed in the vermicompost samples, except for the distilled grape marc (Fig. 8.6d). Generally speaking, processing of dead plant material through vermicomposting resulted in increases in specific metabolic processes potentially beneficial for plant growth and development. Indeed, antibiotic production by beneficial bacteria has been proposed as a plausible mechanism by which vermicompost addition may confer disease

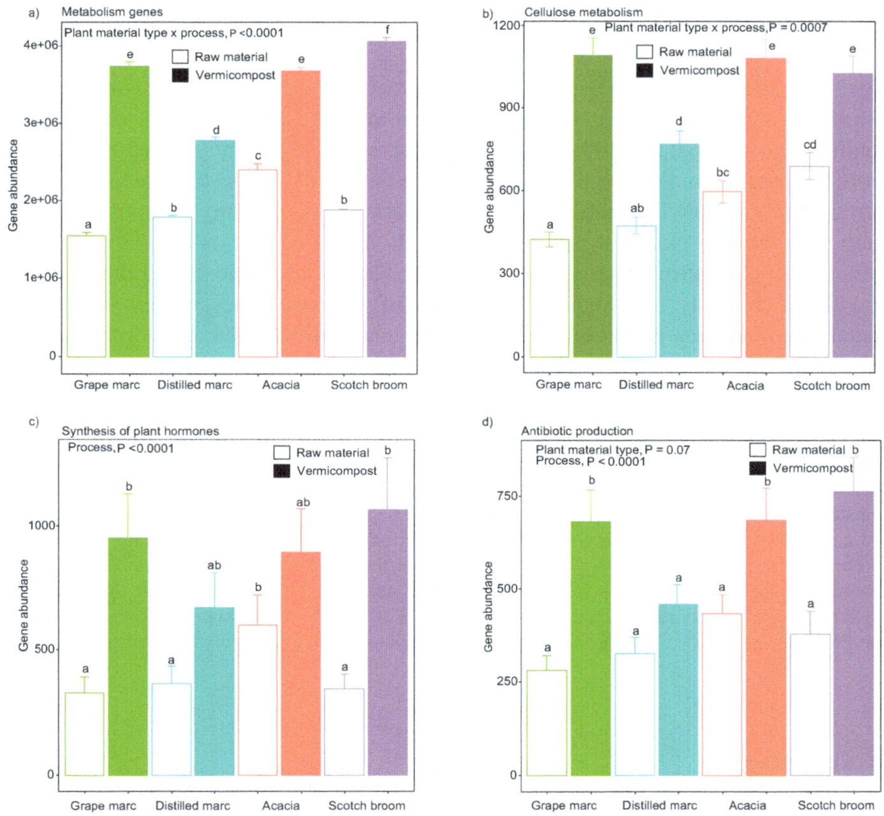

Fig. 8.6 Changes in gene abundance of PICRUSt-predicted KEGG orthologies classified as (**a**) "metabolism," (**b**) cellulose metabolism, (**c**) plant hormone synthesis, and (**d**) antibiotic production of the fresh plant materials (Scotch broom, acacia, raw and distilled grape marcs) and the respective vermicomposts (abbreviated as "Vc" in the x-axis). For each starting material, different letters indicate significant differences between the fresh plant material and the resulting vermicompost (Tukey HSD test, FDR corrected)

resistance in some plants (Lazcano and Domínguez 2011). In this sense, Gopalakrishnan et al. (2011) identified five actinomycete isolates from the genus *Streptomyces* in herbal vermicomposts that contained antagonistic potential against *Fusarium* wilt of chickpea under both greenhouse and field conditions. The selected *Streptomyces* strains were also shown to act as plant growth-promoting agents (Gopalakrishnan et al. 2013). Song et al. (2015) also found that adding vermicompost enhanced the beneficial effects of plant growth-promoting rhizobacteria on both soil and crop, but the extent of this promotion varied with the dose of vermicompost and the crop type.

8.5 Conclusions

The present chapter provides an in-depth characterization of bacterial communities inhabiting green vermicomposts from a taxonomic and functional perspective. Our findings showed that the studied vermicomposts were biologically different not only from the parent plant materials but also among them with regard to the abundance of certain bacterial phyla. Such differences in bacterial community composition may help to define the potential usefulness of each vermicompost as a soil conditioner and/or as a plant growth promoter. Moreover, the vermicomposts exhibited higher functional diversity and higher abundances of putative genes related to specific metabolic processes potentially beneficial for plant growth and development. This reinforces the role of vermicomposts as bioactive organic materials and provides evidence of microbial functions that may explain the positive influence of vermicompost on soil and plants.

Acknowledgements The authors would like to thank Hugo Martínez, Alberto Da Silva and Natalia Ribao for their help with the vermicomposting process, sample collection and DNA extraction. The authors also acknowledge Paul Fraiz for his valuable help regarding the English wording and style. This study was supported by the Spanish Ministerio de Economía y Competitividad (grant numbers CTM2013-42540-R and AGL2017-86813-R) and the Xunta de Galicia (grant numbers ED431F 2018/05 and ED431B 2019/038). MGB acknowledges support by the Programa Ramón y Cajal (RYC-2016-21231; Ministerio de Economía y Competitividad).

References

Aira M, Lazcano C, Gómez-Brandón M, Domínguez J (2010) Ageing effects of casts of Aporrectodea caliginosa on soil microbial community structure and activity. Appl Soil Ecol 46:143–146

Aira M, Domínguez J (2011) Earthworm effects without earthworms: inoculation of raw organic matter with worm-worked substrates alters microbial community functioning. PLoS One 6: e16354

Aira M, Bybee S, Pérez-Losada M, Domínguez J (2015) Feeding on microbiomes: effects of detritivory on the taxonomic and phylogenetic bacterial composition of animal manures. FEMS Microbiol Ecol 91:fiv117

Blouin M, Barrere J, Meyer N, Lartigue S, Barot S, Mathieu J (2019) Vermicompost significantly affects plant growth. A meta-analysis. Agron Sustain Dev 39:34

Cai L, Gong X, Sun X, Li S, Yu X (2018) Comparison of chemical and microbiological changes during the aerobic composting and vermicomposting of green waste. PLoS One 13:e0207494

Callahan BJ, McMurdie PJ, Rosen MJ, Han AW, Johnson AJ, Holmes SP (2016) DADA2: high-resolution sample inference from Illumina amplicon data. Nat Methods 13:581–583

Callahan BJ, McMurdie PJ, Holmes SP (2017) Exact sequence variants should replace operational taxonomic units in marker-gene data analysis. ISME J 11:2639–2643

Chew KW, Chia SR, Yen H-W, Nomanbhay S, Ho Y-C, Show PL (2019) Transformation of biomass waste into sustainable organic fertilizers. Sustainability 11:2266

Domínguez J, Aira M, Gómez-Brandón M (2010) Vermicomposting: earthworms enhance the work of microbes. In: Insam H, Franke-Whittle I, Goberna M (eds) Microbes at work: from wastes to resources. Springer, Berlin, pp 93–114

Domínguez J, Edwards CA (2011) Biology and ecology of earthworms' species used for vermicomposting. In: Edwards CA, Arancon NQ, Sherman RL (eds) Vermiculture technology: earthworms, organic waste and environmental management. CRC Press, Boca Raton, pp 1–14

Domínguez J, Aira M, Kolbe AR, Gómez-Brandón M, Pérez-Losada M (2019) Changes in the composition and function of bacterial communities during vermicomposting may explain beneficial properties of vermicompost. Sci Rep 9:9657

Faith DP (1992) Conservation evaluation and phylogenetic diversity. Biol Conserv 61:1–10

Fernández-Gómez MJ, Nogales R, Insam H, Romero E, Goberna M (2010) Continuous-feeding vermicomposting as a recycling management method to revalue tomato-fruit wastes from greenhouse crops. Waste Manag 30:2461–2468

Fierer N, Nemergut D, Knight R, Craine JM (2010) Changes through time: integrating microorganisms into the study of succession. Res Microbiol 161:635–642

Gómez-Brandón M, Aira M, Lores M, Domínguez J (2011) Epigeic earthworms exert a bottleneck effect on microbial communities through gut associated processes. PLoS One 6:e24786

Gómez-Brandón M, Lores M, Domínguez J (2012) Species-specific effects of epigeic earthworms on microbial community structure during first stages of decomposition of organic matter. PLoS One 7:e31895

Gómez-Brandón M, Domínguez J (2014) Changes of microbial communities during the vermicomposting process and after application to the soil. Crit Rev Environ Sci Technol 44:1289–1312

Gómez-Brandón, M., Vela, M., Martínez-Toledo, M.V., Insam, H., & Domínguez, J. (2015). Effects of compost and vermicompost teas as organic fertilizers. In: Sinha S, Pant K (Eds.) Fertilizer technology. Synthesis, vol. 1. Studium Press LLC, Houston, pp. 301–318

Gómez-Brandón M, Aira M, Kolbe AR, de Andrade N, Pérez-Losada M, Domínguez J (2019) Rapid bacterial community changes during vermicomposting of grape marc derived from red winemaking. Microorganisms 7:473

Gong X, Li S, Sun X, Wang L, Cai L, Zhang J, Wei L (2018) Green waste compost and vermicompost as peat substitutes in growing media for geranium (*Pelargonium zonale* L.) and calendula (*Calendula officinalis* L.). Scientia Horticultura 236:186–191

Gopalakrishnan S, Pande S, Sharma M, Humayun P, Kiran B-K, Sandeep D, Vidya M-S, Deepthi K, Rupela O (2011) Evaluation of actinomycete isolates obtained from herbal vermicompost for the biological control of *Fusarium wilt* of chickpea. Crop Prot 30:1070–1078

Gopalakrishnan S, Srinivas V, Vidya M-S, Rathore A (2013) Plant growth-promoting activities of *Streptomyces* spp in sorghum and rice. Springerplus 2:574

Goswami M, Bhattacharyya P, Mukherjee I, Tribedi P (2017) Functional diversity: an important measure of ecosystem functioning. Adv Microbiol 7:82–93

Ho A, Di Lonardo DP, Bodelier PLE (2017) Revising life strategy concepts in environmental microbial ecology. FEMS Microbiol Ecol 93:fix006

Kanehisa M, Sato Y, Furumichi M, Morishima K, Tanabe M (2019) New approach for understanding genome variations in KEGG. Nucleic Acids Res 47:590–595

Kolbe AR, Aira M, Gómez-Brandón M, Pérez-Losada M, Domínguez J (2019) Bacterial succession and functional diversity during vermicomposting of white grape marc (*Vitis vinifera* v. Albariño). Sci Rep 9:7472

Kozich JJ, Westcott SL, Baxter NT, Highlander SK, Schloss PD (2013) Development of a dual-index sequencing strategy and curation pipeline for analyzing amplicon sequence data on the MiSeq Illumina sequencing platform. Appl Environ Microbiol 79:5112–5120

Langille MG, Zaneveld J, Caporaso JG, McDonald D, Knights D, Reyes JA, Clemente JC, Burkepile DE, Vega Thurber RL, Knight R et al (2013) Predictive functional profiling of microbial communities using 16S rRNA marker gene sequences. Nat Biotechnol 31:814–821

Lazcano C, Domínguez J (2011) The use of vermicompost in sustainable agriculture: impact on plant growth and soil fertility. In: Miransari M (ed) Soil nutrients. Nova Science Publishers, New York, pp 230–254

Lazcano C, Gómez-Brandón M, Revilla P, Domínguez J (2013) Short-term effects of organic and inorganic fertilizers on soil microbial community structure and function: a field study with sweet corn. Biol Fertil Soils 49:723–733

Love MI, Huber W, Anders S (2014) Moderated estimation of fold change and dispersion for RNA-seq data with DESeq2. Genome Biol 15:550

Lozupone C, Knight R (2005) UniFrac: a new phylogenetic method for comparing microbial communities. Appl Environ Microbiol 71:8228–8235

Mas-Carrió E, Dini-Andreote F, Brossi MJ, Salles J, Olff H (2018) Organic amendment under increasing agricultural intensification: effects on soil bacterial communities and plant productivity. Front Microbiol 9:2612

Medina-Sauza RM, Álvarez-Jiménez M, Delhal A, Reverchon F, Blouin M, Guerrero-Analco JA, Cerdán CR, Guevara R, Villain L, Barois I (2019) Earthworms building up soil microbiota, a review. Front Environ Sci 7:81

Price MN, Dehal PS, Arkin AP (2010) FastTree 2—approximately maximum-likelihood trees for large alignments. PLoS One 5:e9490

Quast C, Pruesse E, Yilmaz P, Gerken J, Schweer T, Yarza P, Peplies J, Glockner FO (2013) The SILVA ribosomal RNA gene database project: improved data processing and web-based tools. Nucleic Acids Res 41:D590–D596

Sánchez-Hernández JC, Domínguez J (2019) Dual role of vermicomposting in relation to environmental pollution: detoxification and bioremediation. In: Sánchez-Hernández J (ed) Bioremediation of agricultural soils. CRC Press, Boca Raton, pp 217–230

Shade A, Handelsman J (2012) Beyond the venn diagram: the hunt for a core microbiome. Environ Microbiol 14:4–12

Song X, Liu M, Wu D, Griffiths BS, Jiao J, Li H, Hu F (2015) Interaction matters: synergy between vermicompost and PGPR agents improves soil quality, crop quality and crop yield in the field. Appl Soil Ecol 89:25–34

Tilman D (1999) The ecological consequences of changes in biodiversity: a search for general principles. Ecology 80:1455–1474

Wang Q, Garrity GM, Tiedje JM, Cole JR (2007) Naive Bayesian classifier for rapid assignment of rRNA sequences into the new bacterial taxonomy. Appl Environ Microbiol 73:5261–5267

Yang W, Yang Z, Guan Y, Zhai C, Shi D, Chen J, Wang T, Gu S (2019) Dose-dependent effect of compost amendment on soil bacterial community composition and co-occurrence network patterns in soybean agroecosystem. Arch Agron Soil Sci. https://doi.org/10.1080/03650340.2019.1651450

Yakushev AV, Bubnov IA, Semenov AM (2011) Estimation of the effects of earthworms and initial substrates on the bacterial community in vermicomposts. Eurasian Soil Sci 44:1117–1124

Chapter 9
Vermicomposting Treatment of Fruit and Vegetable Waste and the Effect of the Addition of Excess Activated Sludge

Wenjiao Li, Sartaj Ahmad Bhat, Yongfen Wei, and Fusheng Li

Abstract Fruit and vegetable waste (FVW) is generated in large quantities during production, processing, and consumption. Several conventional methods are used for treating the FVW, such as incineration, landfill, anaerobic digestion, and composting. Compared to these methods, vermicomposting has the advantages of effective stabilization of the organic wastes like FVW through joint action of earthworms and microorganisms in decomposition of the easily decaying organic constituents and of higher reuse value for its final product as organic fertilizer. However, treating fresh FVW directly by vermicomposting is difficult to be conducted because of higher water content and C/N ratio of FVW. Excess activated sludge (EAS), as the main by-product of sewage treatment process, is also an organic waste consisting of microorganisms that can be treated by vermicomposting. Vermicomposting of fresh FVW with the addition of EAS is not only reported as a feasible option to solve the problem caused by the higher water content and C/N ratio of FVW but also can enhance the treatment process. The rich content of microorganisms, nitrogen, and phosphorus in EAS can promote the growth of earthworms and enhance the microbial activity, thus improving the decomposition efficiency of FVW and improving the utilization value of the final product as fertilizer. The present chapter summarizes the several conventional treatment methods of FVW and proposes a more feasible and sustainable treatment method of FVW by vermicomposting with the addition of EAS.

W. Li
Graduate School of Engineering, Gifu University, Gifu, Japan

S. A. Bhat · Y. Wei
River Basin Research Center, Gifu University, Gifu, Japan

F. Li (✉)
Graduate School of Engineering, Gifu University, Gifu, Japan

River Basin Research Center, Gifu University, Gifu, Japan
e-mail: lifs@gifu-u.ac.jp

S. A. Bhat et al. (eds.), *Earthworm Assisted Remediation of Effluents and Wastes*,
https://doi.org/10.1007/978-981-15-4522-1_9

Keywords Fruit and vegetable waste · Excess activated sludge · Vermicomposting · Earthworms · Microbial activity

9.1 Introduction

With the rapid economic development, the lack of resources and energy and environmental issues have become more serious. The concept of recovery, recycling, and reuse from waste resources has been widely accepted for the sustainable treatment of wastes in large quantities and attracted the attention of many researchers. The fruit and vegetable waste (FVW), as one of the major components of wastes, is generated in large quantities during production, processing, and consumption. The FVW is characterized by high water content and rich biodegradable organic substances, normally with water content of about 90 and 85% corresponding to organic matters (Edwiges et al. 2018). These characteristics indicate that the FVW may contribute to negative environmental issues without proper treatment (e.g., leachate and greenhouse gas) and provide a great potential for reuse, recycling, and recovery (Hartmann and Ahring 2006; Plazzotta et al. 2017). Therefore, it is important to find a sustainable method for the treatment and recycling of FVW.

Several conventional methods are used for treatment and recycling of FVW, such as incineration, landfill, anaerobic digestion, and composting. Compared to these conventional methods, vermicomposting has the advantage of effective stabilization of the organic wastes through the joint function of earthworms and microorganisms in decomposition of easily decaying organic constituents and of high utilization value for its final product as fertilizer. On the other hand, excess activated sludge (EAS), as the main by-product of sewage treatment process, is an organic waste mainly consisted of microorganisms which can be treated by vermicomposting. Li et al. (2020) proposed that the wide variety of microorganisms presented in EAS can enhance the microbial activity and positively contribute to the decomposition of FVW since the complex microbial communities play a key role during the vermicomposting (Chen et al. 2018a, b). In addition, the rich content of nitrogen and phosphorus in EAS can promote the growth of earthworms and specific bacterial species, thus improving the stabilization efficiency of FVW and will also improve the utilization value of the final product.

Accordingly, the main objectives of this chapter are to summarize several conventional treatment methods for FVW and to propose a more feasible and sustainable method for treating FVW with the addition of EAS by vermicomposting.

9.2 Fruit and Vegetable Waste

9.2.1 Generation and Characteristics of Fruit and Vegetable Waste

About 1.3 billion tons of food that accounts for 32% of the total food produced for human consumption across the whole food supply chain is lost and wasted every year in the world (Du and Li 2016). The FVW, which account for the largest proportion (45%) of the total lost and wasted food (Abubackar et al. 2019a), are also generated in large quantities along with the entire food supply chain, from production, processing to the consumption. According to the statistical data published by the Food and Agriculture Organization (FAO), about 1.8, 6.5, 32, and 15 million tons of FVW are produced each year in India, the Philippines, China, and the United States, respectively (FAO 2013). From the same information source, it is reported that at least 15% of fruits and 25% of vegetables are wasted at the bottom of the food supply chain globally (FAO 2014). Diverting our attention to the FVW from households, approximately ten million tons of FVW are generated every year in Japan alone (Ministry of Agriculture, Forestry and Fisheries, Japan 2012). All the statistical data indicate that the FVW are a very important class of wastes since they are produced in great amounts from all wholesale markets and other activities in the world (Scano et al. 2014).

The characteristics of FVW can be highly variable depending on its source and are closely dependent on the eating habits of consumers (Cerda et al. 2018). In general, the FVW are characterized by high water content and rich biodegradable organic substances (e.g., carbohydrates, lipids, and organic acids), normally with water content about 90 and 85% (Edwiges et al. 2018; Li et al. 2020). Thi et al. (2015) also reported that the FVW contains 74–90% of water and has a volatile solid to total solid ratio (VS/TS) of 80–97% and a C/N ratio of 14.7–36.4. Edwiges and Frare (2017) investigated the physical and chemical characteristics of FVW sampled monthly during 1 year in Brazil, as summarized in Table 9.1. These characteristics indicate that the FVW has a great potential for reuse, recycling, and recovery (Plazzotta et al. 2017) and should be treated with proper methods.

9.2.2 Treatment Methods of Fruit and Vegetable Waste

The FVW is well-known as liquid and/or semi-liquid waste due to high water content. These kinds of wastes are conventionally disposed by non-scientific methods such as transported to landfill or incineration plants mixed with other municipal solid waste (Liu et al. 2012). In most industrialized countries, incineration is widely used to treat burnable municipal solid waste including FVW (Du and Li 2017). The exhaust gas and ash generated in incineration plant during burning have serious dangerous effects on the human health and the environment (Cangialosi et al.

Table 9.1 Physical and chemical characteristics of fruit and vegetable waste

	Min	Max	Mean
pH	3.9	4.5	4.2
Higher calorific value (MJ/kg)	14.8	21.2	16.5
Total solids (%)	7.2	13.8	9.5
Volatile solids ($\%_{-TS}$)	89.9	93.4	92.0
Proteins ($\%_{-VS}$)	9.6	25.5	15.9
Lipids ($\%_{-VS}$)	1.0	22.3	4.5
Cellulose ($\%_{-VS}$)	13.8	26.9	17.1
Hemicellulose ($\%_{-VS}$)	3.1	15.3	9.4
Lignin ($\%_{-VS}$)	3.0	12.0	6.4
Non-lignocellulosic carbohydrates ($\%_{-VS}$)	20.1	60.1	46.7

TS total solids, *VS* volatile solids

2008). Moreover, the incineration plant is also recognized as a pollution source of heavy metals for the surrounding environment. In many developing countries, however, landfill and direct dumping are still practiced. The landfill and dumping can deteriorate the soil and water environments. Direct dumping to urban rivers or garbage collection stations near domestic areas may cause foul smelling and leachate that affect the living environment of humans (Du and Li 2017). Therefore, it is necessary to overcome the relevant defects of landfill or incineration for treatment and recycling of FVW, concurrently developing an environment-friendly method which can convert FVW to high value products.

Currently, anaerobic digestion and composting are recognized as two efficient and environmentally friendly methods to recycle/recover available resources from organic waste and are used extensively worldwide (Cerda et al. 2018). Diverting organic waste like FVW from landfill or incineration to anaerobic digestion or composting has many environmental benefits, such as reduction of greenhouse gas emission and improvement of soil properties through the application of compost (Bernstad Saraiva Schott et al. 2016). Anaerobic digestion is a biological process in which stabilization of organic waste is achieved by microorganisms in the absence of oxygen. The main products of this process are biogas (a mixture of CH_4, H_2, and other gases) and the digested sludge as the main by-product (Moukazis et al. 2018). The biogas can be used to generate heat and/or electricity or can be used as transport fuel or injection into the natural gas system (Singh et al. 2010). However, it is reported that the required range of pH, carbon to nitrogen (C/N) ratio, and moisture for anaerobic digestion are around 7.0 (Khalid et al. 2011), 20–30, and 70–80% (Hernández-Berriel et al. 2008), respectively. Therefore, it is necessary to introduce a pretreatment process for FVW before the anaerobic digestion. Composting, as one of the most sustainable options for the treatment of organic waste including the FVW, can not only cut down the volume of organic waste but also produce a useful product as fertilizer (Lou and Nair 2009; Cerda et al. 2018). Since the composting is an aerobic biochemical process via the function of thermophilic microorganisms, it is important to maintain the requirements for the growth of microorganisms. Some environmental factors that may influence the microbial activity, like temperature, oxygen concentration, pH, water content, C/N ratio, and particle size, need to be adjusted by a pretreatment process before composting (An Ceustermans et al. 2009). Besides these two methods, some other sustainable treatment methods are being developed (e.g., fermentation, microbial fuel cell). Table 9.2 summarizes some published investigations on the treatment of FVW by using different methods.

Vermicomposting is attracting researchers' attention in recent years owing to the reason that it can degrade organic wastes like FVW and can recycle and convert the valuable nutrients into organic fertilizer (Li et al. 2020). Compared to composting, vermicomposting has more effective functions of biodegradation and stabilization of the organic wastes through the joint action of earthworms and microorganisms (Domínguez 2004). In general, there are two operation systems that are widely used for vermicomposting, namely, mixed system and separated system. The substrate and bed material are mixed together in the mixed system, while the substrate and bed material are simply separated into two layers or separated by using a mesh with holes in the separated system (Li et al. 2020). Many researchers investigated the vermicomposting of FVW with different operation systems and conditions which are summarized in Table 9.3. It is worth to note that these previous studies used dry FVW or the addition of other bulking materials for earthworms. The pretreatment for drying the fresh FVW usually takes 7–21 days before vermicomposting process (Li et al. 2020). This kind of pretreatment not only increases the whole time needed for vermicomposting but also leads to the loss of significant amounts of nutrients through leachate (Huang et al. 2012). Moreover, limited literatures related to the vermicomposting of fresh FVW reported that the earthworms cannot live in the fresh FVW due to the high water content and high electrical conductivity of the generated leachate (Gunadi and Edwards 2003; Huang et al. 2012; Li et al. 2020). Therefore, it is necessary to enhance the efficiency of the treatment of FVW by vermicomposting; owing to that, the sustainable treatment methods are regarded as one of the key approaches to achieve the urban sustainability by recycling resources and recovering energy (Wang et al. 2020).

The excess activated sludge (EAS) is a kind of organic waste generated from wastewater treatment process and can be treated by vermicomposting. However, it is reported that the high content of nitrogen in EAS can give rise to a higher ammonia concentration environment, thus leading to the death of earthworms (Fu et al. 2015; Li et al. 2020). To solve this problem, a pretreatment is required to adjust the C/N ratio by mixing some bulking materials with rich carbon content, such as paper mulch (Ndegwa et al. 2000), straw (Contreras-Ramos et al. 2005), and sawdust (Zhao et al. 2018). Adding these kinds of bulking materials can not only improve the living environment for earthworms by increasing the C/N ratio but also cause a longer decomposition process (non-degradable substrate) and lower utilization value (lower nitrogen content) of final product as fertilizer (Li et al. 2020). Moreover, other pretreatment methods like air drying, airing, creating pellet, and pre-composting could also increase the whole period of vermicomposting. On the other hand, Li et al. (2020) suggested that the high content of nitrogen in EAS not only has negative effects but also has some positive effects if co-vermicomposted with the FVW. The

Table 9.2 Different methods for treating fruit and vegetable waste and their final products

Treatment method	Types of FVW	Pretreatment	Treatment volume	Treatment time (days)	Final products	References
Anaerobic digestion	Mixture of different fruits and vegetables	Inoculation of pig manure	3550 kg (wet basis)	174	Biogas (216 m^3)	Scano et al. (2014)
Anaerobic digestion	FVW from municipal central supply	Drying at 60 °C Ground to a diameter smaller than 2 mm	ND	32	Biogas (CH_4: 377 ± 67 L/kg-VS)	Edwiges and Frare (2017)
Anaerobic digestion	FVW from municipal market	Without any pretreatment for FVW, while the digester was initially loaded with cow rumen	108 kg/day (wet basis)	365	Biogas (CH_4: 0.11–0.15 m^3/kg-VS)	Martí-Herrero et al. (2019)
Composting	Vegetable waste	Mixed with hyacinth/garden prune/sawdust Inoculation of cow dung Adjusted C/N ratio Chopped to a size of 1–2 cm	100 kg	30	Fertilizer (N increased by 14.83%)	Rich et al. (2018)
Composting	Household food waste	Mixed with air-dried and cut cornstalks Mixed with mature compost as bulking agent	ND	35	Fertilizer	Yang et al. (2019)
Fermentation (mesophilic condition)	Mixture of different fruits and vegetables	Ground or cut into small pieces of size <5 cm	ND	8–14	Biogas (H_2: 10.77–23.53 NmL/g-VS)	Abubackar et al. (2019a)
Fermentation (thermophilic condition)		Autoclaved at 120 °C for 20 min Inoculation of sludge from biogas plant	6.5 kg (wet basis)	9–10	Biogas (H_2: 20.81–27.19 NmL/g-VS)	Abubackar et al. (2019b)
Microbial fuel cell (two chambers)	Potato	Boiled and cut into small cubes (length of 5 mm) Inoculation of anaerobic bacterial consortia	2.5 g (wet basis)	52	Electricity (highest current density: 160.1–253.9 mA/m^2)	Du and Li (2016)

FVW fruit and vegetable waste, *ND* no data

EAS can promote the growth of earthworms and specific bacterial species, thus improving the decomposition efficiency, and could improve the nutritional content of the final products as fertilizer (Li et al. 2020). Moreover, the EAS consists of microorganisms with large population and diversity, for example, various nitrogen-fixing bacteria and the phosphate-accumulating bacteria, which can enhance the microbial activity and the decomposition process of FVW, thus accelerating the nitrification and mineralization process and leading to higher content of nitrogen and phosphorus in the final product (Li et al. 2020). This novel idea has been proven to be a more feasible and sustainable method for treating FVW and EAS at the same time.

9.3 Effects of Excess Activated Sludge on Vermicomposting of Fruit and Vegetable Waste

To clarify the effects of EAS on vermicomposting of FVW, comparative experiment was conducted by Li et al. (2020). A novel vermireactor consisting of substrate and bed compartment was used for treating five types of FVW (banana peels, cabbage, lettuce, carrot, and potato) with and without the addition of EAS. The operation condition could be found in Table 9.3, and the vermireactor conditions are displayed in Table 9.4.

The addition of EAS obviously increased the total nitrogen, total phosphorus, copy numbers of 16S rDNA, and the dehydrogenase activity of substrates (FVW) before starting the vermicomposting, as could be seen in Table 9.5. It is well-known that the content of nitrogen and phosphorus in the initial substrate strongly affects the nutrient value of the final product as fertilizer (Huang et al. 2012; Li et al. 2020). Furthermore, the great number and high activity of bacteria in the initial substrate inoculated through the addition of EAS could enhance the vermicomposting since the complex microbial communities are reported to play a key role during vermicomposting (Chen et al. 2018a, b). On the other hand, Zhao et al. (2018) reported that the healthy growth status of earthworms directly reflects a successful vermicomposting since the earthworms are considered as crucial drivers of the process. Body weight changes of the earthworms were monitored over vermicomposting, as displayed in Fig. 9.1. In general, earthworms in the treatment of FVW with the addition of EAS showed a better capability of weight gain compared to the treatment of FVW alone. Moreover, Li et al. (2020) also reported that the cocoon production of earthworms was also promoted by adding EAS.

The total carbon in the substrate compartment of the treatment with addition of EAS was lower than that of the treatment for FVW alone (except for cabbage), as could be seen in Fig. 9.2. The possible reasons of lower total carbon are explained as the better growth of earthworms and higher microbial activity caused by the addition of EAS. However, no significant changes in the total carbon were demonstrated in bed compartments (except for cabbage). The mass reduction rate of substrate in each

Table 9.3 Operation condition and final product value of vermicomposting for fruit and vegetable waste

Types of FVW	Pretreatment	Earthworm species and density	Operation system	Temperature and time	Initial content of N and P	Final content of N and P	References
Vegetable market solid waste	Dried at 60 °C in hot air oven Chopped and sieved (<2 mm) Mixed with cow dung/biogas slurry/wheat straw Thermal stabilization for 3 weeks	*Eisenia fetida*; 26.7 worms/kg-dry (258–278 mg/worm)	Mixed system	26.9 ± 0.36 °C; 105 days	5.82–17.1 g-N/kg 2.73–5.74 g-P/kg	8.24–30.6 g-N/kg; 3.92–8.9 g-P/kg	Suthar (2009)
Vegetable greenhouse waste	Over-dried at 25 °C and chopped Mixed with cow dung or straw Aerated for a week	*Eisenia andrei*; 100 worms/kg-dry (0.17–0.31 g/worm)	Mixed system	24 °C; 12 weeks	15.0–22.8 g-N/kg 3.9–5.8 g-P/kg	14.1–23.0 g-N/kg; 6.9–8.4 g-P/kg	Fernández-Gómez et al. (2010b)
Tomato-fruit wastes	Pre-composting of bed material for 15 days	*Eisenia fetida*; 50 g/kg-wet	Separated system Sheep manure was used as bed material	25 °C 210 days	Sheep manure: 9.6 g-N/kg; 2.6 g-P/kg	12.9 g-N/kg; 2.04 g-P/kg	Fernández-Gómez et al. (2010a)
Vegetable waste	Mixed with cow dung Pre-composting for 21 days	*Eisenia fetida*; 20 worms/kg-dry (adult earthworms)	Mixed system	105 days	Vegetable waste: 17.1 ± 0.46 g-N/kg, 4.4 ± 0.17 g-P/kg; Cow dung: 8.7 ± 0.12 g-N/kg, 7.5 ± 0.12 g-P/kg	15.4–18.5 g-N/kg; 9.5–11.7 g-P/kg	Garg and Gupta (2011)

Banana peels; cabbage; lettuce; potato; watermelon peels	Chopped and wetted by tap water	*Eisenia fetida*; 200 worms/kg-wet (juvenile earthworms)	Separated system FVW and vermicompost as a bed layer was separated by a plastic mesh with holes	25 ± 2 °C; 28 days	6.6–21.6 g-N/kg; 2.7–5.5 g-P/kg	5.8–7.7 g-N/kg; 3.8–5.2 g-P/kg	Huang et al. (2012)
Apple pomace	Mixed with chopped wheat straw (soaked in water for 1 month) Pre-composting at 25 °C for 14 days under aerobic condition	*Eisenia*; 150 worms/L	Mixed system	4 months	13–19 g-N/kg; 5.53–6.43 g-P/kg	N increased by 58%; 10.18–17.79 g-P/kg	Hanc and Chadimova (2014)
Vegetable waste	Mixed with cattle manure Pre-composting for 20 days	*Eisenia fetida*; 65 worms/kg (adult earthworms)	Separated system Chopped hay, banana pulp, and tree leaves were used as bed material	25 °C; 45 days	13–19 g-N/kg; 5.31–6.30 g-P/kg	29–31 g-N/kg; 7.95–9.33 g-P/kg	Varma et al. (2015)
Vegetable market waste	Chopped into pieces (1 cm × 1 cm) Mixed with rice straw	*Eisenia fetida* and *Perionyx excavatus*; 10 worms/kg-wet	Mixed system	2 months	100.8–129.7 mg-N/kg; 35.5–56.6 mg-P/kg	103.3–784.0 mg-N/kg; 88.1–227.5 mg-P/kg	Hussain et al. (2016)
Kitchen vegetable waste	Mixed with paddy straw Pre-composting for 5 days	*Eisenia fetida*, *Eudrilus eugeniae*, and *Perionyx excavates*; 10 worms/kg-wet (juvenile earthworms)	Separated system Cow dung was used as bed material	27–30 °C; 120 days	ND	N increased by 5.86–6.6-fold; P increased by fivefold in maximum	Hussain et al. (2018)

(continued)

Table 9.3 (continued)

Types of FVW	Pretreatment	Earthworm species and density	Operation system	Temperature and time	Initial content of N and P	Final content of N and P	References
Kitchen waste	Dried at 65 °C Smashed and passed 10 mesh sieves Mixed with rice straw	*Eisenia fetida*; 5 worms/kg-wet (300 mg/worm)	Mixed system	20–25 °C; 45 days	19.2–22.2 g-N/kg; 1.36–1.75 g-P/kg	Both N and P decreased	Zhi-wei et al. (2019)
Banana peels; cabbage; lettuce; carrot; Potato	Cut into pieces with a width of about 1 cm	*Eisenia fetida*; 100 worms/kg-wet (adult earthworms)	Separated system. A novel reactor with two compartments (substrate and bed compartments) was used	25 °C; 12–26 days	10.0–40.9 g-N/kg; 1.3–7.0 g-P/kg	16.8–57.8 g-N/kg; 2.0–9.2 g-P/kg	Li et al. (2020)

FVW fruit and vegetable waste, *N* nitrogen, *P* phosphorus

Table 9.4 Vermireactor conditions for investigating the effect of excess activated sludge on vermicomposting of fruit and vegetable waste

	Substrate compartment			Bed compartment	Earthworms	
No.	Composition	Mixing ratio (FVW:EAS)	Weight (g-wet basis)	Weight (g-wet basis)	Individual weight (mg)	Numbers (worms)
1	Banana peels		100	100	350–500	10
2	Banana peels + EAS	3:2	100	100	350–500	10
3	Cabbage		100	100	350–500	10
4	Cabbage + EAS	3:2	100	100	350–500	10
5	Lettuce		100	100	350–500	10
6	Lettuce + EAS	3:2	100	100	350–500	10
7	Carrot		100	100	350–500	10
8	Carrot + EAS	3:2	100	100	350–500	10
9	Potato		100	100	350–500	10
10	Potato + EAS	3:2	100	100	350–500	10
11	EAS		100	100	350–500	10

FVW fruit and vegetable wastes, *EAS* excess activated sludge

vermireactor was calculated for better understanding of the decomposition state of substrates. The calculation equation is described as the following (Li et al. 2020):

$$\text{Mass reduction rate} = \frac{\begin{array}{r}\text{Initial dry weight of substrate} - \text{Final dry weight of substrate}\\ +\text{Sampled dry weight of substrate}\end{array}}{\text{Experimental days}} \ (\text{g/day}).$$

A significantly higher mass reduction rate was recorded in the treatment with the addition of EAS (Fig. 9.3). The conversion of substrates into earthworms cast/excreta or body and the production of CO_2 emission closely related to the higher mass reduction (de Lima Rodrigues et al. 2017; Li et al. 2020). It is reported that the dehydrogenase activity (DHA), as a parameter reflecting total microbial activity, showed an increasing trend in the beginning followed by a decreasing trend at the end of vermicomposting in substrate compartments. Overall, the DHA in substrate compartment of the treatment for FVW with the addition of EAS was markedly higher than the treatment for FVW alone (except for carrot). The enrichment of microbial population and activity by adding EAS can lead to a rapid turnover of microorganisms and the nitrogenous substrates or the encouraged grooming of

Table 9.5 Effect of excess activated sludge on initial properties of fruit and vegetable waste

	Total carbon (g/kg)	Total nitrogen (g/kg)	C/N ratio	Total phosphorus (g/kg)	16S rDNA ($\times 10^{12}$ copies/kg)	Dehydrogenase activity [mg-TF/(kg ·h)]
Banana peels	413.5 ± 1.3	10.5 ± 0.1	39.6 ± 0.3	1.3 ± 0.0	1.3 ± 0.1	0.0 ± 0.0
Banana peels + EAS	424.2 ± 1.1	33.8 ± 0.7	12.6 ± 0.2	6.0 ± 0.4	22.5 ± 3.6	1568.8 ± 113.3
Cabbage	386.6 ± 0.5	28.8 ± 0.0	13.4 ± 0.0	3.0 ± 0.1	9.8 ± 2.2	31.5 ± 7.4
Cabbage + EAS	405.8 ± 0.1	52.7 ± 0.3	7.7 ± 0.0	6.3 ± 0.9	11.1 ± 0.2	1036.5 ± 35.5
Lettuce	376.7 ± 6.9	33.9 ± 0.4	11.1 ± 0.3	4.5 ± 0.1	18.6 ± 0.2	73.1 ± 7.4
Lettuce + EAS	405.4 ± 0.4	58.5 ± 0.3	6.9 ± 0.0	9.7 ± 0.1	25.6 ± 0.3	1447.1 ± 195.6
Carrot	384.5 ± 1.1	15.0 ± 0.3	25.7 ± 0.4	2.3 ± 0.1	17.2 ± 1.2	0.0 ± 0.0
Carrot + EAS	404.7 ± 0.5	46.1 ± 0.5	8.8 ± 0.1	8.0 ± 0.0	32.7 ± 0.05	1567.1 ± 11.0
Potato	417.0 ± 1.5	10.0 ± 0.2	41.9 ± 0.6	1.3 ± 0.0	15.1 ± 0.1	0.0 ± 0.0
Potato + EAS	418.2 ± 0.1	28.2 ± 1.0	14.9 ± 0.5	4.5 ± 0.4	21.3 ± 0.1	861.4 ± 55.7
EAS	422.1 ± 2.7	70.7 ± 0.8	6.0 ± 0.1	12.9 ± 0.3	24.4 ± 2.2	3642.8 ± 213.1

FVW fruit and vegetable wastes, *EAS* excess activated sludge
Data are shown as means ± standard deviation, $n = 3$

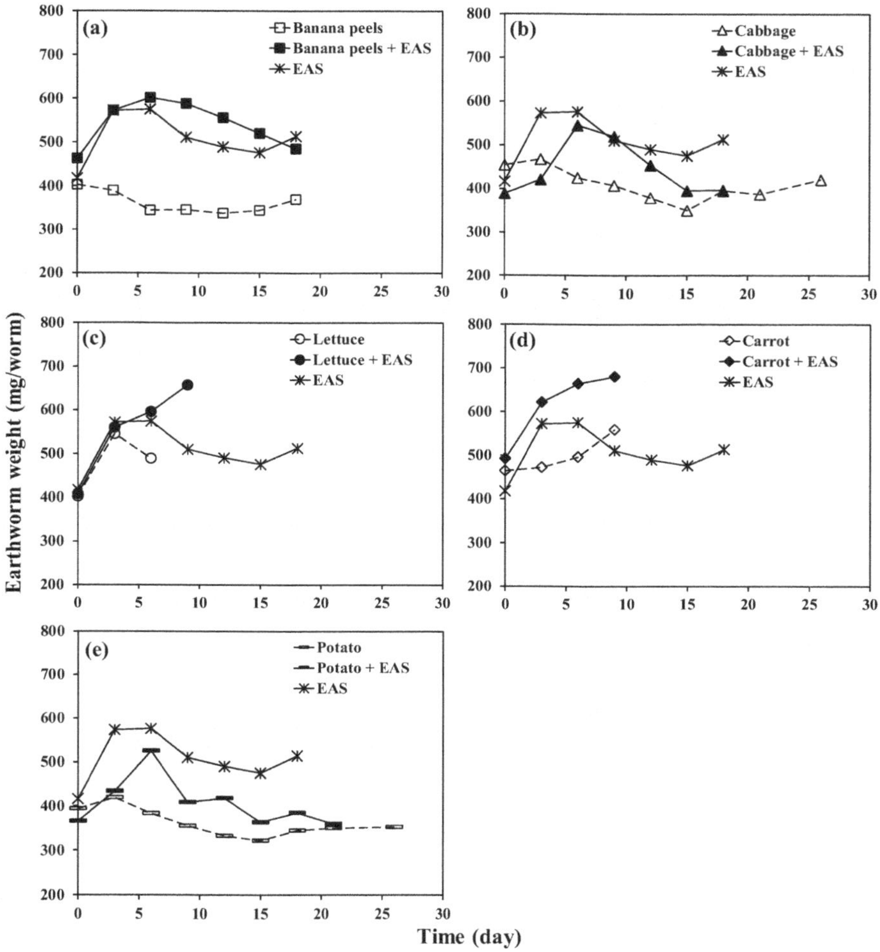

Fig. 9.1 Body weight changes of earthworm during vermicomposting of five types of FVW including (**a**) banana peels, (**b**) cabbage, (**c**) lettuce, (**d**) carrot, and (**e**) potato with and without the addition of EAS. *FVW* fruit and vegetable waste, *EAS* excess activated sludge

earthworms (e.g., mucus or casts), thus enhancing the decomposition efficiency of FVW (Li et al. 2020).

As displayed in Fig. 9.4, the addition of EAS significantly increased the content of total nitrogen in both substrate and bed compartments, except the bed compartment of the treatment for carrot. Similar results of the total phosphorus were also revealed. The mineralization of the higher content of nitrogen and phosphorus in substrate brought about by the addition of EAS, faster mass reduction rate of substrate with the addition of EAS, the higher microbial activity, etc. are the main reasons for the higher fertilizer value of the final product (Li et al. 2020).

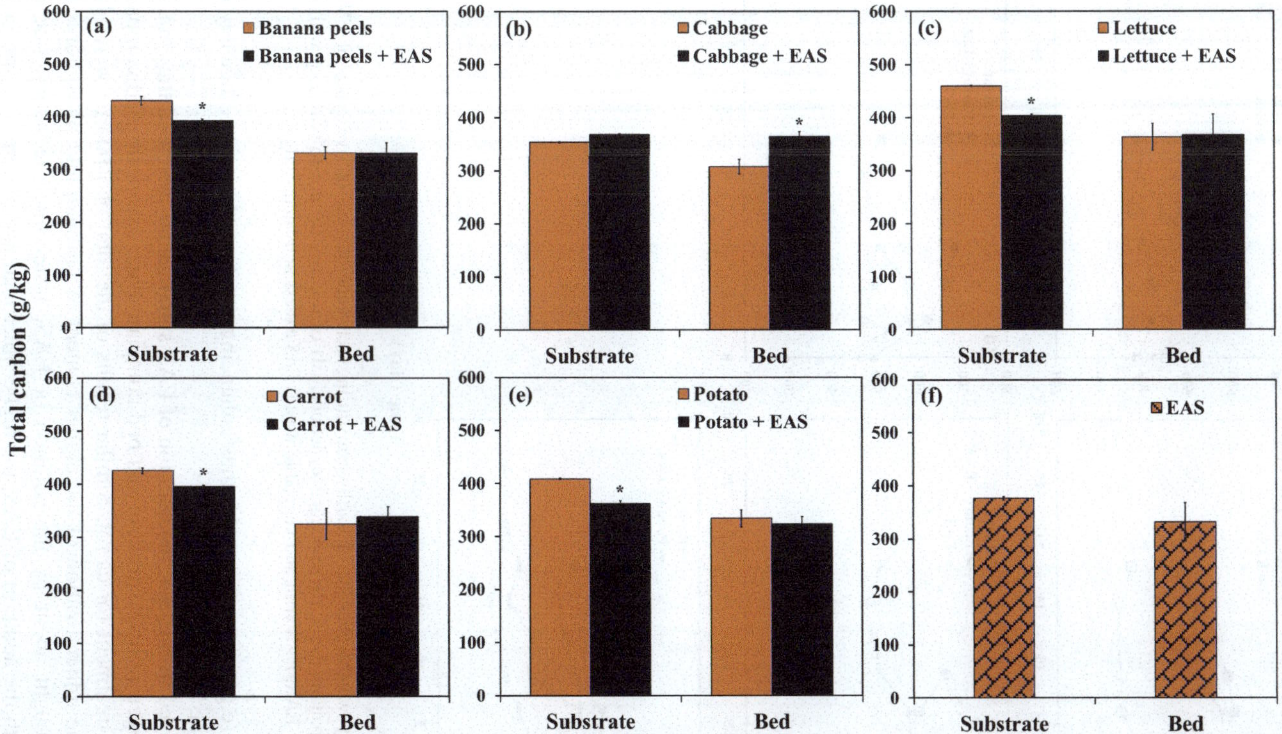

Fig. 9.2 Total carbon in substrate and bed compartments after vermicomposting of FVW and FVW + EAS: (**a**) banana peels; (**b**) cabbage; (**c**) lettuce; (**d**) carrot; (**e**) potato; (**f**) EAS. Data are presented as mean and standard deviation, $n = 3$. The asterisk ($*$) denotes the difference between vermicomposting of FVW and FVW + EAS is statistically significant at 0.05 level (Statistics 21 software). *FVW* fruit and vegetable waste, *EAS* excess activated sludge

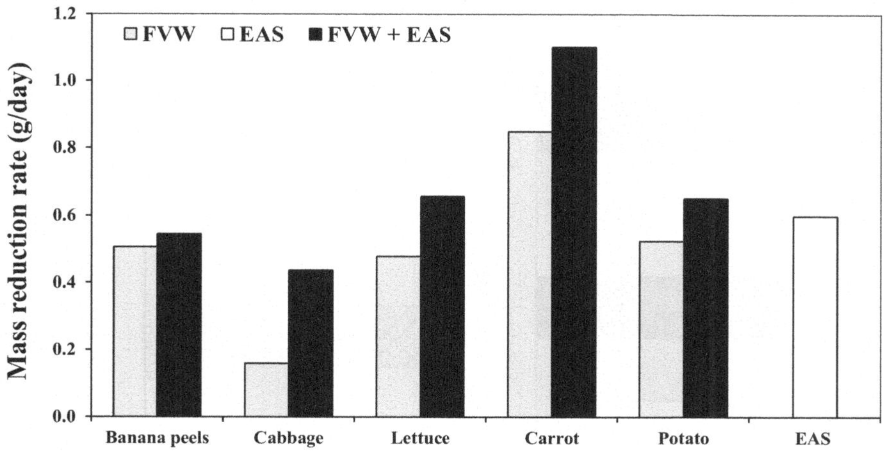

Fig. 9.3 Mass reduction rate of substrates used for vermicomposting. *FVW* fruit and vegetable wastes, *EAS* excess activated sludge

9.4 Conclusion

The large amounts of FVW generated during the whole food supply chain should be treated, recycled, and recovered properly. The vermicomposting is considered as one of the most sustainable methods for the treatment of FVW. However, the vermicomposting of fresh FVW without any pretreatment still has some problems due to the higher water content and C/N ratio of FVW. The addition of EAS could be a feasible option to solve the problems and enhance the vermicomposting of FVW. The EAS can promote the growth and cocoon production of earthworms and the microbial activity, thus enhancing the decomposition efficiency of FVW. Moreover, the content of nitrogen and phosphorus in final product as fertilizer can also be improved by the addition of EAS.

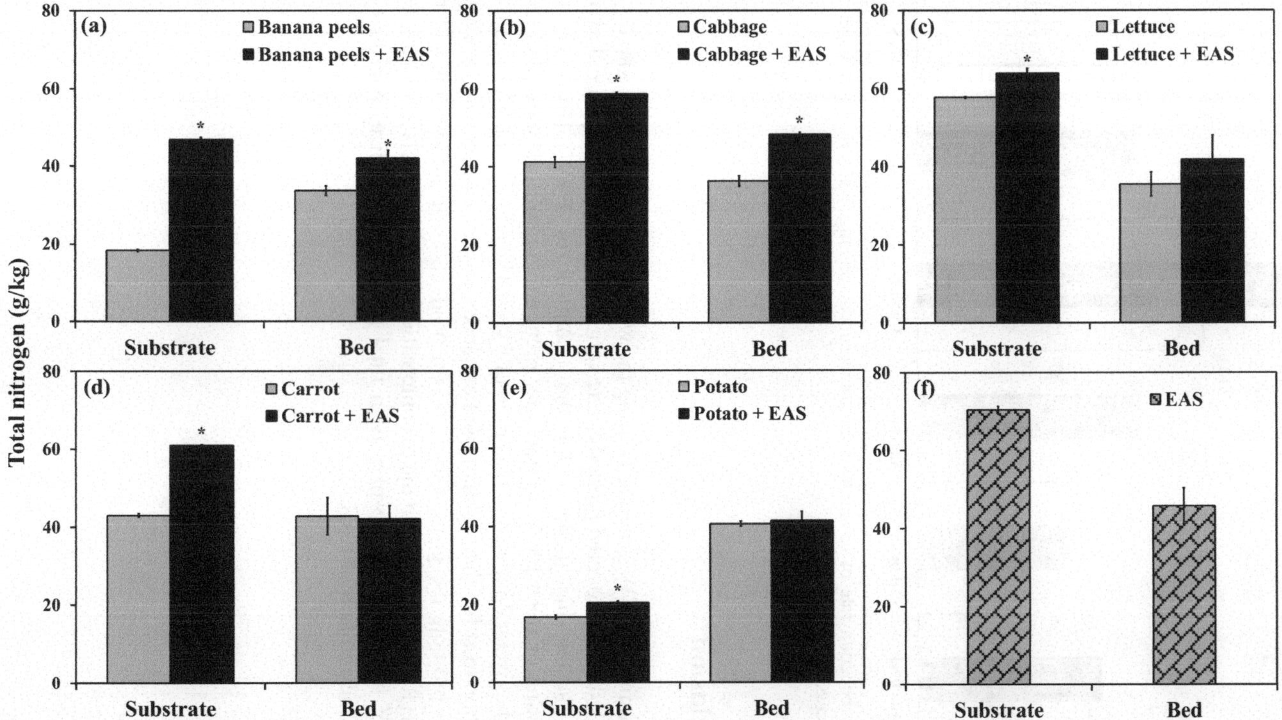

Fig. 9.4 Total nitrogen in substrate and bed compartments after vermicomposting of FVW and FVW + EAS: (**a**) banana peels; (**b**) cabbage; (**c**) lettuce; (**d**) carrot; (**e**) potato; (**f**) EAS. Data are presented as mean and standard deviation, $n = 3$. The asterisk (∗) denotes the difference between vermicomposting of FVW and FVW + EAS is statistically significant at 0.05 level (Statistics 21 software). *FVW* fruit and vegetable waste, *EAS* excess activated sludge

References

Abubackar HN, Keskin T, Arslan K, Vural C, Aksu D, Yavuzyılmaz DK, Ozdemir G, Azbar N (2019a) Effects of size and autoclavation of fruit and vegetable wastes on biohydrogen production by dark dry anaerobic fermentation under mesophilic condition. Int J Hydro Energy 44:17767–17780. https://doi.org/10.1016/j.ijhydene.2019.05.106

Abubackar HN, Keskin T, Yazgin O, Gunay B, Arslan K, Azbar N (2019b) Biohydrogen production from autoclaved fruit and vegetable wastes by dry fermentation under thermophilic condition. Int J Hydro Energy 44:18776–18784. https://doi.org/10.1016/j.ijhydene.2018.12.068

Bernstad Saraiva Schott A, Wenzel H, La Cour Jansen J (2016) Identification of decisive factors for greenhouse gas emissions in comparative life cycle assessments of food waste management—an analytical review. J Clean Prod 119:13–24. https://doi.org/10.1016/j.jclepro.2016.01.079

Cangialosi F, Intini G, Liberti L, Notarnicola M, Stellacci P (2008) Health risk assessment of air emissions from a municipal solid waste incineration plant—a case study. Waste Manag 28:885–895. https://doi.org/10.1016/j.wasman.2007.05.006

Cerda A, Artola A, Font X, Barrena R, Gea T, Sánchez A (2018) Composting of food wastes: status and challenges. Bioresour Technol 248:57–67. https://doi.org/10.1016/j.biortech.2017.06.133

Ceustermans An, Coosemans J, Ryckeboer J (2009) Compost microbial activity related to compost stability. Nature. https://doi.org/10.1007/9783642040436

Chen C, Mitchell NJ, Gratz J, Houpt ER, Gong Y, Egner PA, Groopman JD, Riley RT, Showker JL, Svensen E, Mduma ER, Patil CL, Wu F (2018a) Exposure to aflatoxin and fumonisin in children at risk for growth impairment in rural Tanzania. Environ Int 115:29–37. https://doi.org/10.1016/j.envint.2018.03.001

Chen Y, Chang SKC, Chen J, Zhang Q, Yu H (2018b) Characterization of microbial community succession during vermicomposting of medicinal herbal residues. Bioresour Technol 249:542–549. https://doi.org/10.1016/j.biortech.2017.10.021

Contreras-Ramos SM, Escamilla-Silva EM, Dendooven L (2005) Vermicomposting of biosolids with cow manure and oat straw. Biol Fertil Soils 41:190–198. https://doi.org/10.1007/s00374-004-0821-8

de Lima Rodrigues AS, Mesak C, Silva MLG, Silva GS, Leandro WM, Malafaia G (2017) Organic waste vermicomposting through the addition of rock dust inoculated with domestic sewage wastewater. J Environ Manag 196:651–658. https://doi.org/10.1016/j.jenvman.2017.03.072

Domínguez J (2004) State-of-the-art and new perspectives on vermicomposting research. In: Earthworm Ecology, 2nd edn. CRC Press LLC, Boca Raton, pp 401–424. https://doi.org/10.1201/9781420039719

Du H, Li F (2016) Effects of varying the ratio of cooked to uncooked potato on the microbial fuel cell treatment of common potato waste. Sci Total Environ 569–570:841–849. https://doi.org/10.1016/j.scitotenv.2016.07.023

Du H, Li F (2017) Enhancement of solid potato waste treatment by microbial fuel cell with mixed feeding of waste activated sludge. J Clean Prod 143:336–344. https://doi.org/10.1016/j.jclepro.2016.12.104

Edwiges T, Frare LM (2017) Use of mathematical models to fast predict biochemical methane potential of fruit and vegetable waste, pp 2–6

Edwiges T, Frare L, Mayer B, Lins L, Mi Triolo J, Flotats X, de Mendonça Costa MSS (2018) Influence of chemical composition on biochemical methane potential of fruit and vegetable waste. Waste Manag 71:618–625. https://doi.org/10.1016/j.wasman.2017.05.030

FAO (2013) Utilization of fruit and vegetable wastes as livestock feed and as substrates for generation of other value-added products

FAO (2014) Food wastage footprint: fool cost-accounting

Fernández-Gómez MJ, Nogales R, Insam H, Romero E, Goberna M (2010a) Continuous-feeding vermicomposting as a recycling management method to revalue tomato-fruit wastes from greenhouse crops. Waste Manag 30:2461–2468. https://doi.org/10.1016/j.wasman.2010.07.005

Fernández-Gómez MJ, Romero E, Nogales R (2010b) Feasibility of vermicomposting for vegetable greenhouse waste recycling. Bioresour Technol 101:9654–9660. https://doi.org/10.1016/j.biortech.2010.07.109

Fu X, Huang K, Chen X, Li F, Cui G (2015) Feasibility of vermistabilization for fresh pelletized dewatered sludge with earthworms Bimastus parvus. Bioresour Technol 175:646–650. https://doi.org/10.1016/j.biortech.2014.11.007

Garg VK, Gupta R (2011) Optimization of cow dung spiked pre-consumer processing vegetable waste for vermicomposting using Eisenia fetida. Ecotoxicol Environ Saf 74:19–24. https://doi.org/10.1016/j.ecoenv.2010.09.015

Gunadi B, Edwards CA (2003) The effects of multiple applications of different organic wastes on the growth, fecundity and survival of Eisenia fetida (Savigny) (Lumbricidae). Pedobiologia (Jena) 47:321–329. https://doi.org/10.1078/0031-4056-00196

Hanc A, Chadimova Z (2014) Nutrient recovery from apple pomace waste by vermicomposting technology. Bioresour Technol 168:240–244. https://doi.org/10.1016/j.biortech.2014.02.031

Hartmann H, Ahring BK (2006) Strategies for the anaerobic digestion of the organic fraction of municipal solid waste: an overview. Water Sci Technol 53:7–22. https://doi.org/10.2166/wst.2006.231

Hernández-Berriel MC, Márquez-Benavides L, González-Pérez DJ, Buenrostro-Delgado O (2008) The effect of moisture regimes on the anaerobic degradation of municipal solid waste from Metepec (México). Waste Manag 28:14–20. https://doi.org/10.1016/j.wasman.2008.03.021

Huang K, Li F, Li J, Helard D, Hirooka K (2012) Rapid vermicomposting of fresh fruit and vegetable wastes using earthworm Eisenia Foetida. J Japan Soc Civ Eng Ser G (Environmental Res. 68, III_113-III_120). https://doi.org/10.2208/jscejer.68.iii_113

Hussain N, Singh A, Saha S, Venkata Satish Kumar M, Bhattacharyya P, Bhattacharya SS (2016) Excellent N-fixing and P-solubilizing traits in earthworm gut-isolated bacteria: a vermicompost based assessment with vegetable market waste and rice straw feed mixtures. Bioresour Technol 222:165–174. https://doi.org/10.1016/j.biortech.2016.09.115

Hussain N, Das S, Goswami L, Das P, Sahariah B, Bhattacharya SS (2018) Intensification of vermitechnology for kitchen vegetable waste and paddy straw employing earthworm consortium: assessment of maturity time, microbial community structure, and economic benefit. J Clean Prod 182:414–426. https://doi.org/10.1016/j.jclepro.2018.01.241

Khalid A, Arshad M, Anjum M, Mahmood T, Dawson L (2011) The anaerobic digestion of solid organic waste. Waste Manag 31:1737–1744. https://doi.org/10.1016/j.wasman.2011.03.021

Li W, Ahmad S, Li J, Cui G, Wei Y, Yamada T (2020) Effect of excess activated sludge on vermicomposting of fruit and vegetable waste by using novel vermireactor. Bioresour Technol 302:122816. https://doi.org/10.1016/j.biortech.2020.122816

Liu X, Gao X, Wang W, Zheng L, Zhou Y, Sun Y (2012) Pilot-scale anaerobic co-digestion of municipal biomass waste: focusing on biogas production and GHG reduction. Renew Energy 44:463–468. https://doi.org/10.1016/j.renene.2012.01.092

Lou XF, Nair J (2009) The impact of landfilling and composting on greenhouse gas emissions—a review. Bioresour Technol 100:3792–3798. https://doi.org/10.1016/j.biortech.2008.12.006

Martí-Herrero J, Soria-Castellón G, Diaz-de-Basurto A, Alvarez R, Chemisana D (2019) Biogas from a full scale digester operated in psychrophilic conditions and fed only with fruit and vegetable waste. Renew Energy 133:676–684. https://doi.org/10.1016/j.renene.2018.10.030

Ministry of Agriculture, Forestry and Fisheries, Japan (2012). https://www.caa.go.jp/policies/policy/consumer_policy/information/food_loss/conference/pdf/121005kaigi2_1.pdf

Moukazis I, Pellera FM, Gidarakos E (2018) Slaughterhouse by-products treatment using anaerobic digestion. Waste Manag 71:652–662. https://doi.org/10.1016/j.wasman.2017.07.009

Ndegwa PM, Thompson SA, Das KC (2000) Effects of stocking density and feeding rate on vermicomposting of biosolids. Bioresour Technol 71:5–12. https://doi.org/10.1016/S0960-8524(99)00055-3

Plazzotta S, Manzocco L, Nicoli MC (2017) Fruit and vegetable waste management and the challenge of fresh-cut salad. Trends Food Sci Technol 63:51–59. https://doi.org/10.1016/j.tifs.2017.02.013

Rich N, Bharti A, Kumar S (2018) Effect of bulking agents and cow dung as inoculant on vegetable waste compost quality. Bioresour Technol 252:83–90. https://doi.org/10.1016/j.biortech.2017.12.080

Scano EA, Asquer C, Pistis A, Ortu L, Demontis V, Cocco D (2014) Biogas from anaerobic digestion of fruit and vegetable wastes: experimental results on pilot-scale and preliminary performance evaluation of a full-scale power plant. Energy Convers Manag 77:22–30. https://doi.org/10.1016/j.enconman.2013.09.004

Singh A, Smyth BM, Murphy JD (2010) A biofuel strategy for Ireland with an emphasis on production of biomethane and minimization of land-take. Renew Sust Energ Rev 14:277–288. https://doi.org/10.1016/j.rser.2009.07.004

Suthar S (2009) Vermicomposting of vegetable-market solid waste using Eisenia fetida: impact of bulking material on earthworm growth and decomposition rate. Ecol Eng 35:914–920. https://doi.org/10.1016/j.ecoleng.2008.12.019

Thi NBD, Kumar G, Lin CY (2015) An overview of food waste management in developing countries: current status and future perspective. J Environ Manag 157:220–229. https://doi.org/10.1016/j.jenvman.2015.04.022

Varma VS, Yadav J, Das S, Kalamdhad AS (2015) Potential of waste carbide sludge addition on earthworm growth and organic matter degradation during vermicomposting of agricultural wastes. Ecol Eng 83:90–95. https://doi.org/10.1016/j.ecoleng.2015.05.050

Wang D, Tang YT, Long G, Higgitt D, He J, Robinson D (2020) Future improvements on performance of an EU landfill directive driven municipal solid waste management for a city in England. Waste Manag 102:452–463. https://doi.org/10.1016/j.wasman.2019.11.009

Yang F, Li Y, Han Y, Qian W, Li G, Luo W (2019) Performance of mature compost to control gaseous emissions in kitchen waste composting. Sci Total Environ 657:262–269. https://doi.org/10.1016/j.scitotenv.2018.12.030

Zhao C, Wang Y, Wang Y, Wu F, Zhang J, Cui R, Wang L, Mu H (2018) Insights into the role of earthworms on the optimization of microbial community structure during vermicomposting of sewage sludge by PLFA analysis. Waste Manag 79:700–708. https://doi.org/10.1016/j.wasman.2018.08.041

Zhi-wei S, Tao S, Wen-jing D, Jing W (2019) Investigation of rice straw and kitchen waste degradation through vermicomposting. J Environ Manag 243:269–272. https://doi.org/10.1016/j.jenvman.2019.04.126

Chapter 10
Eco-management of Industrial Organic Wastes Through the Modified Innovative Vermicomposting Process: A Sustainable Approach in Tropical Countries

Ram Kumar Ganguly and Susanta Kumar Chakraborty

Abstract Organic farming demands a great role in soil maintenance, crop productivity and thereby triggers a new trend towards the green economy. Compost generated from vermicomposting technology contains essential nutrients required for the development of plants and hence can be used as bio-fertilizer for sustainable development. The concept of organic farming is still at its infancy in most of the tropical countries, mostly due to lack of consumer awareness. Various experimental studies, which are being conducted in order to boost the improvement and higher production of the crop through nutrient enrichment, rely considerably on the vermicomposting process. Production of vermicompost utilizing different organic decomposable industrial wastes and suitable species of earthworms tend to solve many human problems such as pollution, poverty (through livelihood generation), proper agro-ecosystem development, and environmental sustainability alongside ensuring proper eco-management of industrial wastes. Therefore, this study tries to focus on different types of techniques adapted during vermicomposting and reflects earthworm performance in an ecological box using FTIR and GC MS technology. Furthermore, the study combines the concept of microbial and chemical ecology in context of vermicomposting and provides a complete evidence of its diverse application in agriculture.

Keywords Organic farming · Bio-fertilizer · Vermicomposting · Nutrient enrichment · Eco-management

---

R. K. Ganguly (✉) · S. K. Chakraborty
Department of Zoology, Vidyasagar University, Midnapore, West Bengal, India

S. A. Bhat et al. (eds.), *Earthworm Assisted Remediation of Effluents and Wastes*,
https://doi.org/10.1007/978-981-15-4522-1_10

10.1 Introduction

With regard to sustainable development, the leeway of natural squanders from family unit, farming, and modern sources have caused extreme ecological and financial issues, and with the approach of science, a few advances have been made to address and battle each of those issues. Vermitechnology including vermiculture, and the vermicomposting procedure has given rise to an exceptionally appropriate, easy-to-use and savvy eco-innovation for legitimate natural waste administration. It is a biodegradation procedure where biochemical debasement of the natural waste materials as a substrate through fracture, molding alongside its adjustment by the joint activity of worms and microorganisms (Rakkini et al. 2017). Despite the fact that the procedure includes microbial debasement, earthworms are the real drivers of this innovation (Venkatesh and Eevera 2008). Over the course of innovation, basic supplements, for example, nitrogen, potassium, phosphorus, zinc, calcium, and so on, of waste materials while going through the gut of the earthworm are homogenized in bacterial-rich condition into a profoundly mineralized substance which are significantly more accessible to the plants when utilized as compost (Ndegwa and Thompson 2001). The utilization of organic matter, for example fertilizers, human waste, food squanders, yard squanders, sewage mucks and manures, has been perceived in horticulture as advantageous for plant development and maintenance of soil fertility. New ways to deal with the utilization of natural changes in cultivation have demonstrated vermitechnology to be viable methods for improving soil structure, upgrading soil richness, and expanding crop yields. Organic amendments are phenomenal wellspring of plant-accessible supplements and their addition to soil helps in development of high microbial populaces with expanded estimations of biomass C, basal respiration, biomass C: total organic C proportion, and metabolic quotient (qCO_2). Noteworthy yield due to utilizing mulches from coffee husks along with manures increase the productivity to a considerable extent. Their significant role in the soil and their conceivably constructive outcome on crop yields have made organic compost an important part of homestead treatment in context of farm utilization which leads to a huge development in agribusiness.

Vermicomposting being a multi-stage process incorporates mesophilic, thermophilic, and development stages. Vermicomposts are finely isolated pear-like materials with high porosity, aeration, with good amount of water-holding capacity. They have extraordinarily expanded surface areas, giving more microsites to microbial decaying life forms along with solid adsorption and maintenance of supplements in vermicompost. Furthermore, nutrients in vermicompost are present in much more elemental forms which are supposed to be readily acceptable for plants in terms of carbon, nitrogen, phosphorous, and different soluble elements such as potassium, calcium, magnesium, and so on.

Utilization of earthworm for wastes management, organic matter stabilization, and vermicompost generation has been well documented in literature (Ganguly and Chakraborty 2019; Sharma and Garg 2018). The utility of epigeic earthworm for fruitful debasement of waste is well demonstrated in various enterprises, for example, paper and pulp mash (Ganguly and Chakraborty 2018); dairy sugar preparing (Bhat et al. 2018); winery and refinery (Moorthi 2016); wood and wood chips (Miller et al. 2019); and textile indsutries (Kaushik and Garg 2003). Thermophilic mode of composting gains its importance over the other methods of waste management as it eliminates the damaging effects of organic amendments in soil. Composting proves to be efficient in proper environmental management of organic wastes as it is pathogen-free, odour-free, cheaper, and supports livelihood. In terms of composting, earthworm ingests the organic materials which upon passing through gizzard get grinded and acted upon by different microbial species in derivation of nutrients. This process accelerates the process of decomposition and leads to an alteration in physical and chemical factor such as enrichment of humic acids, good C/N ratio resulting into proper stabilization of organic wastes.

The study reveals the conversion of complex organic substance such as paper mill waste into a usable compound and also intensifies the earthworms' ability in formation of black gold through enrichment of humic substances. The study gives a comprehensive idea over different frameworks of composting and reflects its broad utility in the context of modern farming.

10.2 Advantages of Vermicomposting Over Composting and Other Chemical Processes

- Shorter time to complete the process.
- The rate of humification and bio-conversion of organic wastes are very high.
- Reduction of pathogenic microbes.
- Greater reduction in bioavailable heavy metals than in composting.
- The product is microbially enhanced and has good crumb structure.
- High nutrient status and also contains plant-hormonal agents.
- Supports livelihood generation in rural villages.

10.3 Evolution of this Process of Biotechnology

This can be categorized into four functional domains in the tropical countries as follows:

- Use of exotic earthworm species in vermicomposting
- Application of indigenous breed along with the nutrient supplement
- Modification of these techniques using microbial inoculation as well as isolating different microbial biomass for biomedical applications (Ganguly et al. 2018)
- Further modifications through amendment of bio-reactor for enhancement in compost production

10.4 Aerobic Composting

Vermicomposting is a type of aerobic composting, that is, composting under aerobic condition or in presence of oxygen under apppropriate environmental conditions. The process is divided into three phases: (a) Degradation phase, (b) Transformation phase, and (c) Maturation phase.

10.4.1 Degradation Phase

Degradation phase starts shortly after the addition of organic amendments to the compost bed in the presence of microorganism such as bacteria, actinomycetes, protozoa. This phase is characterized by moderate increase of temperature, or 'mesophilic phase', which readily degrades the soluble compounds. Such phase exist for a couple of days at the beginning of composting. Microorganisms grow exponentially and generate heat during the oxidative process which increases the temperature of composting above 40 °C also termed as 'thermophilic phase'. This phase is characterized by intense biodegradation of carbohydrates, complex lignin, etc. along with production of simple C, N, CO_2 and so on. During this phase, almost all the pathogens get destroyed under high temperature and therefore require proper aeration for correct breakdown of complex organics. After about 3–4 weeks, temperature get lowered which favors the growth of different fungi and acts as a major contributing factor for biodegradation of complex protein and cellulose.

10.4.2 Transformation Phase

At around the second month, temperature drops and the compost enters the transformation phase which is characterized by different micro arthropods such as mites, bettles, ear wigs, etc. However, at this time, compost is not totally matured, and hence it application to flora can have detrimental effects.

10.4.3 Maturation Phase

Maturation phase is characterized by the enrichment of humic acid-like substances or nutrient-rich humus. It also indicates stability or maturity and is ready for use within 3–4 month of composting.

10.5 Microbial Commitment of Vermicomposting

Charles Darwin depicted earthworms as the 'unheralded soldiers of mankind' and Aristotle called them as the 'digestive system of earth', as they could process a wide assortment of organic materials (Sinha et al. 2009). With the advent of science, a synergy between earthworm microflora, microfauna, and soil physiochemical parameters are viewed together as a tool of biotechnology as 'drillosphere' (Kostina et al. 2011). Vermicomposting is being developed as a non-thermophilic natural oxidation process in which toxic natural materials are converted into organic peat-like materials which comprise enriched microbial activities, high porosity, and water-holding capacity (Pathma and Sakthivel 2012). Being a mesophilic process, vermicompost proved to be an efficient incubator for the vast majority of microbial biodiversity and activity in comparison to traditional thermophilic composting. Earthworms ingest soil along with different plant growth advancing microbes such as Pseudomonas, Rhizobium, Bacillus, Azosprillium, Azotobacter, and so on. Alongside rhizospheric soil (Pathma et al. 2019) and on arrival of perfect microenvironment, these bacterial isolates become functional and facilitate the process (Sinha et al. 2009, Ganguly and Chakraborty 2018). This accumulation of microbes animates and improves overall plant growth through solubilization of supplements (Ravindran et al. 2008) and helps in development of different enzymatic activities such as isozymes of arylamidases (Ganguly and Chakraborty 2018) activating nitrogen fixation and furthermore counteracting the development of pathogenic parasite through the secretion of enzymes, specifically chitinases and glucanases (Ravindran et al. 2008).

Recent experimental studies on microbial contribution of vermicomposting through Vitek 2 system revealed a major role of Bacillus in upregulation of different enzymatic activities (Ganguly and Chakraborty 2018).

The gut of a Chinese worm *Eisenia foetida* contains a decent measure of expanded anaerobic N_2-fixing microscopic organisms, for example *Clostridium butyricum*, *C.beijerinckii,* and *C.paraputrificum* (Zhang et al. 2019). Furthermore, atomic screening has uncovered the presence of α-Proteobacteria, β-Proteobacteria, Actinobacteria, Planctomycetes, Firmicutes, and Bacteroidetes (Yasir et al. 2009) and their path of succession along the course of composting through the usage of PLFA investigation, DGGE (Denaturing Gradient Gel Electrophoresis), and other PCR strategies (Fernández-Gómez et al. 2012). This expanded pace of microbial populace alongside the ingested material has allotted the status of worm as a helpful bio-hatchery for the productive development of microorganisms.

Another dynamic segment associated with the procedure of biodegradation and transformation during vermicomposting is fungi. They can debase complex biomolecules and are thus applied as an apparatus for bio-remediation for a wide scope of pollutants (Kästner and Mahro 1996). Ascomycetes and basidiomycetes were seen as increasingly copious and enhanced in vermicompost. This might be due to the special demolition characteristic of quick developing fungi (e.g., zygomycetes and mitosporic organisms) which render a low competition rate and provide a suitable environment for the development of basidiomycetes and ascomycetes (Aira et al. 2006).

In such circumstance, the advanced change of vermicomposting swings around the inoculation of advantageous microorganisms for the productive biodegradation and transformation of natural squanders so as to remediate the most dangerous toxins. Attributable to such serious microbial activities, both by earthworm and external microbial inoculation, a great measure of waste can be changed over to useful, compelling compost in a brief timeframe, and consequently generates employment. Furthermore, the microbial population obtained from the vermicompost has gained importance in context of biomedical applications. Vermicompost also proved to be an incubator for most microbes, and the bioactive molecules thus obtained show marked difference in anticancer and antioxidant properties (Ganguly et al. 2018).

10.6 Large-scale Manufacturing of Vermicompost

By and large, the most widely recognized framework used in tropical nations is beds or bins, which are most appropriate for little-scale vermicomposting. Presently, to meet the global demand in organic farming every country is trying to expand its large-scale composting. It involves two primary frameworks, for example, windrow and flow-through system with certain modifications.

10.6.1 Windrow Framework

For producing huge volume of manure, wastes are gathered and after appropriate blending with cow-dung, they are put in lines or windows. In such construction, length is less important as it depends on the availability of the ground area. The floor is commonly comprised of concrete on which earthworm bed is set up with soil, straw, and the feed is placed at 3 ft high. During the span of composting, PVC (Polyvinyl chloride) sheets are utilized to cover it up so as to avert predators and other ecological barriers. It requires a huge tract of land along with regular supply of labor for the maintenance of livelihood. In tropical countries like India, the windrows are covered with sheets to protect the earthworm from getting scorched in heat, while during rainy season, the windrows are adjusted for proper run off of water (Fig. 10.1). The detriment of this procedure is terrible smell and spread of microscopic organisms, overwhelming metal defilements throughout vermicomposting. Some precautionary measures are taken to prevent excessive wetting of beds and easy drainage of water supply. However, complete processing of organic wastes often takes around 1 year (depending upon the usage amount), but the compost suffers from severe nutrient loses due to volatilization and nutrient leaching. Modification of the windrow framework is a wedge framework in which organic feed is

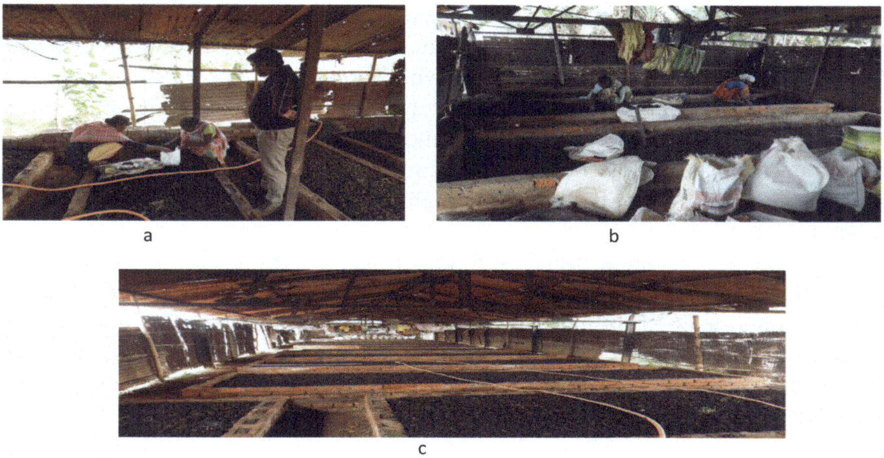

Fig. 10.1 (**a**, **b**) represent vermicomposting as livelihood generation in rural villages. (**c**) represent windrows mode of vermicomposting

added at an angle of 45° which makes a 'wedge' to the existing windrow. Worms for the most part move along the sides overburdened by the load of organic wastes. In this system, earthworm's move from the deeper layers of organic wastes to the fresh organic matter at the surface of wedge, and therefore entire population of earthworms are found to be concentrated approximately one foot below the angled surface. In the long run, following 3–7 months, fertilizer is gathered from the feed depending upon the nature of waste.

10.6.2 Continuous Flow Through the System

The system comprises of large containers mostly raised above the ground, and the organic wastes are added in thin layers at regular intervals. Vermicompost are collected mechanically from the bottom. For the most part, a pit is built with concrete surface to counteract the entrance of predators. This strategy is for the most part utilized for indoor set up. Here, the feed is dumped in one corner of the pit and given consistently once a week. Earthworm populations in such reactors reach a population density of about 9–10 kg/m^2. The procedure gains its importance over windrows in several aspects such as quick returns of capitals, low labor cost, high rate of waste turnover, and most importantly less leaching of nutrients.

Since vermicomposting is a non-thermophilic process no significant changes in temperature are done, only a course of adequate air circulation is performed. This facilitates the earthworm to flourish well inside the vermireactor. Consistent move through the vermireactor is commonly made with poly-vinyl chloride, and bedding is finished with the piece of paper or cardboard at a greatest thickness of 30 cm. The slim layer bedding gives sufficient air circulation and forestalls anaerobic absorption of wastes, which controls the creation of pungent odour. Vermicasts were gathered from the base of vermireactor secured with wire mesh. The produce acquired through this procedure is vastly improved both in quality and quantity when contrasted with different frameworks. In modern day, industrial vermitechnology highlights automated beds where the two contiguous moving surfaces cause finished product at the base of the bed to fall through its perforated base and fresh waste is added to the top.

For small-scale manufacturing, perforated bags are used for proper maintenance of aerobic chamber. In such case, feed is given once and the vermicasts are collected after a definite interval. In most of the cases, Box Crate Systems are practiced in boxes, containers, or trays in different outdoors and indoors settings. Although the technique requires less space, it is labor-intensive and very hard to maintain the adequate moisture content during the process of vermicomposting.

10.7 Development of Vermicompost in Rural Regions of India

Economical agribusiness and natural waste management are twin paths for sustainability in the coming years. In a nation like India where 70% of the population lives in rural areas and where business spins around agriculture related processes, needed an advancement to sustain a better livelihood in those regions. Physically, ethically, and financially, a robust rural community is the major contributors for advancement of a nation. In the general populace, women comprise almost half of the population, and women bear a lot of responsibility than the men in horticulture work like upkeep of domesticated animals, transplanting of seedlings, weeding, reaping, cleaning, and storage of grains.

Recently, to make rural women monetarily autonomous and to inspire their economic wellbeing, the Government of India along with different NGOs have pioneered numerous projects. One such program is ‘women strengthening’ through ‘self-improvement gatherings’. These gatherings are formed to involve women in various vermicompost related projects to make them self-sufficient financially. In rural scenario, vermicompost is produced under proper environmental conditions, and the produce or vermicast are sold in market as fertilizers, while the residual earthworms of composting are sold in market as a rich source of animal feed (Fig. 10.2).

Almost 500 species of earthworm are identified across different regions of India out of the total 2500 species of the world. These earthworms are supplemented with organic wastes and are cultured in crates, tanks, pits, or containers. However, organic wastes are either sundried or pre-digested through anaerobic phase and are supplemented using cow-dung for fast development of earthworm. This is the process used for conversion of organic substance into black gold.

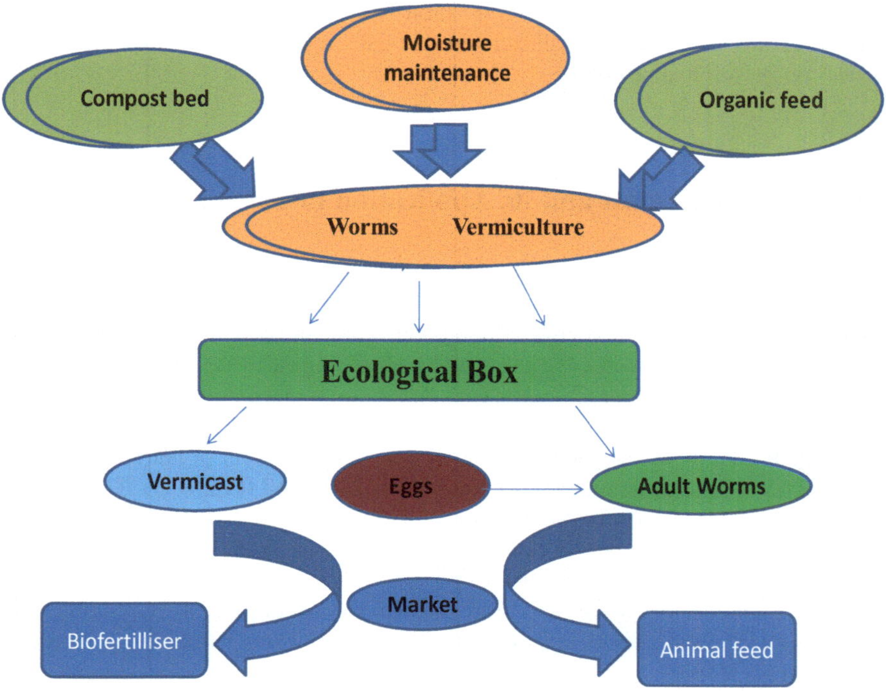

Fig. 10.2 Overview of vermicomposting and its importance in market in most of tropical countries

In rural area, pits are generally built in a concealed region, either under a tree or under a shed. Pits are developed under desirable dimensions (10 ft × 5 ft × 1 ft) with proper ventilation. Essentially two kinds of pits are developed relying on the necessity and accessibility in space of the composters: (1) four-tank framework and (2) two-tank framework. For the simplification of loading procedure, four-tank vermireactors are used in rural areas as it requires low organic wastes and is mostly based on cow-dung composting mixed with different wastes produced at cattle sheds, farm houses, litters, and different types of industrial wastes such as paper mill wastes. A tank having dimension of 6 m × 5 m × 1 m is divided equally into four chambers under shade of tree to prevent from the scorching sun. This is specifically designed for the conversion of small amount of wastes (cattle) of around 30 kg/day. However, two tank systems were mainly built for the handling of small amount of wastes. These systems are used to streamline the waste. Four-tank frameworks utilize the combination of bio dung composting technique alongside vermitechnology. Two-tank frameworks are essentially founded on household garbage and have few natural wastes. These methodologies are highly labor-intensive and henceforth able to generate livelihood for the rural population. However, huge amount of waste water gets generated during the time of composting and cleaning of verm beds. This water can be reused for the new compost to balance the moisture as well as in the enhancement of decomposition process. Waste water generated from this process is mixed with small amount of fresh water and thus enhancement of water conservation is promoted.

10.8 Vermicompost and its Application in Farming

With the expansion in globalization and per capita income rate, it is hard to meet the needs of humankind using limited assets explicitly in tropical nations where the soils are poor in organic content owing to high temperature and huge microbial invasions. In such context, vermitechnology seems, by all accounts, to be a suitable biotechnological tool as a provider for the essential measure of macronutrients such as C, N, P, K and so on (Zhang et al. 2000). Application of vermicompost increases not only the nutrient quality but also enhances the biodiversity, turnover of nutrients, humidification factor, and different enzymatic activities of the soil (Fig. 10.3). The process has gained importance owing to hazardous effect of inorganic mutilation to the soil. Many literatures evidenced the risk of nitrates leaching in soil owing to long-term application of nitrate fertilizer. However, presence of good amount of enzymes, hormones, humic acids proves it to be an efficient soil conditioner by enhancing seed germination, growth of seedling with the increase of plant productivity. Furthermore, vermicompost 'teas' or aqueous extract prepared from vermicompost are successful in suppressing different plant pathogens such as *Verticillium*, *Rhizoctonia*, *Plectosporium* and several parasitic nematodes such as *Meloidogyne and Heterodera* (Arancon et al. 2007).

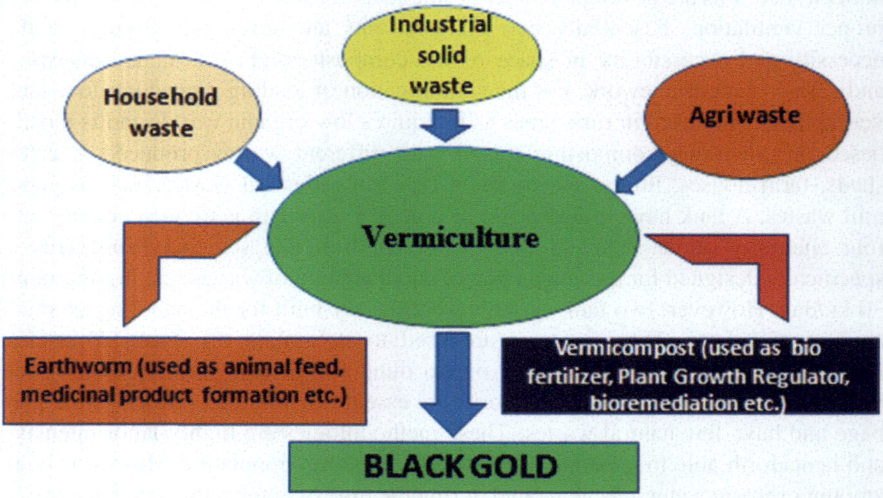

Fig. 10.3 Different types of solid waste used mostly in rural areas

Correlation of fertilizer and vermicompost puts fertility back in soil. Greater yield of potato, spinach, and turnip was recorded upon vermicompost application (Ahirwar and Hussain 2015). Use of vermicompost on various harvests like wheat, paddy, and sugarcane enhanced the yield as accounted previously in literatures by Ansari et al. (2016) and Saranraj and Stella (2012). An improvement in rooting, time of blooming, expansion of leaf region, advancement and stretching of internodes were seen earlier (Padmavathiamma et al. 2008). Few more investigations like tallness of rice plant (Shukla and Singh 2010), spreading of leaf region in maize plant (Joshi et al. 2015), expansion of root knobs in mung bean (Singh and Sharma 2003) are also reflection of utilization of manure in modern-day farming. Further expansion of such application goes to increment in development and yield of some nursery plants (Arancon and Solarte 2019). Spectroscopical information such as FT-IR and GC-MS shows the nearness of a few humic acids like substances which are likely to function as natural conditioner of soil (Ganguly and Chakraborty 2019). However, enhanced humic acid content was found in vermicompost in contrast to conventional composting. Manivannan et al. (2009) likewise demonstrated that the freely accessible minerals such as N, P, K, Zn, Mg, Ca, Na were exponentially expanded in soil enriched with vermicompost. Utilization of vermicompost expanded the mineralizable nitrogen and accessible phosphorus content in the soil (Ganguly and Chakraborty 2019). Therefore, mineral enrichment and lowering of C/N ratio demonstrate its utility in waste management. Furthermore, nitrate accumulation was also lowered upon full-term vermicomposting which lowers the risk of nitrogen leaching from organic wastes.

Mineralization of organic matter and squalor of complex aromatics (lignin, polyphenols) by earthworms into simpler compounds (carbohydrates, lipids) can be analysed by FT-IR spectroscopy. FT-IR spectroscopy method point towards compost mellowness or firmness and has emerged to be an efficient technique for the recognition of functional groups in the decomposing wastes during vermicomposting. Different spectral bands correspond to the absence or presence of certain functional groups which in turn envisage the processes of degradation or stabilization and therefore act as an excellent tool in establishing maturity and stability of the sample. FT-IR spectra during different times (0th and 60th days) of composting for paper mill sludges using *Eisenia fetida* have displayed an intense wide band in between 3300 and 3500 cm^{-1} which indicates an –OH stretching of acid, phenols, and alcohols group (Hussain et al. 2016) (Fig. 10.4). The intensity of the band declines with the progression of vermicomposting for both types of paper mill sludge. Spectral bands around 2920 cm^{-1} correspond to C–H stretching of aliphatic compounds mainly symbolized by alkenes (Ganguly and Chakraborty 2019). A sharp turn down of this band to a substantial point reflects evidence of intense biodegradation coupled with compost maturity. Furthermore, a decrease in the band intensity at 2531 cm^{-1} demonstrates a huge break down of complex –SH groups of paper mill squanders. A surreptitious decrease of band intensity at 2359 cm^{-1} reflects the breakdown of inorganic silanes which are used as coupling agents in paper mill industries. Increase in band intensity at 1790 cm^{-1} indicates C=O stretching of different esters and anhydrides. Decrease in band intensity at

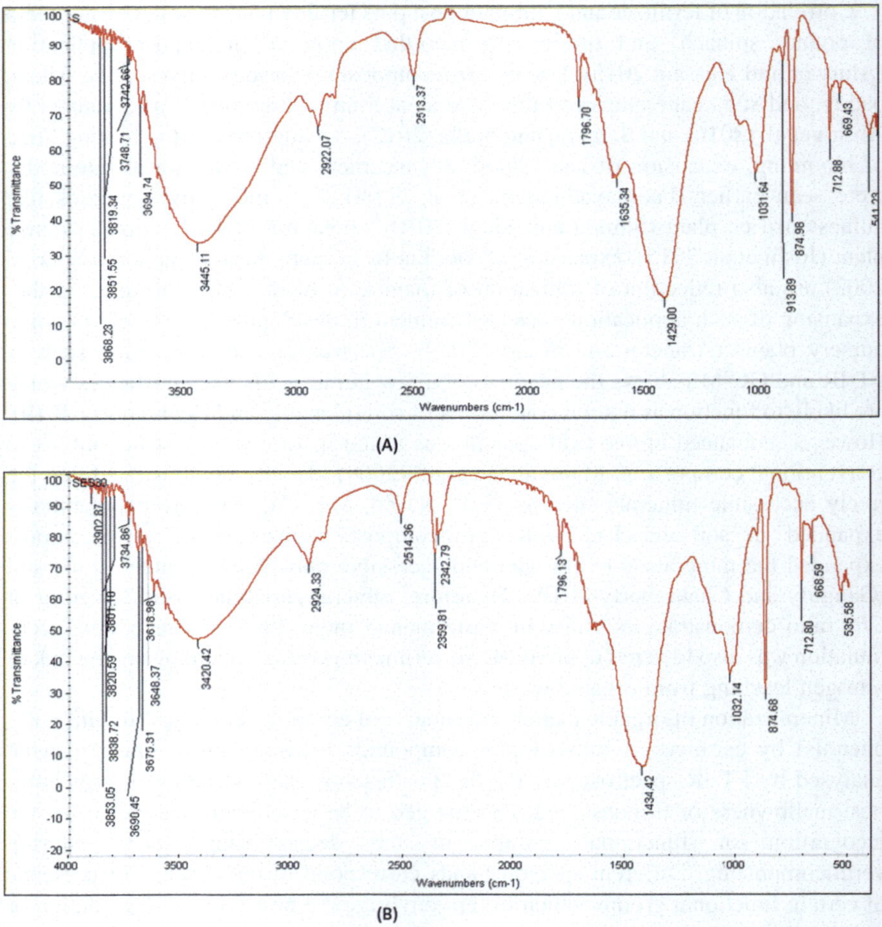

Fig. 10.4 FTIR curves representing the changeover of different functional groups during (**a**) 0th day and (**b**) 60th day of vermicomposting (Ganguly and Chakraborty 2019)

1430 cm^{-1} reflects degradation of organic lignin. So inclusion of all such changes with varying intensity of different bands demonstrate an increment of aromatic C to aliphatic C which reflects enhanced humification during composting of organic sludge by *Eisenia fetida.* The GC-MS study reveals the chemical footprint of the samples in terms of vermicomposting (Fig. 10.5) (Tables 10.1 and 10.2). During composting of primary sludge, 18 peaks were recorded, with maximum peak area for benzoldicarbonsaeure. However, after the end of vermicomposting, several peaks for humic acids were found such as Octadecanoic acid, heptanoic acid, Benzene dicarboxylic acid (Ganguly and Chakraborty 2019). Increase in number of peaks demonstrates the extent of biodegradation of organic wastes by earthworm, such as

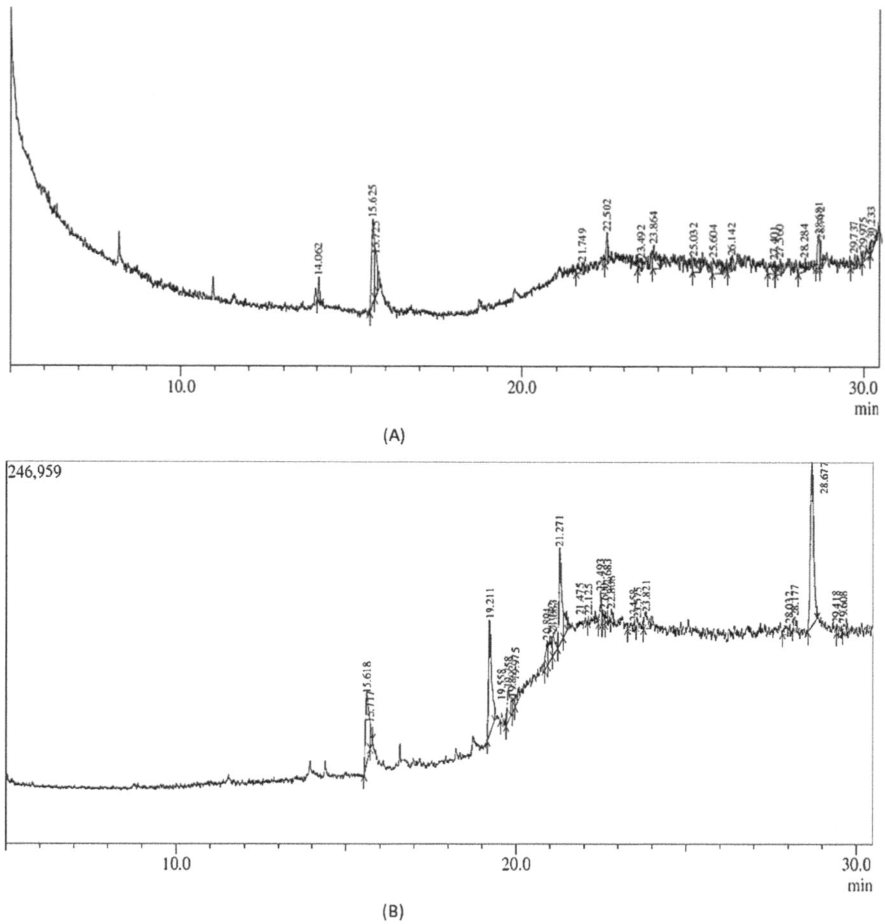

Fig. 10.5 GC MS spectra at (**a**) 0th day and (**b**) 60th day of vermicomposting representing turnover of different humic acid like substances (Ganguly and Chakraborty)

Perionyx excavates. The enrichment of humification factor also determines the stability and extent of maturity of vermicomposting. Therefore, GC-MS study highlights the conversion of organic modification to a beneficial material enriched with Plant Growth Regulators (PGRs), which can be used as biofertilizer.

The enhancement of humification ascertains the microbial degradation taking place inside earthworms gut (Dominguez and Edwards 2004; Ganguly and Chakraborty 2019). These outcomes are shown in the yields of corn, oats, soya bean, peanuts, clover, chicory plants, and other tropical harvests (Devi et al. 2014; Mahendra and Narendra 2012).

Table 10.1 GC-MS spectra revealing list of compounds at 0th day of vermicomposting of a paper mill sludge (Ganguly and Chakraborty 2019)

Peak	R. Time	Area %	Compound name
1	14.062	2.92	Pentadecane
2	15.625	22.20	1,2-Benzoldicarbonsaeure, di-(hex-1-en-5-yl-ester)
3	15.725	7.81	2-Propenoic acid, propyl ester
4	21.749	3.75	Silane, trimethyl(1-methyl-1-phenylethoxy)-
5	22.502	4.65	Isopropyl(2-isopropyl-5-methylcyclohexyl) Phosphinous chloride
6	23.492	2.90	Cyclohexanol, 4-[(tert-butyldimethylsilyl)oxy]-
7	23.864	6.88	Benzenepropanoic acid, tert-butyldimethylsilyl ester
8	25.032	6.00	[(*T*-butyl)dimethylsilyl] phenylacetate
9	25.604	7.46	Caprolactimether, (nb)-*o*-[(diethylboryloxy)(ethyl)boryl]-
10	26.142	2.92	Cyclohexanol, 3-[(tert-butyldimethylsilyl)oxy]-, trans-
11	27.401	4.90	1H-Purine-8-propanoic acid, alpha-amino-2,3,6,7-tetrahydro-1,3,7-Trimethyl-2,6-dioxo
12	27.560	3.66	5 h-Dibenz[b,f]azepine, 5-(2,2-diethoxyethyl)-10,11-dihydro-
13	28.284	3.93	2-Butanone, 1-(2-furanyl)-
14	28.681	5.68	Phosphorous acid, tris(2-ethylhexyl) ester
15	28.742	3.90	1-Iodo-4-hexyne
16	29.737	2.77	Valerophenone, 4′-(trimethylsiloxy)-
17	29.975	4.65	2-Butyn-1-al diethyl acetal
18	30.233	3.01	1 h-Purine-8-propanoic acid alpha-amino-2,3,6,7-tetrahydro-1,3,7-trimethyl-

aR.Time is Retention Time

10.9 Conclusion

Advancement of human activities cannot conquer nature. Nature incorporates human, as a holder, and therefore harmony is important to orchestrate the enthusiasm of both man and nature. Farming is no exception to this. A strong collaboration always exists between organism and multitude of abiotic factors for the maintenance of healthy ecosystem in any given ecosystem. Comprehension of nature is mandatory to establish sustainability.

Waste management systems are considerably predisposed to different environmental, socio-economic factors, population growth, and advancement in waste disposal systems. Henceforth, many countries have shifted their age-old techniques in disposing wastes in land filling and have developed novel eco-sustainable methods for recycling of wastes. In order to meet the demands of sustainable development, 'zero waste' concept is being accentuated globally to create a platform for reutilization of different wastes, such as agricultural, household, and other organic wastes, for the betterment of future generations. The most fundamental and integral part of agriculture is the quality of the soil. Use of different chemical fertilizers poses a great threat to human health and exploits the natural microfauna of

Table 10.2 GC-MS spectra revealing list of compounds at 60th day of vermicomposting of a paper mill sludge (Ganguly and Chakraborty 2019)

Peak	R. Time	Area %	Compound name
1	15.618	10.47	1,2-benzenedicarboxylic acid, diethyl ester
2	15.717	0.81	2(s)-fluoro-gamma-butyrolactone
3	19.211	15.82	Heptadecanoic acid, 16-methyl-, methyl ester
4	19.558	0.81	Trisiloxane, 1,1,1,5,5,5-hexamethyl-3,3-bis[(trimethylsilyl)oxy]-
5	19.758	4.06	1,2-benzenedicarboxylic acid, dibutyl ester
6	19.892	1.21	Hydroxyurea, N,N′,O-trimethyl-
7	19.975	0.78	Benzofuro[3,2-b]pyridine-1(2 h)-carboxylic acid, 3-cyano-4-methyl-, ethyl ester
8	20.894	2.30	Isophthalic acid, pentadecyl pent-4-enyl ester
9	21.042	3.85	Cardinalin 12
10	21.083	2.17	1,1,1,3,5,5,7,7,7-Nonamethyl-3-(trimethylsiloxy)tetrasiloxane
11	21.271	11.83	Octadecanoic acid, methyl ester
12	21.475	2.82	*N*-Methoxy-*N*-benzoyl-1,1-dimethyl-2-carbomethoxyethylamine
13	22.125	0.88	Imidazole-5-carboxylic acid, 2-amino-
14	22.493	2.40	Propanedioic acid, hexyl-, diethyl ester
15	22.600	1.21	1,7-Di(2,5-dimethylphenyl)-2,2,4,4,6,6-hexamethyl-1,3,5,7-tetraoxa-2,4,6-trisilaheptane
16	22.683	0.77	Cyclohexanecarboxamide, *N*-furfuryl-
17	22.808	0.77	Benzeneacetic acid, .alpha.,3,4-tris[(trimethylsilyl)oxy]-, trimethylsilyl
18	23.458	1.35	Benzoflex
19	23.575	1.19	Tetrasiloxane, decamethyl-
20	23.821	2.90	Decane, 1,2-dibromo-
21	28.032	2.02	Glutaric acid, di(hex-4-yn-3-yl) ester
22	28.177	1.54	1,1,3,3,5,5,7,7,9,9,11,11-dodecamethyl-hexasiloxane
23	28.677	25.85	1,2-benzenedicarboxylic acid, dioctyl ester
24	29.418	1.11	Silane, (4-ethylphenyl)trimethyl-
25	29.608	1.09	Tris(tert-butyldimethylsilyloxy)arsane

aR.Time is Retention Time

soil and its quality. Therefore the use of different organic augmentations with vermitechnology can serve as a tool to ensure sustainable agro-ecosystem for betterment of crop productivity and soil quality. This technology can also be adopted for proper waste management, because bacterial carrier in the gut wall of earthworm provides an essential microenvironment for appropriate degradation of large polymers such as lignin, cellulose, etc. Improvement in the rate of composting or compost quality is done through microbial inoculation, nanoparticle utilizations, and regulated surveillance during composting. To sum up, it can be claimed that for robust maintenance of ecological health of soil, crop quality, and least chemical bio-magnification, this traditional methodology offers a new solution for sustainable agriculture management and human health.

Acknowledgments This study acknowledges Vidyasagar University for supporting all sorts of laboratory and computer facilities.

Conflict of Interest: Authors have no conflict of interest.

References

Ahirwar CS, Hussain A (2015) Effect of vermicompost on growth, yield and quality of vegetable crops. Int J Appl Pure Sci Agric 1(8):49–56

Aira M, Monroy F, Domínguez J (2006) Eisenia fetida (Oligochaeta, Lumbricidae) activates fungal growth, triggering cellulose decomposition during vermicomposting. Microb Ecol 52(4):738

Ansari AA, Jaikishun S, Islam MA, Kuri SK, Fiedler K, Nandwani D (2016) Principles of vermitechnology in sustainable organic farming with special reference to Bangladesh. In: Organic farming for sustainable agriculture. Springer, Cham, pp 213–229

Arancon NQ, Solarte Z (2019). Vermiculture in greenhouse plants, field crop production, and hydroponics. In: Oxford research encyclopedia of environmental science. https://oxfordre.com/environmentalscience/view/10.1093/acrefore/9780199389414.001.0001/acrefore-9780199389414-e-295

Arancon NQ, Edwards CA, Yardim EN, Oliver TJ, Byrne RJ, Keeney G (2007) Suppression of two-spotted spider mite (Tetranychus urticae), mealy bug (Pseudococcus sp) and aphid (Myzus persicae) populations and damage by vermicomposts. Crop Prot 26(1):29–39

Bhat SA, Singh S, Singh J, Kumar S, Vig AP (2018) Bioremediation and detoxification of industrial wastes by earthworms: vermicompost as powerful crop nutrient in sustainable agriculture. Bioresour Technol 252:172–179

Devi U, Singh KP, Kumar S, Sewhag M (2014) Effect of nitrogen levels, organic manures and Azotobacter inoculation on yield and economics of multi-cut oats. Forage Res 40:36–43

Dominguez J, Edwards CA (2004) Vermicomposting organic wastes: a review. In: Soil zoology for sustainable development in the 21st century. Cairo, pp 369–395

Fernández-Gómez MJ, Nogales R, Insam H, Romero E, Goberna M (2012) Use of DGGE and COMPOCHIP for investigating bacterial communities of various vermicomposts produced from different wastes under dissimilar conditions. Sci Total Environ 414:664–671

Ganguly RK, Chakraborty SK (2018) Assessment of microbial roles in the bioconversion of paper mill sludge through vermicomposting. J Environ Health Sci Eng 16(2):205. https://doi.org/10.1007/s40201-018-0308-4

Ganguly RK, Chakraborty SK (2019) Assessment of qualitative enrichment of organic paper mill wastes through vermicomposting: humification factor and time of maturity. Heliyon 5(5):e01638

Ganguly RK, Midya S, Chakraborty SK (2018) Antioxidant and anticancer roles of a novel strain of Bacillus anthracis isolated from vermicompost prepared from paper mill sludge. Bio Med Research International, Article ID 1073687, 7p, https://doi.org/10.1155/2018/1073687

Hussain N, Abbasi T, Abbasi SA (2016) Transformation of the pernicious and toxic weed parthenium into an organic fertilizer by vermicomposting. Int J Environ Stud 73(5):731–745

Joshi R, Singh J, Vig AP (2015) Vermicompost as an effective organic fertilizer and biocontrol agent: effect on growth, yield and quality of plants. Rev Environ Sci Biotechnol 14(1):137–159

Kästner M, Mahro B (1996) Microbial degradation of polycyclic aromatic hydrocarbons in soils affected by the organic matrix of compost. Appl Microbiol Biotechnol 44(5):668–675

Kaushik P, Garg VK (2003) Vermicomposting of mixed solid textile mill sludge and cow dung with the epigeic earthworm Eisenia foetida. Bioresour Technol 90(3):311–316

Kostina NV, Bogdanova TV, Umarov MM (2011) Biological activity of the coprolites of earthworms. Moscow Univ Soil Sci Bull 66(1):18–23

Mahendra S, Narendra K (2012) Effect of FYM, vermicompost, vermiwash and NPK on growth, microbial biomass and yield of soybean. Soybean Res 10:60–66

Manivannan S, Balamurugan M, Parthasarathi K, Gunasekaran G, Ranganathan LS (2009) Effect of vermicompost on soil fertility and crop productivity—beans (*Phaseolus vulgaris*). J Environ Biol 30:275–281

Miller JJ, Owen ML, Drury CF, Chanasyk DS (2019) Short-term legacy effects of feedlot manure amendments on earthworm abundance in a clay loam soil. Can J Soil Sci 99(0):1–11

Moorthi M (2016) Vermi-technology of organic solid waste with using earthworm Eudrilus Eugeniae. J Zool Stud 3(4):48–51

Ndegwa PM, Thompson SA (2001) Integrating composting and vermicomposting in the treatment and bioconversion of biosolids. Bioresour Technol 76(2):107–112

Padmavathiamma PK, Li LY, Kumari UR (2008) An experimental study of vermi-biowaste composting for agricultural soil improvement. Bioresour Technol 99(6):1672–1681

Pathma J, Sakthivel N (2012) Microbial diversity of vermicompost bacteria that exhibit useful agricultural traits and waste management potential. Spring Plus 1(1):26

Pathma J, Raman G, Sakthivel N (2019) Microbiome of rhizospheric soil and vermicompost and their applications in soil fertility, pest and pathogen management for sustainable agriculture. In: Soil fertility management for sustainable development. Springer, Singapore, pp 189–210

Rakkini VM, Vincent S, Kumar AS, Baskar K (2017) An overview: organic waste management by earthworm. J Civil Eng Environ Sci 3(1):013–017

Ravindran B, Dinesh SL, Kennedy LJ, Sekaran G (2008) Vermicomposting of solid waste generated from leather industries using epigeic earthworm Eisenia foetida. Appl Biochem Biotechnol 151(2-3):480–488

Saranraj P, Stella D (2012) Vermicomposting and its importance in improvement of soil nutrients and agricultural crops. Novus Nat Sci Res 1(1):14–23

Sharma K, Garg VK (2018) Vermicomposting: a green technology for organic waste management. In: Waste to wealth. Springer, Singapore, pp 199–235

Shukla R, Singh K (2010) Effect of vermicompost of different organic wastes on the germination, growth and productivity of certain crops. J Solid Waste Technol Manag 36(3):170–181

Singh A, Sharma S (2003) Effect of microbial inocula on mixed solid waste composting, vermicomposting and plant response. Comp Sci Utilizat 11(3):190–199

Sinha RK, Herat S, Chauhan K, Valani D (2009) Earthworms: the 'unheralded soldiers of mankind' and 'farmer's friend' working day and night under the soil: reviving the dreams of sir Charles Darwin for promoting sustainable agriculture. Am Eurasian J Agric Environ Sci 5:5–13

Venkatesh RM, Eevera T (2008) Mass reduction and recovery of nutrients through vermicomposting of fly ash. Appl Ecol Environ Res 6(1):77–84

Yasir M, Aslam Z, Kim SW, Lee SW, Jeon CO, Chung YR (2009) Bacterial community composition and chitinase gene diversity of vermicompost with antifungal activity. Bioresour Technol 100(19):4396–4403

Zhang BG, Li GT, Shen TS, Wang JK, Sun Z (2000) Changes in microbial biomass C, N, and P and enzyme activities in soil incubated with the earthworms Metaphire guillelmi or Eisenia fetida. Soil Biol Biochem 32(14):2055–2062

Zhang W, Wang C, Liu M, Yu Y (2019) Integrated reclamation of saline soil nitrogen transformation in the hyphosphere by earthworms and arbuscular mycorrhizal fungus. Appl Soil Ecol 135:137–146

Chapter 11
Growth and Reproductive Biology of Earthworms in Organic Waste Breakdown Under the Indian Condition

Priyasankar Chaudhuri and Susmita Debnath

Abstracts Environmental improvement is an accepted national goal. Under the present condition of the acute energy crisis and environmental degradation with rise in population, it has become essential to develop an appropriate technology for the recovery of energy from non-conventional sources like organic wastes. India produces 350 million tonnes of organic wastes from different agricultural sources. Out of these, several million tonnes of plant nutrients are produced. Earthworm (verm) is known to be a good biological source for the recovery of vermifertilizer and vermiprotein from the organic wastes to be used in agro-ecosystem and aquaculture and poultry, respectively. Vermicomposting is a process of conversion of organic waste through the synergistic actions of epigeic earthworm and bacteria. When earthworms (endogeic or anecic) are applied for breakdown and stabilization of municipal solid wastes, the process is termed 'vermistabilization'. The ultimate goal of both the processes is the production of plant-available nutrients and earthworm biomass (vermiprotein). Earthworm species to be selected for the process (1) should have great adaptability with respect to environmental factors, (2) should be capable of inhabiting high percentage of organic matter, (3) should be a prolific breeder (high fecundity) with high hatching success, and (4) should have short life cycle. Recent studies indicated that under Indian conditions epigeic species, such as *Perionyx excavatus*, *Perionyx ceylanensis*, *Eudrilus eugeniae*, *Eisenia fetida*, *Eisenia andrei,* can complete their life cycle in organic waste and produce vermicompost. *Lampito mauritii*, *Polypheretima elongata*, *Drawida nepalensis*, *Drawida willsi*, *Metaphire posthuma*, *Metaphire houlleti* can degrade organic matter in municipal sewage sludge which contains considerable amounts of sand particles. This chapter deals with the growth and reproduction of these organic waste degrading earthworms. Based on the study, suitable species for organic waste degradation are being suggested.

P. Chaudhuri (✉) · S. Debnath
Department of Zoology, Earthworm Research Laboratory, Tripura University, Suryamaninagar, West Tripura, India

S. A. Bhat et al. (eds.), *Earthworm Assisted Remediation of Effluents and Wastes*,
https://doi.org/10.1007/978-981-15-4522-1_11

Keywords Earthworm · Growth rate · Cocoon · Fecundity · Vermicomposting · Vermistablization

11.1 Introduction

Charles Darwin (1881) in his book *The Formation of Vegetable Mould through the Action of Worms* mentioned earthworm's role in the disappearance of overground leaf litter. Since then, more than a century has passed to evaluate the role of earthworms in the breakdown of organic wastes, plant litter, and their redistribution and incorporation in the soils through their feeding, burrowing, and casting activities (Lee 1985). Earthworm is physically an aerator, crusher, mixer, and chemically a degrader and biologically a stimulator in the decomposer system (Mitchell 1978). The burrowing activities of earthworm enhance aeration and porosity of the soil, thereby improving its water-holding capacity.

Earthworms are metamerically segmented, vermiform, skin-breathing coelomate invertebrates belonging to the class Oligochaeta under phylum Annelida. Their locomotion is aided by the hydrostatic function of the coelome and segmentally arranged chitinous setae. They possess a simple straight alimentary canal with terminal mouth and anus. The circulatory system is of closed type. Central nervous system is simple with paired cerebral ganglia (Brain), paired subesophageal ganglia, circumesophageal connectives joining the cerebral with subesophageal ganglia and ventral nerve cord that originates from subesophageal ganglia. They have no discrete endocrine glands, instead the highly vascularized central nervous system along with enormous number of neurosecretory cells act as the neuroendocrine system. The neurosecretory cells are the sole source of hormones (neurohormone) in Oligochaetes that control all the physiological events such as growth, reproduction, cocoon laying, regeneration, thermal acclimation, osmoregulation, etc.

Earthworms are found in the soils of forest, grassland, pasture, agricultural lands, and also in the organic wastes. There are more than 8300 species of Oligochaeta of which about 50% are terrestrial earthworms (Reynolds and Wetzel 2004). According to Julka (2014), presently, 420 species of earthworms under 70 genera are found in the Indian soil. Endemism, both at the genus and species level, is very high. About 71% of genera and 89% of species are endemic (Julka 2014).

Earthworm scientists often recognize three broad ecological groups of earthworms: epigeic, endogeic, and anecic. The epigeic earthworms (e.g. *Perionyx excavatus*) inhabit degrading organic matter, such as cattle dung, litter heaps, etc. They are generally small to medium-sized, deeply pigmented, flat ventrally, tolerant to disturbance and have short life cycle, higher fecundity and regeneration capacity (Bhattacharjee and Chaudhuri 2002; Banik and Chaudhuri 2017). Anecic earthworm (*Metaphire houlleti*) lives in deep vertical burrows in the soil, and they emerge at night to draw dead leaves and other organic decaying matter down into their burrows. Anecics are phytogeophagous. They are generally large, moderate to heavily pigmented and intolerant to disturbance. Endogeic species (*Pontoscolex corethrurus*) rarely come to the surface. They inhabit mineral soil horizons, feed more on soils than organic matter and tolerant to some disturbance. Both endogeic and anecic have a long life cycle with limited regeneration capacity (Banik and Chaudhuri 2017). Unlike temperate regions, majority of earthworm species of tropics are geophagous, i.e. endogeic, and only a few are detritivorous (Lavelle 1978). Both endogeic and anecic earthworms are called soil ecosystem engineers and provide soil health for over ground crop development. Epigeic earthworms increase the nutrient content of top soils by soil casting activities.

The dramatic rise in the human population and increased agricultural activities have generated a large amount of organic wastes, the disposal of which is one of the major hazards for upkeeping the environment. In recent times, vermicomposting technology is being developed to recycle organic wastes through the synergistic action of epigeic earthworms and bacteria to form the organic fertilizer called vermicompost. The term 'vermistabilization' is used when instead of organic wastes of plant origin municipal solid wastes are acted upon by earthworms to form nutrient-rich substrates. During the process of vermicomposting, earthworm reproduces, grows, and produces biomass depending upon the quality of organic wastes. The ultimate product of vermicomposting is organic fertilizer, called vermicompost, a plant growth medium, rich in plant-available nutrients and earthworm biomass chiefly containing protein (vermiprotein).

This chapter aims to report on the organic waste degrading earthworms, selective characteristic of composting earthworms, with short notes on their growth and reproductive biology.

11.2 Composting Earthworms and their Selective Features

11.2.1 Selective Features of Composting Earthworms

1. Earthworms should have adaptability to environmental factors.
2. Earthworms should be capable of inhabiting high percentage of organic matter.
3. Short incubation period of cocoon.
4. More than one hatchling.
5. They should be prolific breeders (high fecundity) with high hatching success.
6. Earthworms should have quick maturation time (short life cycle).

Earthworm species with potential for waste management under Indian conditions include *Perionyx excavatus* (Indian blue), *Perionyx ceylanensis*, *Perionyx sansibaricus*, *Eudrilus eugeniae* (African night crawler), *Eisenia fetida* (tiger worm), *Eisenia andrei* (European red worm), *Lampito mauritii*, *Polypheretima elongata*, *Drawida nepalensis*, *Drawida willsi*, *Metaphire posthuma*, *Dichogaster bolaui*, *Dichogaster curgensis*, *Dichogaster modiglianii*, *Amynthus cortices*, *Bimastos parvus*, etc. Among these earthworms, *E. fetida*, *E. andrei*, *P. excavatus*, and *E. eugeniae* are widely used in India. Recently, Debnath and Chaudhuri (2019) reported 11 earthworm species in the waste deposit sites of West Tripura out of which only *P. excavatus* is a typical vermicomposting species while others have affinity for organic wastes and play some role in waste degradation and vermistablization.

11.3 Growth and Reproduction in Earthworms Used for Vermicomposting and Vermistabilization

Earthworms are hermaphrodite with separate testes and ovaries. Most of the earthworm species reproduce by cross-fertilization, although some species produce cocoons parthenogenetically or by self-fertilization. Following copulation, which may take as long as an hour, the worms separate and each clitellum produces a secretion that eventually hardens over its surface. When it gets hard, the partners move backward, so drawing the tube over its head, and when the worms are completely free, the ends of the tube close to form the cocoon which is roughly a lemon-shaped structure but with a shape typical of each species (Fig. 11.1) (Edwards and Bohlen 1996). Fertilization which occurs in the cocoon is thus external. From the cocoon, after a certain period of development (incubation period), baby worm emerges out (Fig. 11.1 and 11.2).

According to Hyman (1940), earthworms grow continuously throughout their lives by the addition of segments proliferated from a 'growing zone' in front of the anus. Edwards and Bohlen (1996) supported the view of Evans (1946) who opined that some earthworm species possess the adult number of segment on hatchling, whereas other species add further segments during post-emergence growth. In earthworms, there is a pre-reproductive growth phase followed by a phase of reproduction after attainment of sexual maturity. Many workers have confirmed a sigmoid growth curve for *E. fetida* and in other species that grow in organic wastes (Edwards and Neuhauser 1988). Rate of growth and reproduction of some epigeic earthworm species utilized in waste degradation are shown in Table 11.1.

11.3.1 Perionyx excavatus *(Perrier 1872)*

Potential of *P. excavatus* for breaking down of organic waste was first reported by Kale et al. (1982). This tropical earthworm species is commonly found over a large area of tropical South Asia (Stephenson 1930; Gates 1972) although it has also been transported to Europe and North America. This native epigeic species lives solely in organic wastes (preferably cattle dung) with adequate moisture content (50–60%) and organic matter (4.5–12%) (Bhattacharjee and Chaudhuri 2002). This earthworm species has high fecundity (150 cocoons worm^{-1} year^{-1}) with a short life cycle of 40–50 days. Its cocoon development time is 13 days (Bhattacharjee and Chaudhuri

Fig. 11.1 and 11.2 Cocoons of earthworms: (a) *Perionyx excavatus*, (b) *Metaphire posthuma*, (c) *Lampito mauritii*, (d) *Dichogaster bolaui*; Emergence of hatchling from the cocoon on *Lampito mauritii*

2002). Sexual maturity as indicated by the appearance of clitellum is attained within 20–28 days. Growth rate of *Perionyx excavatus* in cow dung is 2.86 mg day^{-1}, but addition of straw to cow dung enhances its growth rate significantly as high as 4.75 mg day^{-1}, while it is 4.02 mg day^{-1} when bamboo leaf litter acts as an additive (Chaudhuri and Bhattacharjee 2002). Likewise highest rate of reproduction in *P. excavatus* was also noticed in cow dung-straw diet (14.0 juveniles adult^{-1} week^{-1}) which was followed by cow dung-bamboo leaf litter (11.7 juveniles adult^{-1} week^{-1}) and cow dung (2.45 juveniles adult^{-1} week^{-1}) (Chaudhuri and Bhattacharjee 2002).

11.3.1.1 *Perionyx ceylanensis* Michaelsen

This species was first reported from Sri Lanka but at present abundantly found in Peninsular India. It has a great similarity with *P. excavatus* and is usually found in leaf litter with high moisture and organic matter content. A recent report by Prakash

Table 11.1 Growth and reproduction rate in some epigeic earthworm species

Earthworm species	Culture materials	Growth rate (mg worm^{-1} day^{-1})	Reproduction rate		References
			Cocoon (worm^{-1} day^{-1})	Juvenile (adult^{-1} week^{-1})	
Perionyx excavatus	Cowdung	22.91	0.25	1.39	Chaudhuri and Debnath (2020)
	Acacia leaf litter-Cowdung	3.47	0.3	3.31	Chaudhuri and Debnath (2020)
	Bamboo leaf litter-Cowdung	12.6	0.23	2.22	Chaudhuri and Debnath (2020)
	Mikania micrantha-Cowdung	9.47	0.13	0.66	Chaudhuri and Debnath (2020)
	Cowdung	2.86	–	2.45	Chaudhuri and Bhattacharjee (2002)
	Cowdung-kitchen wastes	2.47	–	1.37	Chaudhuri and Bhattacharjee (2002)
	Cowdung-straw	4.75	–	14	Chaudhuri and Bhattacharjee (2002)
	Cowdung-leaf litter	4.02	–	11.7	Chaudhuri and Bhattacharjee (2002)
	Rubber leaf litter	5.04	–	0.2	Chaudhuri et al. (2003)
	Pressmud-Cowdung	4.81	0.79	–	Birundha et al. (2013)
Perionyx ceylanensis	Cowdung	1.34	0.94	–	Karmegam and Daniel (2009)
Perionyx sansibaricus	Kitchen waste-*Mangifera indica* leaf litter	3.77	0.25	–	Suthar (2007)
	Cattle solid wastes	8.00	0.22	–	Suthar (2009)

Eisenia fetida	Cowdung	16.3	0.39	–	Garg et al. (2005)
	Goat waste	16.5	0.32	–	Garg et al. (2005)
	Sheep wastes	26.2	0.44	–	Garg et al. (2005)
	Rubber leaf litter	6.2	–	1.3	Chaudhuri et al. (2003)
Eudrilus eugeniae	*Diospyrosa meanoxylon* leaf litter	68.00	0.54	–	Kadam (2015)
	Rubber leaf litter	28.8	–	1.4	Chaudhuri et al. (2003)
Lumbricus rubellus	Cowdung	8.02	0.77	–	Elvira et al. (1996)
Dendrobaena rubida	Cowdung	3.84	0.2	–	Elvira et al. (1996)
Dichogaster modiglianii	Pasture soil	–	0.19	–	Bhattacharjee and Chaudhuri (2002)

et al. (2008), Karmegam and Daniel (2008a, b) confirmed the high vermicomposting potential of *P. ceylanensis*. Its mean growth rate varies from 1.34 to 1.79 mg worm^{-1} day^{-1}. Worms cultured singly always have shown a much better growth rate than those cultured in batches (Karmegam and Daniel 2009). The highest rate of cocoon production ranging from 1.18 to 1.31 cocoon worm^{-1} day^{-1} was recorded during the age of 120–150 days. The average incubation period is 18.5–19 days and hatching success rate varies from 75–83%. Interestingly, copulation is not essential for cocoon production in *P. ceylanensis* (Karmegam and Daniel 2009).

11.3.1.2 *Perionyx sansibaricus* Michaelsen

This epigeic earthworm is found in soil surface with leaf litter containing high moisture, kitchen wastes and drainage, sewage and is believed to be suitable for vermicomposting. A mature worm produces about 70 cocoons per year and one to two juveniles are hatched from one cocoon.

11.3.1.3 *Eudrilus eugeniae* (Kinberg 1866)

This earthworm species is native to Africa, but it has been extensively bred in Canada, the United States, India, and other countries. It is a large earthworm that grows extremely rapidly and is a prolific breeder, and under optimum conditions it is ideal for the production of animal feed protein. Due to its voracious food habit, it is capable of decomposing large masses of organic wastes quickly. Compared to other vermicomposting species, *E. eugeniae* has the preference for rubber leaf litter where its mean growth rate and reproduction rate are 29 mg worm^{-1} day^{-1} and 1.4 juveniles worm^{-1} day^{-1} (Chaudhuri et al. 2003). It has preference for high temperature with maximum biomass production occurring at 25–30 °C. The life cycle of *E. eugeniae* ranges from 50 to 70 days. Sexual maturity is attained within 40–49 days, and a week after this period, the individual starts to lay cocoons at the rate of 0.41–0.51 cocoon day^{-1}. The cocoon incubation period ranges from 12 to 16 days, and hatching success from 75 to 84% with the mean number of earthworm cocoon^{-1} between 2 and 2.7 (Reinecke and Viljoen 1993; Dominguez et al. 2001). Lifespan can be 1–3 years.

11.3.1.4 *Eisenia fetida* (Savigny 1826) and *Eisenia andrei* (Bouche 1972)

These dynamic lumbricid species are commonly utilized in vermicomposting mainly because of their worldwide adaptability. Moreover, these worms can easily be handled and have a wide range of temperature and moisture tolerance. While *E. andrei* corresponds to the uniformly reddish morph, *E. fetida* has lightly yellowish bands around the intersegmental groove. The two species are syntopic, commonly living in mixed colonies in dung and vermicompost heaps, and therefore hybridization is possible (Dominguez and Edwards 2011). Both the species show cross as well as self-fertilization. Life cycle of these are short, number of hatchling cocoon^{-1} is 2.5–3.8 and incubation time (days) is 18–26. Hatching success is comparatively more in *E. fetida* (73–80%) than *E. andrei* (72%). Maturation time (days) is much higher (28–30 days) in *E. fetida* than *E. andrei* (21–28 days). The mean growth rate of *E. fetida* in rubber leaves is 6.02 mg worm^{-1} day^{-1} (Chaudhuri et al. 2003).

11.3.1.5 *Polypheretima elongata* (Perrier 1872)

This earthworm species is exotic to India and has been tested for the treatment of solid organic wastes, including municipal and slaughterhouse waste, human and diary wastes (Dominguez and Edwards 2011). A project in India using this species claimed to have commercially viable facilities for vermistabilization of 8 tonnes of solid organic wastes day^{-1}. They developed a vermifilter (packet of vermicompost and live earthworms) that produces reusable water from sewage sludge and organic wastewater from sewage sludge and from the food processing centre. *P. elongata* has a continuous breeding strategy with fecundity 23 cocoons worm^{-1} year^{-1}, incubation time 49 days, hatchling cocoon^{-1} 1.33, and hatching success 40% (Bhattacharjee and Chaudhuri 2002). Vermistabilization of municipal sewage sludge by *P. elongata* enhancing its nutrient contents was reported by Chaudhuri and Bhattacharjee (2001).

11.3.1.6 *Amynthas cortices* (Kinberg)

A. cortices occur in forest litter, crop fields, pasture, kitchen drainage, and compost heaps. In India, it is commonly found in hilly areas. It feeds on surface litter, decomposing material, and soil. Exhibits peculiar behaviour of thrashing and jumping and breaks off its tail when disturbed. Its preferred temperature varies from 21 to 27 °C. Number of juveniles per cocoon is 1. Incubation period is about 30 days.

11.3.1.7 *Drawida nepalensis* (Michaelsen)

D. nepalensis is not widely utilized vermicomposting although it has an affinity for organic matter and shows some suitable characteristics for vermiculture. This species possesses semi-continuous breeding strategy with cocoon production 29 cocoons worm^{-1} year^{-1}, development time 24 days, hatching success 37.50%, and hatchling cocoon^{-1} 1.80 (Bhattacharjee and Chaudhuri 2002). Kaushal et al. (1994) observed a growth rate of 5.07 mg worm^{-1} day^{-1} for *D. nepalensis* reared on the horse manure. Fed on cow manure and oak litter *D. nepalensis* has a growth rate of 3.52 mg worm^{-1} day^{-1} (Joshi and Dabral 2008).

11.3.1.8 *Drawida willsi* (Michaelsen)

This epigeic worm is reported from crop fields, soil with organic matter, compost pits, kitchen wastes, etc. Favourable growth and cocoon laying occurs in 25 °C in laboratory culture (Dash 2012). Cocoon case is very hard. Cocoon production is about 15 mature worm^{-1} year^{-1}. Number of young per cocoon varies from 2 to 3. Incubation period is 14–18 days during October to December.

11.3.1.9 *Lampito mauritii* Kinberg 1866

L. mauritii is an epianecic or anecic earthworm species occurring in grassland, crop field, forest litter, garden soil, garbage, compost heaps, and sewage. Its origin was India (Palni/ Cardamom hills in the Western Ghats) and presently it is widely distributed in India. It feeds on decaying plant material, soil micro-fungi, nematodes, and soil. Casts are deposited on the soil surface in the form of small globular pallets. This species is suitable for organic waste conditioner for canteen waste, domestic waste, municipal organic waste, agricultural waste, etc. But for its survival in organic waste, it needs soil bedding. This species has semi-continuous breeding strategy, fecundity 43 cocoons worm^{-1} year^{-1}, development time 15 days, hatching success 60%, and 1.67 hatchling cocoon^{-1}(Bhattacharjee and Chaudhuri 2002). Its life cycle in cow dung and press mud is 117 days and 108 days, respectively (Parthasarathi 2007). In press mud, sexual maturity is attained within 51–76 days (Parthasarathi 2007). Growth rate of 14 mg worm^{-1} day^{-1} and 17 mg worm^{-1} day^{-1} in cow dung were recorded (Parthasarathi 2007).

P. elongata, L. mauritii, M. posthuma, D. modiglianii, D. bolaui, and *Metaphire houlleti* have an affinity for organic wastes. These earthworm species may be utilized for breaking down of municipal solid wastes (vermistabilization) which contains a considerable amount of soils. However, these species are not suitable for vermicomposting, because they need soil bedding under the organic wastes for their regular activity.

11.4 Effect of Food Supply on Growth and Reproduction in Earthworms

Earthworms feed on a wide variety of organic material. The kind and amount of food available influences not only the size of the earthworm population but also the species present and their growth rate and fecundity (Edwards and Bohlen 1996). Evans and Guild (1948) observed that earthworms fed on any nitrogen-rich diet grew faster and produce more cocoons than those with little nitrogen available. Earthworms, in general, grow faster when provided with animal dung, leaf litter rich in protein, and soluble sugar (Edwards and Bohlen 1996). Satchell (1967) reported that there was an inverse correlation between palatability of litter and its total polyhydric phenol content and positive correlation with the amount of soluble carbohydrates. In tropics, Tian et al. (1993) reported that earthworm populations were negatively correlated with the ratio of lignin to nitrogen of the plant residue. According to Muys et al. (1992), earthworm biomass diminished under an oak tree (*Quercus palustris*) due to poor quality of the oak litter and acidification of the soil by oak leaves. High contents of polyphenol, flavonoid, and lignin in the litter of 3- and 10-years-old rubber plantations, through their effects on food intake probably resulted in low biomass values of earthworms in those age group of the plantation (Chaudhuri et al. 2013).

11.5 Influence of Environmental Factors on Survival and Growth of Earthworms

Growth of earthworm species, cocoon production, cocoon development, and hatching all are critically influenced by the environmental conditions.

11.5.1 Temperature

Within the temperature range of 25 and 31 °C growth, reproduction and cocoon production occurs. If these limits are greatly exceeded, worms may move to a suitable zone in the wastes or leave it or may even die. Edwards (1988) studied the life cycle and optimal condition for survival growth of *E. fetida*, *E. eugeniae*, and *P. excavatus*. According to Edwards (1988), optimum temperature for *E fetida*, *E. eugeniae*, and *P. excavatus* were 25 °C, and the latter two species died below 9 °C and above 30 °C when kept for longer period. In fact, temperature beyond optimum level acts as cues for neurosecretory activity thus affecting cocoon production (Olive and Clark 1978). Temperature affects the incubation period of cocoons. With the increase in room temperature, incubation period is increased in *P. elongata and D. nepalensis* and decreased in *P. excavatus* and *Dichogaster modiglianii* (Bhattacharjee and Chaudhuri 2002). Reinecke et al. (1992) reported that temperature higher than 25 °C decreased the mean incubation period in the epigeic worms, *P. excavatus* and in *E. eugeniae* and *E. fetida*. The optimum temperature in which the cocoon of an earthworm species hatch is the threshold temperature, which according to Holmstrup et al. (1991) is an adaptation to the particular habitat condition in which the species live. In general, the threshold temperature during hatching is lower than the temperature during the incubation period of the cocoon of earthworm species. Thus, room temperature (°C) during hatching in *P. excavatus* (29.52 ± 0.25), *L. mauritii* (29.60 ± 0.49), *D. modiglianii* (29.88 ± 0.31), *D. nepalensis* (29.56 ± 0.60), and *Eutyphoeus gammiei* (28.0) was lower than the room temperature during the incubation period (31.07 ± 0.5, 29.71 ± 0.27, 30.85 ± 0.14, 30.35 ± 0.24, and 30.46) (Bhattacharjee and Chaudhuri 2002). This may be due to the tendency for the avoidance of "heat shock" by the baby worm following emergence.

11.5.1.1 Moisture Content

Moisture is the limiting factor for earthworm survival, as earthworms can't respire in dry soils. There is a strong relationship between the moisture content of the organic wastes and the growth rate of earthworms (Dominguez and Edwards 2011). Bedding moisture of 70–80% for *E. eugeniae*, *E. fetida*, *P. excavatus*, 28–42% for *D. modiglianii*, and 25–30% for *D. nepalensis*, *P. elongata,* and *L. mauritii* are required for their reproduction (Bhattacharjee and Chaudhuri 2002). *E. andrei* and *E. fetida* grow more rapidly between 80 and 90% moisture in organic wastes (Domínguez and Edwards 1997).

11.5.1.2 pH

Most epigeic earthworm species favour pH level in the acidic side of neutral (Edwards and Bohlen 1996) when subjected to pH gradient 5–9.

11.5.1.3 Aeration

Earthworms have no specialized respiratory organ, and oxygen and carbon dioxide diffuse through their body wall, thus they are highly sensitive to anaerobic conditions.

11.5.1.4 Ammonia and Salts

Earthworms are very sensitive to ammonia and cannot survive in organic wastes having a high level of this cation (e.g. fresh poultry dropping). They also die in wastes with large quantities of inorganic salts (fresh kitchen waste). Both inorganic salt and ammonia have sharp cut-off points between toxic and nontoxic (Edwards 1988). Organic waste contains a high level of ammonia, and salts can become acceptable to earthworms after removal of ammonia by a period of pre-composting or by several washing with water.

11.6 Conclusions

Under Indian conditions, *Perionyx excavatus*, *Perionyx ceylanensis*, *Eudrilus eugeniae*, *Eisenia fetida*, *Eisenia andrei* can be utilized in vermicomposting technology. All these earthworm species possess wide range of tolerance to environmental factors and high growth rate, short life cycle survival in organic wastes. *Lampito mauritii*, *Polypheretima elongata*, *Drawida nepalensis*, etc. can be utilized in municipal solid waste management, because these species survive well in municipal solid organic waste.

Acknowledgments This study was partly funded by DBT (Sanction Order No. BT/PR24972/NER/95/932/2017), New Delhi.

References

Banik D, Chaudhuri PS (2017) Regeneration ability in seventeen top soil and sub soil earthworm species. J Environ Biol 38(3):393–399

Bhattacharjee G, Chaudhuri PS (2002) Cocoon production, morphology, hatching pattern and fecundity in seven tropical earthworm species—a laboratory based investigation. J Biosci 27:283–294

Birundha M, John PJA, Mariappan P (2013) Growth and reproduction of *Perionyx excavatus* in different organic wastes. Int J Curr Microbiol Appl Sci 2(2):28–35

Bouche MB (1972) Lombriciens de France. In: Ecologie et systematique. Annales de Zoologie Ecologie Animale 72, Institut national de la recherche agronomique, pp 1–671

Chaudhuri PS, Bhattacharjee G (2001) Changes in the level of nutrients during processing of municipal sewage sludge by earthworms (*Polypheretima elongata* and *Eutyphoeus gammiei*). Proc Zool Soc 54(2):81–83

Chaudhuri PS, Bhattacharjee G (2002) Capacity of various experimental diets to support biomass and reproduction of *Perionyx excavatus*. Bioresour Technol 82:147–150

Chaudhuri PS, Pal TK, Bhattacharjee G, Dey SK (2003) Rubber leaf litters (*Hevea brasiliensis* var RRM 600) as vermiculture substrate for epigeic earthworm, *Perionyx excavatus*, *Eudrilus eugeniae* and *Eisenia fetida*. Pedobiologia 47:796–800

Chaudhuri PS, Bhattacharjee S, Dey A, Chattopadhyay S, Bhattacharya D (2013) Impact of age of rubber (*Hevea brasiliensis*) plantation on earthworm communities of West Tripura (India). J Environ Biol 34:59–65

Chaudhuri PS, Debnath S (2020) Physico-chemical changes during vermicomposting of a terrestrial weed, Mikania micrantha and leaf litters of Acacia auriculiformis and Bambusa polymorpha mixed with cowdung. J Environ Biol 41(2):178–185

Darwin C (1881) The formation of vegetable moulds through the action of worms, with observations of their habits. Murray, London

Dash MC (2012) Charles Darwin's plough: tools for vermitechnology. IK International Publishing House, New Delhi

Debnath S, Chaudhuri P (2019) Earthworm communities in the waste deposit sites (Cowdung heaps and municipal solid wastes) of West Tripura, India. Int J Ecol Environ Sci 45(1):1–14

Domínguez J, Edwards CA (1997) Effects of stocking rate and moisture content on the growth and maturation of Eisenia andrei (Oligochaeta) in pig manure. Soil Biol and Biochem 29 (3-4):743–746

Dominguez J, Edwards CA (2011) Biology and ecology of earthworm species used for vermicomposting. In: Edwards CA, Arancon NQ, Sherman R (eds) Vermiculture technology. CRC Press, Boca Raton, pp 249–261
Dominguez J, Edwards CA, Ashby J (2001) The biology and population dynamics of *Eudrilus eugeniae* (Kinberg) (Oligochaeta) in cattle waste solids. Pedobiologia 45:341–353
Edwards CA (1988) Breakdown of animal, vegetable and industrial organic wastes by earthworms. Agric Ecosyst Env 21:21–31
Edwards CA, Bohlen PJ (1996) Biology and ecology of earthworms. Chapman and Hall, London
Edwards CA, Neuhauser EF (1988) Earthworms in waste and environmental management. SPB Academic Publishing, The Hague
Elvira C, Dominguez J, Briones MJI (1996) Growth and reproduction of Eisenia andrei and E. fetida (Oligochaeta, Lumbricidae)nin different organic residues. Pedobiologia 40:377–384
Evans AC (1946) Distribution of numbers of segments in earthworms and its significance. Nature (London) 158:98
Evans AC, Guild WJML (1948) Studies on the relationships between earthworms and soil fertility. IV on the life cycles of some British Lumbricidae. Ann Appl Biol 35:471–484
Garg VK, Chand S, Chhillar A, Yadav A (2005) Growth and reproduction of *Eisenia foetida* in various animal wastes during vermicomposting. Appl Ecol Environ Res 3(2):51–59
Gates GE (1972) Burmese earthworms. An introduction to the systematics and biology of Megadrile Oligochaetes with special reference to Southeast Asia. Trans Am Philos Soc 62:1–326
Holmstrup M, Ostergaard IK, Nielsen A, Hansen BT (1991) The relationship between temperature and cocoon incubation time for some Lumbricid earthworm species. Pedobiologia 35:179–184
Hyman LH (1940) Aspects of regeneration in annelids. Am Nat 74:513–527
Joshi N, Dabral M (2008) Life cycle of earthworms *Drawida nepalensis*, *Metaphire houlleti* and *Perionyx excavatus* under laboratory controlled conditions. Life Sci J 5:83–86
Julka JM (2014) Diversity and distribution of exotic earthworms (Annelida, Oligochaeta) in India a review. In: Chaudhuri PS, Singh SM (eds) Biology and ecology of tropical earthworms. Discovery Publishing House, New Delhi, pp 73–83
Kadam DG (2015) Growth and reproduction of *Eudrilus eugeniae* in Tendu (*Diospyros melanoxylon Roxb.*) leaf residues as influenced by feed particle size. Int J Curr Microbiol Appl Sci 4(11):206–210
Kale RD, Bano K, Krishnamoorthy RV (1982) Potential of *Perionyx excavatus* for utilization of organic wastes. Pedobiologia 23:419–425
Karmegam N, Daniel T (2008a) Investigating efficiency of *Lampito mauritii* (Kinberg) and *Perionyx ceylanensis* Michaelsen for vermicomposting of different types of organic substrates. Environmentalist. https://doi.org/10.1007/s10669-008-9195-z
Karmegam N, Daniel T (2008b) Effect of vermicompost and chemical fertilizer on growth and yield of hyacinth bean, *Lablab purpureus* (L.) sweet. Dynam Soil Dynam Plant 2:77–81
Karmegam N, Daniel T (2009) Growth, reproductive biology and life cycle of the vermicomposting earthworm, *Perionyx ceylanensis* Mich. (Oligochaeta: Megascolecidae). Bioresour Technol 100:4790–4796
Kaushal BR, Bisht SBS, Kalia S (1994) Effect of diet on cast production by the megascolecid earthworm, *Amynthas alexandri* in laboratory culture. Biol Fertil Soils 17:14–17
Kinberg JGH (1866) Annulata nova. Ofersigt af Kungliga Svenska Vetenskaps-Akademiens Forhandlingar. Stockholm 23:97–357
Lavelle P (1978) Les Vers de Terre de la savane de Lamto (Côte d'Ivoire): peuplements, populations et fonctions dans l'écosystème. École Normale Supérieure, Laboratoire de Zoologie
Lee KE (1985) Earthworms: their ecology and relationships with soils and land use. Academic, Sydney
Mitchell MJ (1978) Role of invertebrates and microorganisms in sludge decomposition. In: Hartenstein R (ed) Utilization of soil organisms in sludge management. National Technical Information Services, Springfield, pp 35–50. PB286932

Muys B, Lust N, Granval P (1992) Effects of grassland afforestation with different tree species on earthworm communities, litter decomposition and nutrient status. Soil Biol Biochem 24:1459–1466

Olive PJW, Clark RB (1978) Physiology of reproduction. In: Mill PJ (ed) Physiology of annelids. Academic, London, pp 271–368

Parthasarathi K (2007) Life cycle of *Lampito mauritii* (Kinberg) in comparison with *Eudrilus eugeniae* (Kinberg) cultured in different substrates. J Environ Biol 28:803–812

Perrier E (1872) Recherches pour server a '1' histoire des lombriciens terrestris. Nouvelles Archives du Muséum d'Histoire Naturelle, Paris 8:5–198

Prakash M, Jayakumar M, Karmegam N (2008) Physicochemical characteristics and fungal flora in the casts of the earthworm *P. Ceylanensis* Mich reared in *Polyalthia longifolia* leaf litter. J Appl Sci Res 4:53–57

Reinecke AJ, Viljoen SA (1993) Effects of worm density on growth and cocoon production of the African nightcrawler (*Eudrilus eugeniae*) Oligochaeta. Eur J Soil Biol 29(1):29–34

Reinecke AJ, Viljoen SA, Saayman RJ (1992) The suitability of *Eudrilus eugeniae*, *Perionyx excavatus* and *Eisenia fetida* (Oligochaeta) for vermicomposting in Southern Africa in terms of their temperature requirement. Soil Biol Biochem 24:1295–1307

Reynolds JW, Wetzel MJ (2004) Terrestrial Oligochaeta (Annelida: Clitellata) in North America north of Mexico. Megadrilogica 9:71–98

Satchell JE (1967) Lumbricidae. In: Burgess A, Raw F (eds) Soil biology. Academic Press, London, pp 259–322

Savigny E (1826) Enterion chloroticum. Mém Acad Sci Inst France 5:183

Stephenson J (1930) The Oligochaeta. Oxford University Press, Oxford

Suthar S (2007) Vermicomposting potential of Perionyx sansibaricus (Perrier) in different waste materials. Bioresour Technol 98(6):1231–1237

Suthar S (2009) Growth and fecundity of earthworms: Perionyx excavatus and Perionyx sansibaricus in cattle waste solids. The Environmentalist 29(1):78–84

Tian G, Brussaard L, Kang BT (1993) Biological effects of plant residues with contrasting chemical compositions under humid tropical conditions: effects on soil fauna. Soil Biol Biochem 25 (6):731–737

Chapter 12
Vermicomposting of *Parthenium hysterophorus* L.: A Solution to Weed Menace in Terrestrial Ecosystem

Deepshikha Sharma and Anu Bala Chowdhary

Abstract *Parthenium hysterophorus* L. is one of the most harmful and aggressively colonizing terrestrial weed which is invading and degrading the natural ecosystems in India. It is causing a serious impact on the human health, economy and environment. It also threatens the native biodiversity of India. Its immediate management is the need of the hour to maintain the ecological balance in terrestrial habitat. To manage the menace of *Parthenium* weed, a number of management approaches have been used, but unfortunately each has its own limitations. An effective tool for the management of invasive *Parthenium* weed is vermitechnology. This chapter discusses the hazards associated with the *Parthenium* weed and its mitigation with vermicomposting by using locally available epigeic earthworms.

Keywords *Parthenium hysterophorus* L. · Management · Menace · Invasive · Vermicompost

12.1 Introduction

Parthenium weed, *Parthenium hysterophorus* L. (Asteraceae; Heliantheae), is an annual herbaceous plant that grows luxuriantly in terrestrial ecosystem of India, mainly agricultural fields, abandoned lands, railway tracks, forest ecosystems, irrigation canals, etc. It is a native of Northeast Mexico and is endemic to America. It is commonly known as carrot grass, congress grass, star weed, bitter weed, etc. *Parthenium* grows to a height of 0.50–2.0 m. It germinates at any time of the year. Flowering begins at 6–8 weeks after germination in summer and continues until

D. Sharma (✉)
Department of Environmental Sciences, Government Degree College Kathua, Kathua, Jammu and Kashmir, India
A. B. Chowdhary
Department of Botanical and Environmental Sciences, Guru Nanak Dev University, Amritsar, Punjab, India

S. A. Bhat et al. (eds.), *Earthworm Assisted Remediation of Effluents and Wastes*,
https://doi.org/10.1007/978-981-15-4522-1_12

drought or frost kills the plant. Each mature *Parthenium* plant can generate as many as 15,000 seeds (Dhileepan 2012). *Parthenium* is considered as one of the most harmful terrestrial weeds because of its allelopathic effect and photo- and thermo-insensitivity. *Parthenium* weed contains a number of potential allelochemicals within its leaves and roots. It is reported that these chemicals can inhibit the germination and growth of a wide range of plant species, including native plants as well as various crops and pasture species. The allelopathic effect of this weed is evident by various experiments carried out over the time. A study conducted on the effect of *Parthenium* root extract on the germination and shoot growth of maize and barley reported that with the increased concentration of root extract, there is a decrease in the germination and shoot growth. On germination shoot length, root length, seed vigour, tolerance index and fresh and dry weight of bean seedlings are inhibited by the *Parthenium hysterophorus* L. leaves, stem and flower extracts (Tahseen et al. 2015). *Parthenium* weed residues are reported to have a phytotoxic effect upon the number of important field, horticultural, vegetable and agroforestry crops (Adkins and Shabbir 2014). It is spreading in all the ecosystems, suppressing the native vegetations and disturbing the natural habitat of endemic species.

This is the noxious weed which is considered to be one of the worst weeds currently known. This weed is responsible for severe human and animal ailments like asthma, bronchitis, dermatitis, allergies, hay fever, black spots, choking, etc. (Maishi et al. 1998). It is believed that the seeds of this weed came to India with grains imported from the USA in the 1950s under the USPL 480 scheme, also known as 'Food for Peace' which was a food assistance program of the US government (Singh et al. 2008). Since then this weed is causing a great damage to the terrestrial habitat. *Parthenium hysterophorus* L. has a large number of harmful aspects, and its eradication is a challenge because of its epidemic proliferation and strong reproductive potential. Due to its high fecundity, a single plant of *Parthenium* can produce 10,000–15,000 viable seeds, and these seeds can disperse and germinate to cover large areas.

Parthenium hysterophorus L. is a terrestrial weed which is threatening the terrestrial ecosystems of the world mainly agricultural fields. The management of this weed has been tried by many ways including mechanically, manually and by burning, but this type of management is not a sustainable method to control this weed. Moreover it is not having any economic benefit. The objective of this chapter is to suggest one of the most suitable methods for the management of the *Parthenium* weed which is environment-friendly and economical. It also helps in converting the *Parthenium* biomass into a useful product, i.e. vermicompost. This vermicompost can be used in the agricultural fields as a nutrient-rich fertilizer for the crop growth.

12.2 Management of *Parthenium* Weed

Many methods have been devised to manage this weed like chemical, mechanical and biological control methods (Morin et al. 2009), but the results were not so promising due to one or other disadvantages (Garg and Bhardwaj 2000). There is a need to explore beneficial uses of it if any. The biomass of this plant is never put to any use and is simply disposed off in open or burnt. Out of various available options, vermitechnology holds a great promise and is compatible with the weeds' high growth potential (Abbasi and Ramasamy 2001). Yadav and Garg (2011) have reported that *Parthenium hysterophorus* can be effectively converted into vermicompost if mixed in appropriate quantities with animal dung. Composting process has also served as an important method for waste treatment. Today, the introduction of earthworms in the composting process has met with fair success, and the technique is known as vermitechnology. In this process, the biological degradation of organic waste takes place in controlled conditions due to earthworm feeding on the materials. The end-product vermicomposting process is vermicompost which is a stable fine granular organic matter, enriched with higher levels of nutrients (C, N, P, K, Na, Mg and Ca).

In the vermicompost, some of the secretions of worms and the associated microbes act as growth promoters along with other nutrients. When the organic matter passes through the gut of earthworms, the vermicompost acquires enzymes and hormones which can stimulate plant growth (Atiyeh et al. 2000; Zaller 2007; Lazcano et al. 2010; Samrot et al. 2015) and improve yield and quality of fruits (Singh et al. 2008; Doan et al. 2015). The organic carbon in vermicompost releases the nutrients slowly and steadily into the system and enables the plant to absorb these nutrients. The vermicompost influences the growth and yield of crops and causes a positive impact on their overall growth (Ravindran et al. 2013).

Vermicomposting with *Eisenia fetida* of *Parthenium hysterophorus* mixed with cow dung in different ratios (25%, 50% and 75%) in an 18 weeks' experiment shows that in all the treatments, a decrease in pH and C/N ratio but increase in EC, N total, P available, Ca total, K total and heavy metals was recorded. The cocoon production and growth rate were maximum in 100% cow dung. The results indicated that *Parthenium* can be a raw material for vermicomposting if mixed with cow dung in appropriate quantity (Yadav and Garg 2011). The macronutrients (N, P and K) and micronutrients were increased in *Parthenium*-mediated vermicompost and decreased in compost of *Parthenium*-mediated compost. *Parthenium* can be utilized effectively as an organic manure by composting and vermicomposting and thereby control this weed (Sivaraj and Narendhran 2013).

It was observed that *Parthenium hysterophorus* mixed with other organic supplements imparts suitable physicochemical conditions for maximum worm production and large-scale vermicompost production. The future research is required to explore the utilization of *Parthenium hysterophorus* in vermicompost production to enhance crop yield (Anbalagan and Manivannan 2012). *Parthenium hysterophorus* can be used as additive for effective vermicomposting process of sludge. Maximum rate of 80% germination of tomato seeds was observed in *Parthenium* and sludge-mediated vermin extract. The combination of *Parthenium* and cow dung enhanced the nutrient value of the compost and increased the germination of *Arachis hypogaea*. The toxicity of allelochemicals could be minimized though compost (Rajiv et al. 2013). When *Parthenium*-mediated vermicompost was analysed with vermicompost produced from another weed *Argemone mexicana*, the higher concentration of nitrogen was seen in *Parthenium* vermicompost than other weeds. Overall NPK value is also observed in higher concentration in both the weeds (Mistry et al. 2015). In various physical, chemical and biochemical characteristics that were tested, the vermicomposting seemed to be plant-friendly, giving finest results after applied at concentrations of 1.5% in soil (w/w) (Hussain et al. 2015). In a similar way, the experiments can be carried out with other terrestrial weeds which are major constituents of agricultural land and also available in other land covers. The significance of vermicompost prepared by these terrestrial weeds can be analysed and can be initiated for a large scale. There are so many terrestrial weeds and dominating invasive species. The compost prepared out of it can be used as organic fertilizer and soil conditioner.

Adequate moisture content is the major requirement for efficient vermicompost by the worms. Basically the moisture level higher than 50% is required. In the initial characterization for all the plant weed substrates, the moisture level is between 70% and 80%. The C/N ratio is also in the range of 25–30 for all the plant substrate which is suitable for the rapid degradation by the earthworms and makes an excellent environment (Munroe 2007).

Parthenium competition increased with an increase in its dry mass up to 448% and relative competition index up to 52%, and consistent increases in the uptake of N (up to 581%), P (up to 700%) and K (up to 669%) were also recorded; competition period of *Parthenium* weed competition period is 35 DAE decreased of grain yield and produces index of maize (Safdar et al. 2016). Over the competitive displacement, it can cause up to 90% degeneration in the herbaceous constituent of natural plant communities where this weed invades (Mahadevappa 1997). As the growth of *Parthenium* increased, NPK-uptake was also increased. In this weed-crop competition with agriculture crops, it was observed that uptake of *Parthenium* during the different competition period at the range of 2.7–18.4, 0.2–2.4 and 2.3–17.7 N, P and K kg/ha, respectively, at its different competition periods (Safdar et al. 2016).

Also a study conducted on the impact of kitchen waste, water hyacinth and *Parthenium* weed vermicompost on the growth of tomato (*Lycopersicon esculentum* Mill. var. Pusa Ruby) revealed that various vermicomposts in various concentrations have a positive impact on the growth of tomatoes (Rampal and Sharma 2006). The study conducted on the vermicompost production potential of various species of earthworms, viz. *Amynthas diffringens*, *Metaphire houlleti* and *Octolasion tyrtaeum* on *Parthenium* weed, reported that each species successfully vermicomposted the *Parthenium* weed and the vermicompost production of *Octolasion tyrtaeum* was highest among the three species (Rampal and Sharma 2007). Also the physicochemical characteristics of *Parthenium* weed vermicompost produced by three local species were observed to be high in *Parthenium* weed vermicompost (Sharma 2017).

Earlier it was thought that vermicompost produced from *Parthenium* weed may be toxic to soil or plants as *Parthenium* is a toxic weed. Recent studies conducted by Hussain et al. (2016) revealed that vermicompost generated from *Parthenium* by using *Eisenia fetida* has none of the allelopathy and soil or plant or animal toxicity associated with *Parthenium*. It rather has enhanced the plant growth.

12.3 Vermitechnology

Vermitechnology is a simple technology with low investment, which does not need sophisticated infrastructure and can be adopted for the management of organic wastes more effectively.

12.3.1 Vermitechnology Has Two Components

1. *Vermiculture*: It means a scientific method of breeding and rearing the specific variety of earthworms in controlled conditions. Vermiculture was started in the middle of the twentieth century, and the first experiment was set up in Holland in 1970 and then in England and Canada, while other countries that are engaged in this technology are the USA, Mexico, West Germany, Italy, etc. Culturing of earthworms can be done indoors in humid places with proper shelter to avoid direct sunlight or heavy downpour. The culturing can be done in vermary, cattle shed or any other congenial place. Cement tanks, wooden boxes or plastic trays of approximately 1 m × 1 m × 0.3 m size, which accommodate 1000 worms, may be used as containers. These containers should be covered to prevent the escape of worms. Water is to be sprinkled on the bed to get a moisture level of 40–50%. For bedding materials, saw dust or husk, fine sand or layer of garden soil is used. The earthworms are introduced in the pit, and organic waste is added to the vermipit which is converted into vermicasts and vermicompost by the earthworms.
2. *Vermicomposting*: It is the bioconversion of organic waste materials through earthworm consumption. Earthworms are cultured and reared by providing them proper living conditions and then feeding them with organic wastes like postharvest stribbles; sugarcane thrash; coir waste; paper pulp; faecal matter of cow, sheep and, horse; activated sludge; biogas sludges of poultry droppings; slaughter house wastes; municipal solid wastes; etc.

The overall products of vermitechnology, vermicompost and earthworms are utilized in various ways (Fig. 12.1).

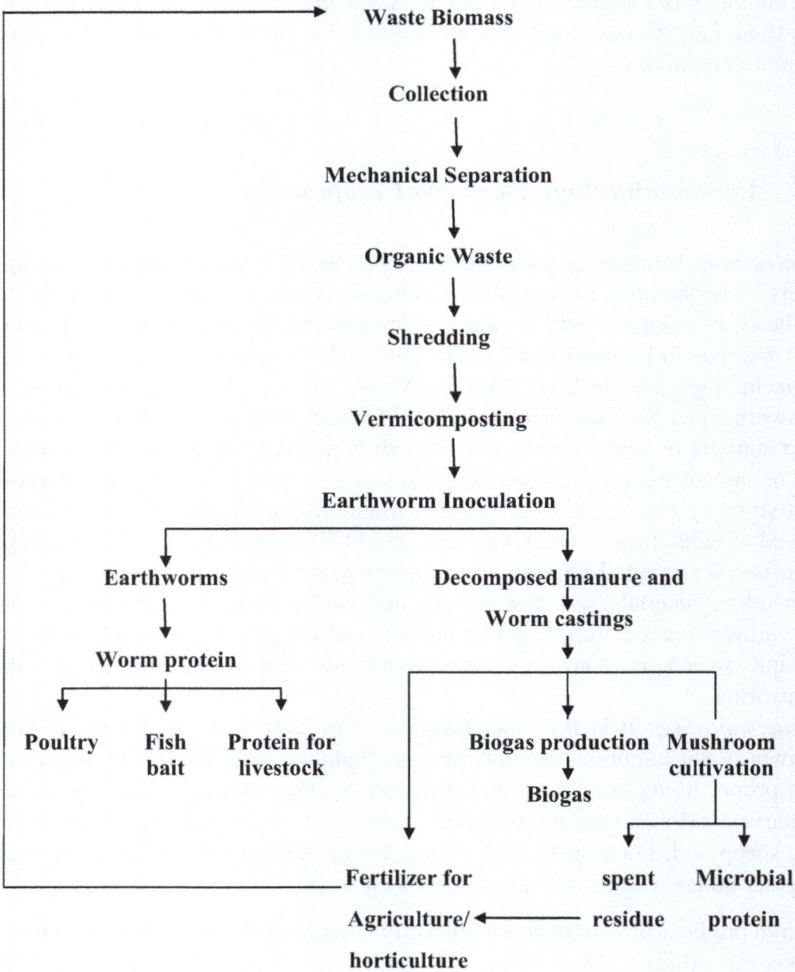

Fig. 12.1 Vermitechnology flow sheet

12.3.2 Methods of Vermicomposting

Vermicompost can be mainly produced by two methods: pit method and windrow method.

1. *Pit method* is commonly used for small-scale production of vermicompost. These steps can be followed:
   (a) Construct a pit of 3 × 2 × 1 m size ($L \times W \times D$) over the ground surface using bricks. Size of the pit may vary as per availability of raw materials.
   (b) Fill the pit with the following five layers:
      - First layer—paddy straw or other crop residues of 5–6 cm which will be used for providing aeration to the pit.
      - Second layer—sand or sandy soil of 20 cm. This layer helps to drain excess water from the pit.
      - Third layer—garden soil up to 20 cm to provide bedding to the earthworms.
      - Fourth layer—15 to 30 days of cattle dung at a thickness of 20 cm. This helps in initiating microbial activity.
      - Fifth layer—predigested organic waste about 25 cm.

   Inoculate earthworm at 1000 worms per square meter area or 10 kg earthworm in 100 kg of organic matter. Spray water on the bed and gunny bag. Maintain 40–60% moisture of the pit by periodical water spraying.
2. *Windrow method* is widely used for large-scale production of vermicompost. These steps can be followed:
   (a) Load the organic wastes in the form of bed (preferably 10 ft L × 3 ft W × 1.5 ft H). Size of the bed may vary as per availability of organic waste.
   (b) After loading, the fresh bed should be covered with jute mat or dry agriculture wastes such as rice bran, banana leaf, maize residue, etc.
   (c) Sprinkle water over the covered vermibed to maintain 40% moisture in bed.
   (d) Bed should be mixed thoroughly to prevent it from becoming compact.
   (e) The first lot of vermicompost is ready for harvesting after 2–2.5 months, and the subsequent lots can be harvested after every 6 weeks of loading.
   (f) Watering of bed should be stopped for at least 2–3 days before harvesting. Earthworms go down in the moist soil, and the compost is collected from the top without disturbing the lower layers of vermibed having earthworm. Vermicompost harvested will be of dark brown colour and free flowing.
   (g) The harvested compost should be stored in a dark and cold place.

12.4 About Earthworms

Earthworms belong to class Annelida, class Clitellata and order Oligochaeta. There are about 1800 species of earthworms belonging to 10 families. The most optimum temperature for earthworms is 20–25 °C, and most optimum moisture content is 50–60%. The preferred pH by earthworm is 7. Earthworms became mature at 10 weeks. They produce two to three cocoons a week. Each cocoon holds two to five young ones, and cocoon takes 3 weeks to hatch. In 6 months 8 earthworms multiply into 1500 if conditions are suitable for their survival. There are three ecological varieties of earthworms, i.e. epigeics, endogeics and anecics. Epigeics are litter- and surface-dwelling species. Endogeics are soil-dwelling species, and anecics are deep-burrowing species. Out of the three types, only epigeic earthworm species are considered most suitable for vermicomposting process. Mainly epigeic species like *Eisenia fetida* (Hartenstein et al. 1979a) and *Eudrilus eugeniae* (Kale and Bano 1988) have been used in converting organic waste into vermicompost.

12.4.1 Suitable Earthworm Species for Vermicomposting

Although there are thousands of earthworm species, only epigeic earthworms are considered most suitable in vermicomposting, and that depends on their replication under a suitable environmental conditions. The earthworms should be able to process large amounts of organic waste under a given condition. Of the nearly 200 species of earthworms identified and reported from various parts of India, relatively few species have been harnessed by man in the promotion of vermiculture for vermicomposting practices. Of the three ecological varieties of earthworms, the epigeics and anecics have been utilized in the vermicomposting process.

Epigeics like *Eisenia fetida* (Hartenstein et al. 1979a, b) and *Eudrilus eugeniae* have been used in converting organic wastes into vermicompost. Though these surface dwellers are capable of working hard on the litter layer and converting all the organic wastes into manure, they are of no significant value in modifying the structure of the soil. The anecics, however, are capable of both organic waste consumption and in modifying the structure of the soil. Burrowing species that are widely used in soil management are *Pheretima elongata* and *Lampito mauritii*.

Of various varieties of earthworms, *Eisenia fetida* and *Eudrilus eugeniae* have widely been used throughout the world in converting organic litter into compost. Though not the natives of India, these earthworms have been introduced in India, and several experiments are in progress. The in situ soil community especially the local variety of earthworms comprises the epigeic and endogeic varieties for the combined process of litter and soil management. The advantages of using local varieties of earthworms provide a cheap eco-friendly technology for the waste management and the generation of organic fertilizers.

12.5 About Vermicomposts

Vermicompost is a stable fine granular organic matter and when added to clay soil loosens the soils and provides the passage for the entry of air. The mucus associated with the cast being hygroscopic absorbs water, prevents water logging and improves water holding capacity. In the vermicompost, some of the secretions of worms and the associated microbes act as growth promoters along with other nutrients. It improves physical, chemical and biological properties of soil in the long run on repeated application. The organic carbon in vermicompost releases the nutrients slowly and steadily into the system and enables the plant to absorb these nutrients. The soil enriched with vermicompost provides additional substances that are not found in the chemicals. The multifarious effects of vermicompost influence the growth and yield of crops.

Vermicompost contains the higher level of C, N, P, K, Na, Mg and Ca than that in the surrounding soil. The production of mucus and nitrogenous excreaments enhance the level of nitrogen, and this helps in bringing down the ratio of carbon to nitrogen which is most essential in humification process (Govindan 1998). It also contains trace elements such as manganese (Mn), copper (Cu), boron (B) and molybdenum (Mo) which are essential for the growth of crop plants. In addition, if the garbage contains cattle dung and human excreta, it may contribute to the presence of soluble sugars, amino acids, nucleic acids and enzymes such as proteinase, urease, phosphatase and invertase which can improve the soil's nutrient status and dehydrogenase and ATPase which can improve the aeration in the soil and the energy status of the soil to which this compost is applied (Sinha and Sinha 2000). The application of worm casts or vermicompost has shown higher yield of wheat, paddy and green leaves ranging from 10% to 200% (Govindan 1998). The application of chemical fertilizer along with vermicompost increased the nutrient uptake and net production of wheat and sugarcane, respectively (Kale et al. 1992). Vermicompost significantly improves the soil fertility, texture and water retaining capacity. The addition of compost to soil helps in erosion control. Vermicompost is rich both in nutrients and in enzymes. Vermicompost having high quality humus can be used as a soil conditioner.

12.6 Methodology for Vermicomposting of Parthenium Weed

12.6.1 Collection and Processing of Parthenium Weed

Parthenium plants can be removed from any terrestrial habitat. Thereafter they should be sundried and chopped into smaller pieces. Chopped *Parthenium* weed is then soaked into water for 24 h prior to its transfer into the vermibeds. This

Parthenium weed can be transferred to the vermibeds slowly in a period of 2–3 days (Rampal and Sharma 2007).

12.6.2 Collection of Earthworms

Locally available epigeic earthworm species can be collected from 3 to 10 cm depth of the soil and transferred to the vermibeds. Epigeics species like *Eisenia fetida* (Hartenstein et al. 1979a) and *Eudrilus eugeniae* (Kale and Bano 1988) are also used for converting *Parthenium* weed into vermicompost.

12.6.3 Preparation of Vermibeds and Production of Vermicompost

Culturing of earthworms can be done indoors in humid places with proper shelter to avoid direct sunlight or heavy downpour. The culturing can be done in kitchen garden, cattle shed or any other congenial place. Cement tanks, wooden boxes or plastic trays of approximately 1 m × 1 m × 0.3 m size, which accommodate 1000 worms, may be used as containers. These containers should be covered with gunny bags to prevent the escape of worms. Water is to be sprinkled on the bed to get a moisture level of 40–50%. For bedding materials, saw dust or husk, fine sand or layer of garden soil is used. Then week old cattle dung is spread over the surface of the soil as a bait for the acclimatization of earthworms. The holes are made at the base of the vermipit to release the extra water from the vermibed, since earthworms cannot survive in waterlogged conditions. The locally available epigeic earthworms can be collected from the loose soils rich in organic matter at a depth of 3–10 cm of soil surface and introduced in the pit, and processed *Parthenium* weed is added to the vermipit which is converted into vermicasts and vermicompost by the earthworms.

After weeks of time when earthworms get acclimatized to the vermibeds, the processed *Parthenium* weed is added slowly to the vermibeds and again covered with the moist gunny bags. Now the vermibeds are kept in the shady place, water is sprinkled regularly depending upon the season and temperature is also monitored in the vermibeds as the optimum temperature for worm function is 20–25 °C although earthworms can tolerate temperature from 5 to 25 °C. Also care should be taken to protect the earthworms from pests and predators.

12.6.4 Collection of Vermicompost

After the formation of vermicompost, the sprinkling of water is stopped to ensure the migration of earthworms to deeper layers of vermibed. Then the topmost loose layer of soil along with worm casts is collected, sundried and crushed. This final product which is harvested from vermibed is a vermifertilizer. This vermicompost is a nutrient-rich fertilizer which can be used for the growth of plants in the agricultural fields.

12.7 Overall Benefits of Vermicomposting

Vermicomposting process conserves water, energy and land resources that are required for waste treatment by aerobic composting. Vermicomposting is fast and has low cost and sustainable method for organic waste treatment by earthworms, with the advantage of getting good-quality compost. The earthworms can hold nutrients for a longer period without impacting the environment. Vermicompost requires no curing like in aerobic composting. The overall time duration and cost for processing the waste are substantially reduced, besides the process has on toxic by-products or waste materials.

12.8 Conclusion

Parthenium weed can be properly managed with large-scale production of vermicompost using epigeic earthworm species. The vermicompost produced from *Parthenium* weed will be nutrient-rich and odour-free. *Parthenium* weed vermicompost can be used as biofertilizers in the agricultural fields which can give a better plant growth without degrading the environment. This would give benefits to the national economy as widespread use of this technology would lead to less import of agrochemicals, reduced wasteland formation and better environmental management of low cost and by-product recovery. Therefore it is recommended that recycling of this weed as composted manure should be done to reduce its toxic and poisonous effects and resource generation.

References

Abbasi SA, Ramasamy EV (2001) Solid waste management with earthworms. Discovery Publishing House, New Delhi

Adkins S, Shabbir A (2014) Biology, ecology and management of the invasive parthenium weed (Parthenium hysterophorus L.). Pest Manag Sci 70(7):1023–1029

Anbalagan M, Manivannan S (2012) Assessment of impact of invasive weed Parthenium hysterophorus L. mixed with organic supplements on growth and reproduction preference of Lampito mauritii (Kinberg) through vermitechnology. Int J Environ Biol 2:88–91

Atiyeh RM, Arancon N, Edwards CA, Metzger JD (2000) Influence of earthworm-processed pig manure on the growth and yield of greenhouse tomatoes. Bioresour Technol 75(3):175–180

Dhileepan K (2012) Reproductive variation in naturally occurring populations of the weed (Asteraceae) in Australia. Weed Sci 60(4):571–576

Doan TT, Henry-des-Tureaux T, Rumpel C, Janeau JL, Jouquet P (2015) Impact of compost, vermicompost and biochar on soil fertility, maize yield and soil erosion in Northern Vietnam: a three year mesocosm experiment. Sci Total Environ 514:147–154

Garg K, Bhardwaj N (2000) Effect of vermicompost of parthenium on two cultivars of wheat. Indian J Ecol 27:177–180

Govindan VS (1998) Vermiculture and vermicomposting. In: Trivedi RK, Kumar A (eds) Ecotechnology for pollution control and environmental management. Enviro Media, Maharashtra, pp 49–57

Hartenstein R, Neuhauser EF, Kaplan DL (1979a) Reproductive potential of the earthworm Eisenia foetida. Oecologia 43(3):329–340

Hartenstein R, Neuhauser EF, Kaplan DL (1979b) A progress report on the potential use of earthworms in sludge management. In: Proceedings of the eighth national sludge conference, Florida, Mar 1979, pp 238–241

Hussain N, Abbasi T, Abbasi SA (2015) Vermicomposting eliminates the toxicity of Lantana (Lantana camara) and turns it into a plant friendly organic fertilizer. J Hazard Mater 298:46–57

Hussain N, Abbasi T, Abbasi SA (2016) Transformation of Toxic and allelopathic lantana into a benign organic fertilizer through vermicomposting. Spectrochim Acta A Mol Biomol Spectrosc (SAA) 163:162–169

Kale RD, Bano K (1988) Earthworm cultivation and culturing techniques for the production of vee COMP83E UAS. Mysore J Agric Sci 2:339–344

Kale RD, Mallesh BC, Kubra B, Bagyaraj DJ (1992) Influence of vermicompost application on the available macronutrients and selected microbial populations in a paddy field. Soil Biol Biochem 24(12):1317–1320

Lazcano C, Sampedro L, Zas R, Domínguez J (2010) Assessment of plant growth promotion by vermicompost in different progenies of maritime pine (Pinus pinaster Ait.). Comp Sci Util 18 (2):111–118

Mahadevappa M (1997) Ecology, distribution, menace and management of Parthenium. In: Proceedings of the first international conference on Parthenium management, Karnataka, India, 6–8 Oct 1997. University of Agricultural Sciences, pp 1–12

Maishi AI, Ali PS, Chaghtai SA, Khan G (1998) A proving of Parthenium hysterophorus, L. Br Homeopath J 87(01):17–21

Mistry J, Mukhopadhyay AP, Baur GN (2015) Status of NPK in vermicompost prepared from two common weed and two medicinal plants. Int J Appl Sci Biotechnol 3(2):193–196

Morin L, Reid AM, Sims-Chilton NM, Buckley YM, Dhileepan K, Hastwell GT et al (2009) Review of approaches to evaluate the effectiveness of weed biological control agents. Biol Control 51(1):1–15

Munroe G (2007) Manual of on-farm vermicomposting and vermiculture, vol 39. Organic Agriculture Centre of Canada, Nova Scotia, p 40

Rajiv P, Narendhran S, Subhash KM, Sankar A, Sivaraj R, Venckatesh R (2013) Parthenium hysterophorus L. compost: assessment of its physical properties and allelopathic effect on germination and growth of Arachis hypogeae L. Int Res J Environ Sci 2(2):1–5

Rampal R, Sharma D (2006) Impact of kitchen waste, water hyacinth and parthenium weed vermicompost on the growth of tomato (Lycopersicon esculentum mill. Var. Pusa ruby). Pollut Res 25(3):577

Rampal R, Sharma D (2007) Vermicompost potential of various local earthworm species of Jammu on Parthenium weed. Indian J Environ Ecoplan 14(3):611–614

Ravindran B, Sravani R, Mandal AB, Contreras-Ramos SM, Sekaran G (2013) Instrumental evidence for biodegradation of tannery waste during vermicomposting process using Eudrilus eugeniae. J Therm Anal Calorim 111(3):1675–1684

Safdar ME, Tanveer A, Khaliq A, Maqbool R (2016) Critical competition period of parthenium weed (Parthenium hysterophorus L.) in maize. Crop Prot 80:101–107

Samrot AV, Vignesh MP, Vignesh V (2015) Utilization of vermicompost of organic waste for plant growth. Int J Pharm Bio Sci 6:B326–B332

Sharma D (2017) Parthenium weed management by vermicomposting using local earthworm species of Jammu. IJAR 3(6):1170–1171

Singh R, Sharma RR, Kumar S, Gupta RK, Patil RT (2008) Vermicompost substitution influences growth, physiological disorders, fruit yield and quality of strawberry (Fragaria × ananassa Duch.). Bioresour Technol 99(17):8507–8511

Sinha RK, Sinha AK (2000) Waste management. INA Shree Publishers, Jaipur, pp 143–157

Sivaraj R, Narendhran S (2013) Ruthenium mediated compost versus Parthenium mediated vermicompost: a comparative study of nutrition status. In: International conference on chemical, agricultural and medical sciences (CAMS-2013), Dec 2013, pp 29–30

Tahseen NK, Kumar H, Jagannath S (2015) Assessment of allelopathic efficacy of Parthenium hysterophorus L. plant parts on seed germination and seedling growth of Phaseolus vulgaris L. Braz J Biol Sci 2(3):85–90

Yadav A, Garg VK (2011) Vermicomposting—an effective tool for the management of invasive weed Parthenium hysterophorus. Bioresour Technol 102(10):5891–5895

Zaller JG (2007) Vermicompost as a substitute for peat in potting media: effects on germination, biomass allocation, yields and fruit quality of three tomato varieties. Sci Hortic 112(2):191–199

Chapter 13
Evaluating Method of Mica Waste Application in Earthworm Cast-Treated Soil for Enhancing Potassium Availability to the Plants with Reference to Tea

Prabhat Pramanik, Chayanika Kalita, Pallabi Kalita, and Anup Jyoti Goswami

Abstract Potassium (K) is an essential plant nutrient, and deficiency in the availability of K limits plant growth. Organic amendment in agricultural fields is required to sustain soil health; however, maintaining K availability is a major challenge in organic matter-treated soil. In this context, mica waste, a high potassium-containing mineral, could be an alternative K source for arable soil. In this study, a series of experiments were conducted to evaluate how mica waste could be applied with vermicompost for improving K availability in organic-amended soil. It was observed that vermicomposting of mica waste-amended organic substrates did not increase total K content in the final vermicompost, though earthworms harbour potassium-solubilizing bacteria in their guts. Hence, application of vermicompost, prepared from mica waste-amended organic substrates, did not increase K availability in soil. However, combined application of conventional vermicompost with mica waste was effective to increase K availability in soil, and that was effective to achieve better plant growth and higher K uptake in tea. In this study, combined application of vermicompost and mica waste (at 10% weight of vermicompost) was effective to increase plant growth and K availability in soil.

Keywords Mica waste · Earthworms · Plant nutrients · Bacteria

P. Pramanik (✉)
Department of Soil Science, Tocklai Tea Research Institute, Tea Research Association, Jorhat, Assam, India
Department of Soil Science, North Bengal Regional R&D Centre, Tea Research Association, Nagrakata, West Bengal, India

C. Kalita · P. Kalita · A. J. Goswami
Department of Soil Science, Tocklai Tea Research Institute, Tea Research Association, Jorhat, Assam, India

S. A. Bhat et al. (eds.), *Earthworm Assisted Remediation of Effluents and Wastes*,
https://doi.org/10.1007/978-981-15-4522-1_13

13.1 Introduction

Potassium (K) is one of the three major nutrients required for plant growth. Potassium plays a crucial role in the number of physiological processes and in stress resistance of plants (Sitienei et al. 2013). Hence, adequate K supplement is necessary to obtain the desired plant growth. Muriate of potash (MoP) is the most recognized K fertilizer in agriculture, and 100% of the consumed MoP is imported in several countries like India due to lack of K reservoir. That makes K fertilization an expensive soil management practice in those countries (Basak and Biswas 2009). Though applied chemical fertilizers immediately increase nutrient content in soil, long-term continuous fertilizer application deteriorates soil health (Biswas and Mukherjee 2000). In years 2014–2015, 37.97 lakh tonnes of MoP were imported in India to be applied as the chemical fertilizer in agricultural soils (MoC&F 2015). Therefore, it is required to evaluate the suitability of some alternative indigenous high K-containing resources for at least partially replacing doses of MoP to supply K in soil. Application of organic substrates especially vermicompost boosts soil health by improving physical, chemical and biological properties of soil (Pramanik et al. 2010). However, sustaining soil K availability is a major challenge during organic farming. Hence, it is necessary to increase K availability in vermicompost-treated soil. A combined utilization of high K-containing minerals and vermicompost could be an alternative to K management. However, the method of application is not standardized before.

The objective of the experiments mentioned in this chapter was to optimize a method to increase K availability in vermicompost-treated soil by utilizing mica waste in soil.

13.2 Mica Waste Application in Soil

Mica waste could be a good alternative resource for K fertilization in agricultural soil. India has the largest mica deposit in the world (Nishanth and Biswas 2008). Cleaning of raw mica ores after their mining generates a large quantity of mica waste containing approximately 10% potash. Those mica waste substrates could be an important K resource in agriculture due to its large availability and low cost, and those mica waste compounds were often used in the past as slow-releasing K amendment in soil during crop cultivation (Singh et al. 2010; Meena et al. 2014). Mica is a high potassium-containing orthoclase-type aluminosilicate mineral (Brady 2003). In mica, K atoms are present in a sandwiched form between two expandable layers of aluminosilicate crystals (Fig. 13.1). Those K atoms of inter-lattice position could be released slowly, and that was able to sustain plant growth under controlled condition (Xiao et al. 2017). However, lower exchangeable K content in mica waste-treated soil as compared to MoP treatment suggested that the released K from mica waste may not be sufficient to supplement K to plants under field condition.

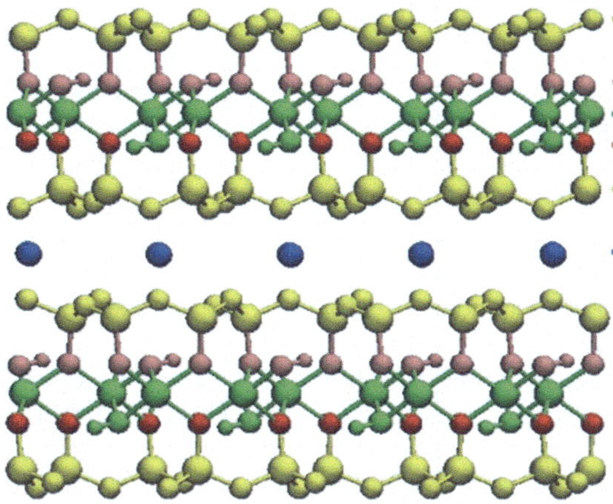

Fig. 13.1 Crystal structure of mica

However, Pramanik et al. (2019) observed that application of mica waste at the recommended dose increased exchangeable K content in soil as compared to control (received no K input); however, the value was lower than that of MoP treatment. The mica waste-treated plants had better growth and higher K uptake than plants with the MoP treatment. Application of mica waste at half of the recommended dose increased K content in the soil above no K-treated soil and recorded comparable growth and K accumulation by plants as MoP (Pramanik and Kalita 2019). Hence, mica waste could be used as K amendment in acidic soil, especially during tea cultivation.

Türkmen et al. (2004) suggested that applied humic acids had a synergistic effect on K availability in soil. Humic acids may have some interaction with bound K compounds for releasing K in ionic form. Humic acids have the ability to release K from inter-lattice space of mica waste (Pramanik et al. 2019). Their SEM analysis and X-ray diffractogram analysis suggested that crystal structure and lattice formation of mica waste was definitely affected due to humic acid treatment. In this study, humic acid-treated mica waste released 43.9 mg K mL^{-1} in solution after 10 days of incubation, while additional 40.6 mg K_2O kg^{-1} were released in soil microcosm after 75 days of incubation under laboratory condition. Therefore, it could be inferred that humic acid treatment was effective to release significantly ($P < 0.05$) higher amount of K in soil as compared to the sole application of mica waste. Therefore, it could be hypothesized that incorporation of mica waste with humic acid-enriched substrates like vermicompost will hasten K release from the minerals.

13.3 Vermicompost and Vermicomposting

Vermicomposting is a biological process to convert organic wastes into humic acid-enriched soil amendments. Vermicomposting is the controlled oxidative decomposition of organic substrates by mutual interaction between earthworm and microorganisms. Cow manure is generally mixed with initial organic wastes to provide easily available energy source and favourable environment to the earthworms (Reinecke et al. 1992). Unlike conventional composting, vermicompost does not contain any pathogens. Pramanik et al. (2010) observed that vermicompost prepared from the organic portion of medical wastes is also free from pathogens. During vermicomposting, organic matter is stabilized by the enhanced decomposition (humification) in the presence of earthworms (Atiyeh et al. 2001) but by a non-thermophilic process (Elvira et al. 1998). Earthworms harbour microorganisms in their intestine, which secretes several exo-cellular enzymes for degrading injected organic substrates and for releasing nutrients in an ionic form from the bound complex structures. Hence, vermicompost has higher concentration of nutrients in ionic plant-available form as compared to initial organic substrates (Fig. 13.2).

Inoculation of suitable cellulolytic and lignolytic strains has been reported to hasten the rate of composting, which in turn leads to the enrichment of nutrients in the final product (Pramanik et al. 2007). Kaushik and Garg (2004) reported that inoculation of N-fixing bacteria during composting increased the nitrogen content of

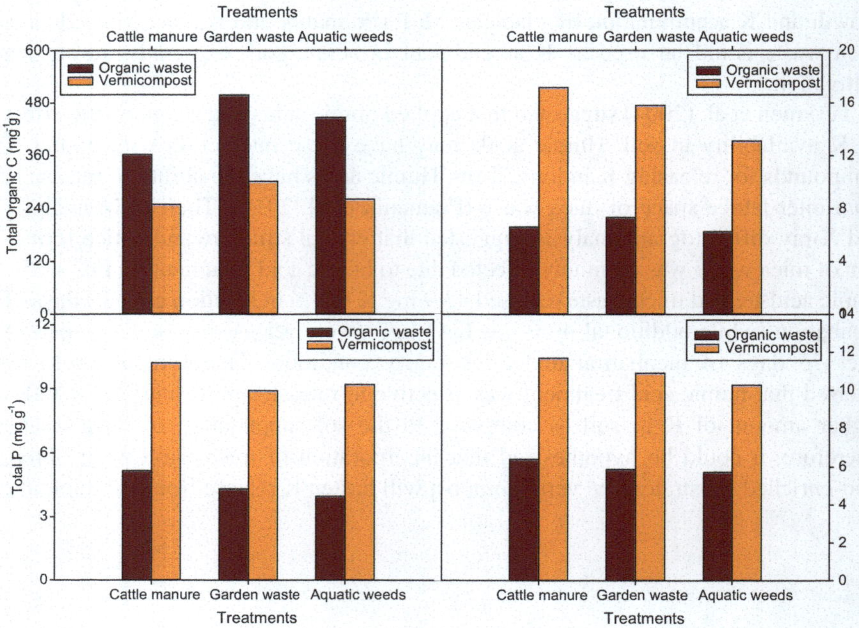

Fig. 13.2 Changes in nutrient content in different organic wastes due to vermicomposting

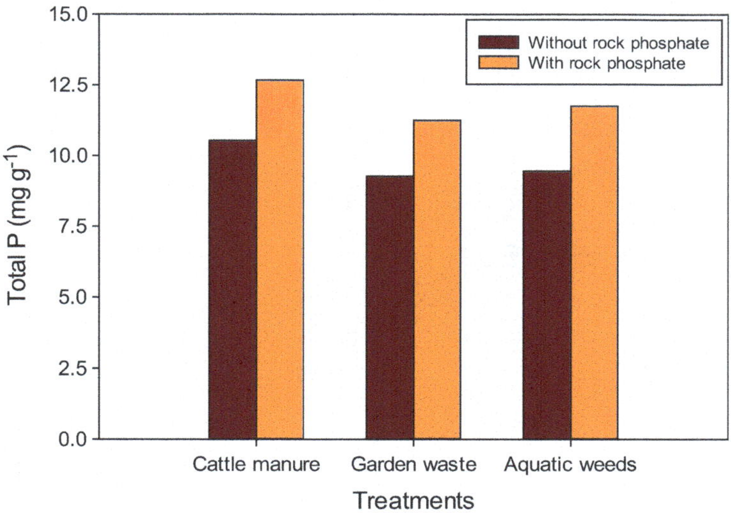

Fig. 13.3 Changes in phosphorus content of vermicompost due to mixing of rock phosphate with initial organic wastes

the stabilized product. It was found that microorganisms, present in the earthworm's intestine, contribute enzymes to the organic substrates during their passage through the gut of earthworms during vermicomposting. Pramanik et al. (2009) revealed that phosphate-solubilizing microorganisms contribute additional isozymes of phosphatase enzyme in the final vermicompost. Vermicomposting of rock phosphate-amended organic substrates significantly increases phosphate ion content in the final vermicompost as compared to that of vermicompost prepared from only organic substrates (Fig. 13.3). Meena et al. (2016) observed that potassium-solubilizing microorganisms are present in the earthworm cast, which in turn increases potassium availability to the plants when applied in soil. Hence, it could be a very interesting study to evaluate the effect of addition of another mineral like mica waste on potassium content in prepared vermicompost.

13.4 Vermicomposting with Mica Waste

Potassium in mica waste is present in a sandwiched form between expandable aluminosilicate lattice structures. Application of humic acid facilitates K release from mica waste. Vermicomposting is the biological process to convert organic substrates into humic acid-enriched organic amendment. The formation of humic acids and presence of potassium-solubilizing bacteria (KSB) are expected to hasten the release of K from mica waste during vermicomposting. However, the fate of mica waste during vermicomposting with organic substrates was not evaluated

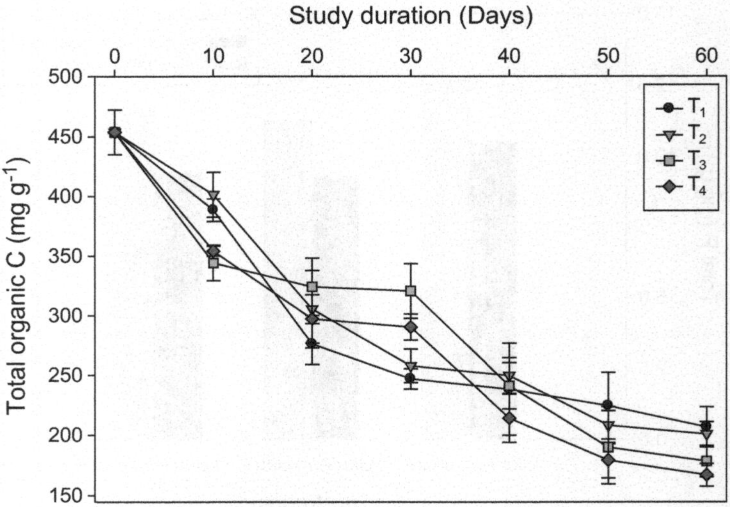

Fig. 13.4 Changes in total organic C content in cattle manure and mica waste combinations during vermicomposting

before. An experiment was conducted with various compositions of cattle manure and mica waste to evaluate changes in potassium content in the final product. During this study, 5 kg of air-dried cattle manure was mixed with mica waste in below-mentioned proportions. Four (4) treatments in this experiment were only cattle manure (T_1), cattle manure + 250 g mica waste (T_2), cattle manure + 500 g mica waste (T_3) and cattle manure + 750 g mica waste (T_4). Initially, cattle manure had total organic C 571.2 $\pm$ 30.8 mg g^{-1} with C/N ratio 110.27 and total K 4.83 $\pm$ 0.33 mg g^{-1}. All the treatments were maintained in triplicate and kept in suitable-size earthen pots for vermicomposting. One hundred fifty *Eudrilus eugeniae* earthworms were introduced in each treatment. Moisture content in these treatments was maintained within 65–75% by periodical sprinkling of water throughout the study period.

Total organic C content in all the treatments was steadily decreased during vermicomposting (Fig. 13.4). Initial total organic C content in cattle manure was 453.82 $\pm$ 18.65 mg g^{-1}, which was decreased up to 206.96–167.15 mg g^{-1} during vermicomposting. The completion of vermicomposting was determined by stabilization of total organic C content. The lowest total organic C was recorded in vermicompost of T_4 treatment; however, total organic C contents of different treatments did not vary significantly during vermicomposting. Therefore, mixing of mica waste did not show any effect on total organic C content of organic substrates during vermicomposting. Vermicomposting is the biochemical alteration of organic wastes into nutrient-rich organic amendment (Pramanik and Chung 2011). Loss of organic C in the form of carbon dioxide leads to the relative increase in concentration of major nutrients in organic substrates during vermicomposting

(Ndegwa and Thompson 2000). In this study, vermicomposting of T_1 treatment leads to 73.6% increase in TK content, which might be attributed to 54.4% reduction in its organic C content. Garg et al. (2006) observed that acid production by microorganisms is accounted for solubilization of insoluble K. Total K content in initial cattle manure was 5.69 ± 0.17 mg g^{-1}. Irrespective of mica waste addition, vermicomposting increased TK content in all the treatments (Fig. 13.5). During vermicomposting, total K content was significantly ($P < 0.05$) increased after 40 days and kept on increasing till 50 days of vermicomposting. At the end of vermicomposting, T_2 recorded significantly ($P < 0.05$) higher total K content as compared to T_1 treatment. However, mixing of 500 g or more mica waste significantly ($P < 0.05$) decreased total K content in organic substrates during vermicomposting. Vermicompost from T_4 treatment recorded the lowest total K content, though the value was statistically at par with that of T_3 treatment.

Previously, researchers mixed chemical substrates like lime, rock phosphate, etc. for increasing N and P contents in the final vermicompost (Pramanik et al. 2007; Domínguez 2004). Addition of mica waste with cattle manure increased (up to 63.2%) loss of organic C during vermicomposting. Increased earthworm count (mature and juvenile) in mica waste-amended organic substrates suggested that earthworms preferred the combination of mica waste with organic substrates in their feeding materials. However, addition of 10% or more (on weight basis) mica waste with organic substrates decreased TK content in the final vermicompost as compared to T_1 (Control) treatment. Exchangeable K and water-extractable K contents in initial cattle manure were 2.06 ± 0.34 and 1.22 ± 0.10 mg g^{-1}, respectively. Values of both exchangeable K and water-extractable K were significantly ($P < 0.05$) increased after 20–30 days of vermicomposting (Fig. 13.5). Vermicomposting of 500 g mica waste-added cattle manure (T_3) recorded the highest EK and WEK contents, though the values were not significantly different from those of T_2 treatment. In spite of adding 750 g mica waste in T_4 treatment, exchangeable K and water-extractable K contents in final vermicompost did not differ from those of T_1 treatment.

13.5 Potassium-Solubilizing Bacteria in the Gut of Epigeic Earthworms

Suthar and Singh (2008) also reported that extractable K content in organic substrates was increased during the vermicomposting and interaction between earthworms and microorganisms plays a major role in releasing nutrients from a bound form by degrading organic matter.

For extracting earthworm gut materials, three matured *E. eugeniae* earthworms, collected from the vermin-bed, were cleaned thoroughly to remove any kind of substance adhered to their body. Later, the specimens were kept on a moist filter paper and allowed to starve overnight for complete gut evacuation (Dey et al. 2017).

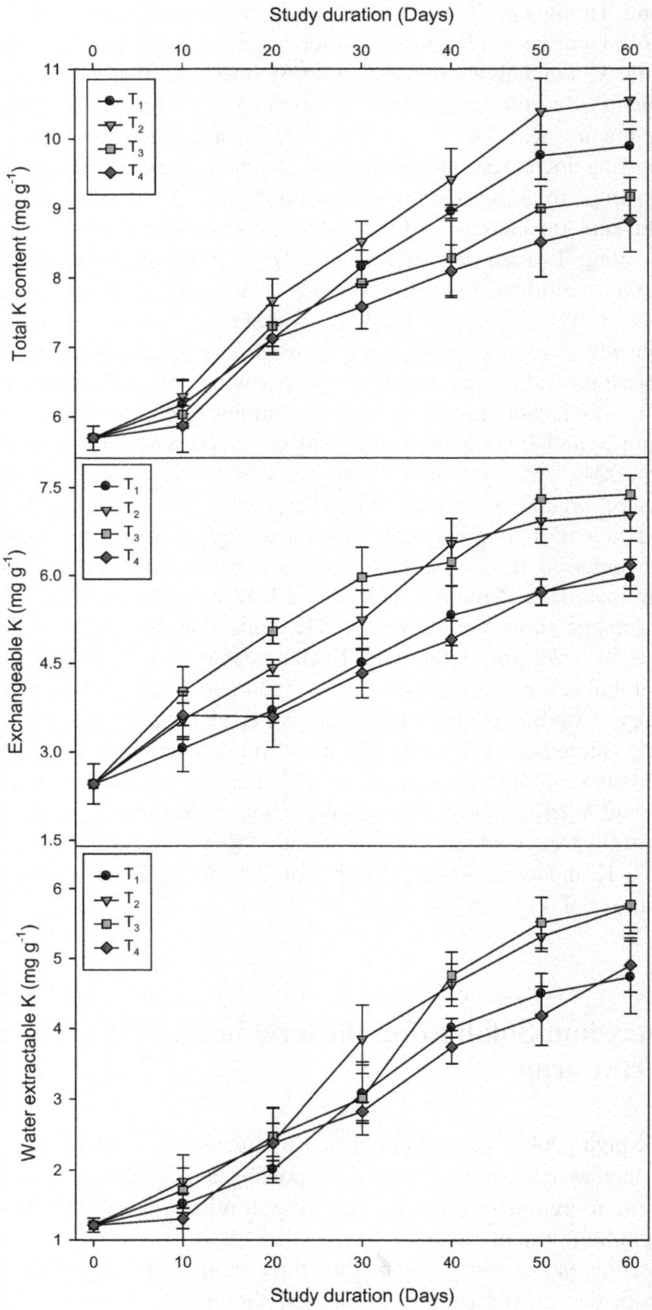

Fig. 13.5 Changes in TK and different K fractions in cattle manure and mica waste combinations during vermicomposting

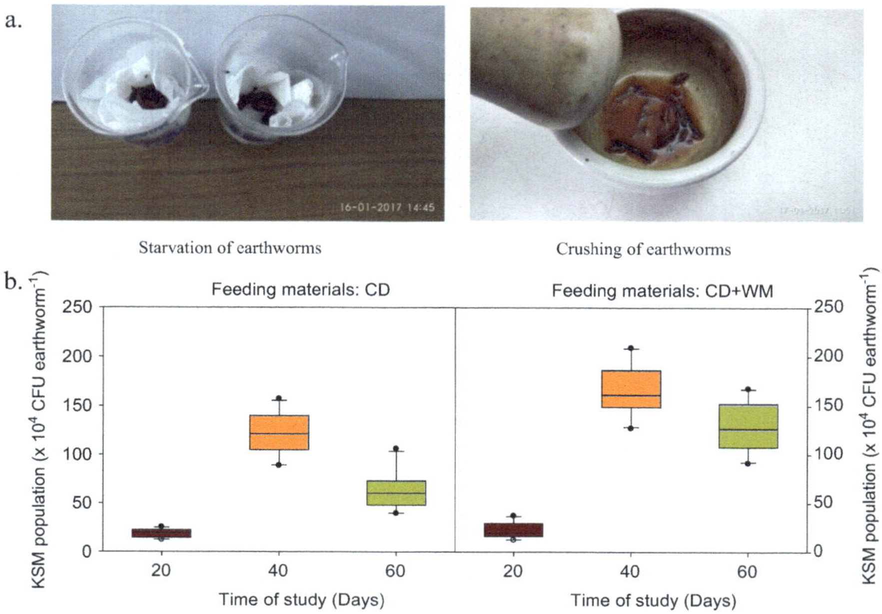

Fig. 13.6 (**a**) Extraction of earthworm gut materials (**b**) Abundance of KSM in earthworm gut during vermicomposting of different feeding materials [CD: cattle manure and WM: mica waste]

Gut-cleaned earthworms were crushed (Fig. 13.6) under aseptic condition, and the homogenate sample was centrifuged at 10,000 rpm for 10 min. Then, the supernatants were collected for estimating KSB population following the above-mentioned protocol.

For measuring KSB count in different parts of earthworms, *E. eugeniae* earthworms were starved overnight for removing ingested organic matter from their guts and intersected into three portions, namely, mouth, midgut and lower intestine. All those sections were crushed aseptically, and KSB in the supernatant was estimated as mentioned before.

In this study, the presence of KSB was also detected in the gut material of *E. eugeniae* earthworm. Total KSB population in earthworm gut material was varied depending on the nature of feeding materials and duration of vermicomposting. Though KSB populations in gut materials of all treatments were same up to 20 days, after 40 days of vermicomposting, KSB population in WM treatment was higher than that of control treatment. After 60 days of vermicomposting, KSB population in WM treatment was significantly ($P < 0.05$) higher than that of without WM treatment.

It was observed that the midgut and lower intestine of *E. eugeniae* generally harboured KSB, and KSB population in those parts was influenced by the presence of bound K materials like WM. Consumption of WM-mixed feeding materials

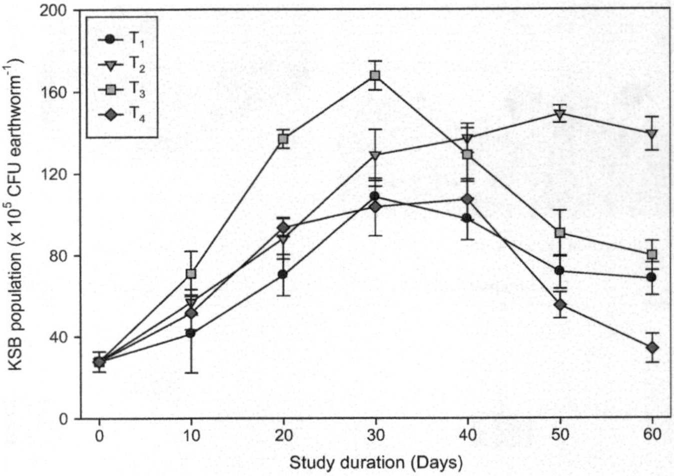

Fig. 13.7 Changes in KSB population in organic substrate during vermicomposting

increased total KSB population in the midgut and lower intestine of *E. eugeniae* by 45.53% and 87.48%, respectively.

This study indicated that KSB present in the gut of earthworms possibly played a major role in solubilizing bound K compounds and that in turn increased water-extractable K and exchangeable K in organic substrates during vermicomposting. The presence of KSB was detected in organic substrates during vermicomposting. Population of KSB in organic substrates was increased up to 40 days of vermicomposting (Fig. 13.7). Mixing of mica waste at 500 g or more with 5 kg of waste materials significantly ($P < 0.05$) increased KSB population in organic substrates up to 40 days of vermicomposting. The highest KSB population was observed in T_3 treatment; however, KSB in T_3 and T_4 treatments were significantly ($P < 0.05$) higher than that of T_1 and T_2 treatments.

Nishanth and Biswas (2008) observed that mica waste-amended compost had higher TK content than the normal compost. The difference between conventional composting and vermicomposting is that organic substrates pass through the earthworm guts in latter case (Aalok and Tripathi 2010). Hence, the variation in K release from mica waste might be attributed to the activity of gut-borne microorganisms during vermicomposting. The guts of earthworms harbour various beneficial microorganisms, which act on organic substrates to alter their chemical and biochemical properties during vermicomposting (Fracchia et al. 2006). The study revealed that KSB are present in the guts of *E. eugeniae*, a species of epigeic earthworm, and that KSB population in the guts was increased due to feeding mica waste-enriched organic substrate. The results indicated that *E. eugeniae* has the capability to solubilize K from the ejected mica waste. Therefore, it could be inferred that earthworms possibly assimilated solubilize K in their biomass which in turn was responsible for reduction in TK content in mica waste-amended vermicompost.

13.6 Application of Vermicompost in Soil

Vermicompost has a good balance of major plant nutrients in an organically bound form and in an inorganic plant-available form. Hence, application of vermicompost as an organic amendment supplies sufficient nutrients to support plant growth, and it also builds up reserve pool of nutrients in soil. Pramanik et al. (2010) revealed that applied organic matter through vermicompost increased the activity of several important enzymes, viz. phosphatase, urease, etc., in soil. Hence, vermicompost application further increases the availability of major nutrients like nitrogen and phosphorus in soil (Pramanik et al. 2007). However, sustaining potassium availability is a major challenge in vermicompost-treated soil.

Vermicompost application also has a positive impact on microbial community in soil. Microbial activity, determined by quantifying carbon dioxide emission from soil, is significantly increased by vermicompost application and that in turn was attributed to increased population of all groups of soil microorganism, viz. bacteria (both Gram-positive and Gram-negative) and fungi (Pramanik et al. 2012). Microorganisms of all specific groups like nitrifying bacteria, phosphate-solubilizing microorganisms and potassium-solubilizing bacteria show higher population in vermicompost-treated soil as compared to untreated control soil. Hence, higher potassium-solubilizing bacterial population could be exploited by mixing mica waste along with vermicompost to enhance and sustain potassium availability in vermicompost-treated soil.

13.7 Application of Mica Waste and Vermicompost in Soil

Application of mica waste is an alternative method to supplement K in soil. Mica waste is a slow K-releasing mineral, which possesses K atoms in the interlayer space of two expandable octahedral lattices. Humic substrates facilitate K release from mica waste by distorting its crystal structure. Since vermicompost contains up to 5% humic substrates, combined application of mica waste and vermicompost is expected to enhance K availability in soil.

An experiment was conducted in the net house of the Tocklai Tea Research Institute (TTRI) (26°41″N and 94°43″E), Tea Research Association, Jorhat, Assam, India, to evaluate the effect of mica waste and vermicompost on K availability in soil and on plant growth. A virgin soil, collected from Borbetta Tea Estate, was used for this experiment. This soil was sandy clay loam in texture and had pH 5.11 ± 0.27, total organic carbon 0.92 ± 0.06 g kg^{-1} and exchangeable K content 108.33 ± 16.47 mg kg^{-1}. Six-month-old tea plants (Tocklai Vegetative (TV)—23 cultivars) were transplanted in pot soil (6.5 kg), and after 1 week of acclimatization, total six treatments were applied in soil. The treatment details are mentioned in Table 13.1. Here, the doses of soil amendments were considered following recommendations of tea-growing soil. Nylon net of the net house restricts about 70% of the

Table 13.1 Details of the treatments tested in soil under controlled condition

Treatments	Details
T_0	Control
T_1	MoP treatment at 60 kg K_2O ha^{-1}
T_2	Vermicompost prepared from 5% (w/w) MW amended organic wastes at 2 t ha^{-1}
T_3	Vermicompost prepared from 10% (w/w) MW amended organic wastes at 2 t ha^{-1}
T_4	Vermicompost at 2 t ha^{-1} along with mica waste at 5% (w/w)
T_5	Vermicompost at 2 t ha^{-1} along with mica waste at 10% (w/w)

incident solar radiation, which protects young tea plants from strong sunshine and heavy precipitation.

In northeastern part of India, about 450–500 mm precipitation is received within the monsoon season (mid-June to August month). Since overhead net cover allows limited rainwater percolation inside the study site, additional water application was given to the plot soil only in later part of August and September months. Plant heights were recorded at 15 days interval, and soil properties were tested at the end of this study.

13.7.1 Plant Growth

Irrespective of treatments, height of tea plants was increased throughout the study duration. Initially, tea plants of almost same height (35.4–37.6 cattle manure) were selected for this study. After 120 days of trial, average plant height of different treatments varied within 43.5–51.3 cm (Fig. 13.8). With 35.4% increase, the highest plant height was recorded in T_5 treatment, while the lowest plant height was observed in T_3 treatment. This study indicated that application of mica waste and vermicompost combination significantly ($P < 0.05$) increased plant height, whereas mica waste-amended vermicompost did not show any effect on changes in the height of tea plants.

13.7.2 Soil K Content

Application of all compost treatments (T_1–T_5) increased exchangeable K content in soil; however, the extent of increase was varied depending on the nature of treatments (Fig. 13.9). The highest exchangeable K content was recorded in the soil of T_5 treatment, though its value was statistically at par with that of T_4 and T_2 treatments. Exchangeable K content of T_1 and T_3 treatments did not vary its increase significantly as compared to control (T_0) treatment.

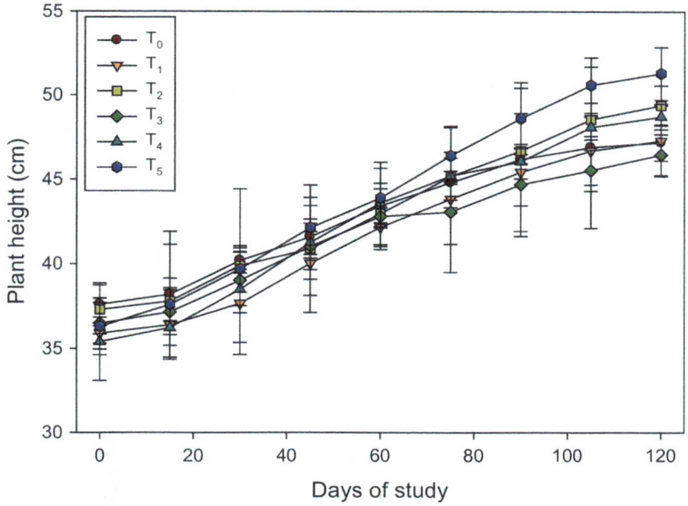

Fig. 13.8 Changes in plant height as affected by different treatments (vertical lines indicated standard error of the corresponding mean vales, $n = 3$)

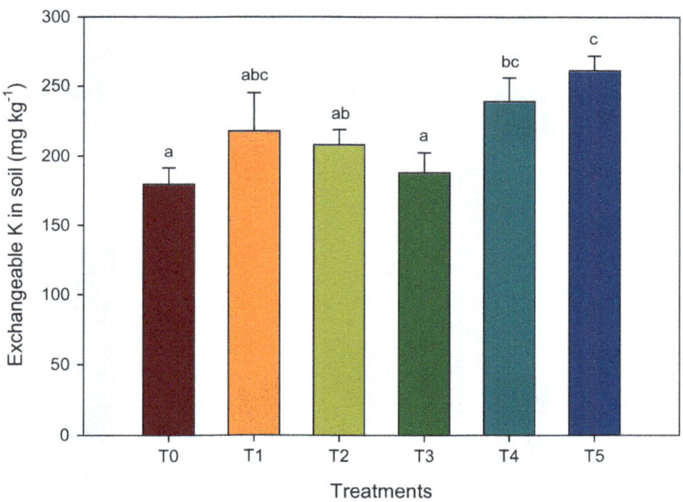

Fig. 13.9 Effect of different treatments on exchangeable K content in soil (different alphabets indicated that the corresponding data are significantly different at $P < 0.05$)

13.8 Biomass Weight and K Uptake by Tea Plants

Biomass weight of tea plants was varied due to application of different treatments (Fig. 13.10). Average biomass weight of tea plants of T_0 treatment was 4.05 ± 0.36 g ($n = 3$). Vermicompost application increased biomass weight (5.11 ± 0.20) of tea plants; however, the increase was not statistically significant as compared to that of T_0 treatment. Application of mica waste in combination with vermicompost significantly increased biomass weight of tea plants (Fig. 13.10).

Total K content in tea plants ranged within 1.06–1.19 mg g^{-1} and did not vary among the treatments. Potassium uptake in tea plants was 4.68 ± 0.69 mg $plant^{-1}$,

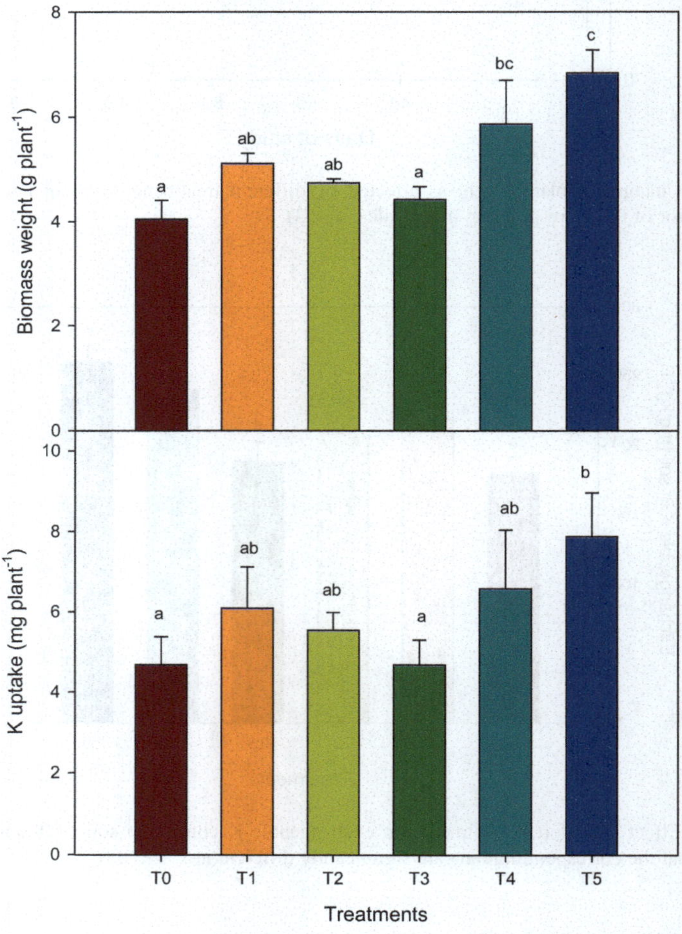

Fig. 13.10 Changes in biomass weight and K uptake of tea plants as affected by different treatments (different alphabets indicated that the corresponding data are significantly different at $P < 0.05$)

which was significantly increased (7.88 ± 1.08 mg plant^{-1}) in T_5 treatment (Fig. 13.10). Though all the vermicompost treatments increased K uptake in tea plants, the difference was not statistically significant for other treatments.

Application of organic matter improves soil health by increasing chemical, biochemical and microbial properties in soil (Pramanik et al. 2010). Low K content in mica waste-amended vermicompost might be attributed to the lower plant growth in these treatments. However, application of mica waste and vermicompost combination had shown significantly better response over vermicompost prepared amending same amount of mica waste. Vermicompost is a humic substrate-enriched organic amendment (Pramanik et al. 2007), and application of mica waste and vermicompost combination released more K in soil as compared to sole vermicompost treatment. Humic substrates alter crystal structure of mica waste and that in turn facilitate release of K from its inter-lattice space (Pramanik et al. 2019). In this study, application of vermicompost in combination with 10% mica waste (weight basis) had the potential to significantly increase plant growth and K availability in soil.

13.9 Conclusion

Addition of 50 g mica waste with each kilogram of waste materials was effective to increase TK content in the final vermicompost; however, mica waste addition at higher rates leads to the reduction in TK content in vermicompost as compared to conventional vermicompost. The gut of *E. eugeniae* harbours KSB, which plays an important role in solubilizing bound K in injected materials during vermicomposting. Application of vermicompost, prepared from mica waste-amended organic substrates, did not increase K availability in soil, whereas combined application of conventional vermicompost with mica waste was effective to increase K availability in soil and that in turn recorded higher plant growth and K uptake in tea. Under controlled condition, vermicompost in combination with 10% mica waste (weight basis) had the potential to increase plant growth and K availability in soil.

Acknowledgement The corresponding author is thankful to the Director of Tea Research Association and Deputy Director of North Bengal Regional R&D Centre, Nagrakata, for kindly providing the facilities to conduct the project work. The work of this chapter was financially supported by the funding of the National Tea Research Foundation (NTRF), Tea Board of India (NTRF: 183/2016).

References

Aalok A, Tripathi AK (2010) Composting-vermicomposting of different types of leaves using earthworm species Eisenia fetida. In: Dynamic soil, dynamic plant, Special Issue 4. Global Science Books Publisher, Ltd., Middlesex, pp 139–144

Atiyeh RM, Arancon NQ, Edwards CA, Metzger JD (2001) The influence of earthworm-processed pig manure on the growth and productivity of marigolds. Bioresour Technol 81:103–108

Basak BB, Biswas DR (2009) Influence of potassium solubilizing microorganism (*Bacillus mucilaginosus*) and waste mica on potassium uptake dynamics by sudan grass (*Sorghum vulgare* Pers.) grown under two Alfisols. Plant Soil 317:235–255

Biswas TD, Mukherjee SK (2000) Text book of soil science, 2nd edn., Sixth reprint. Tata-McGraw Hill Publishing Company Ltd., New Delhi

Brady N (2003) The nature and properties of soil, 13th edn. Colier McMillan, London

Dey MD, Das S, Kumar R, Doley R, Bhattacharya SS, Mukhopadhyay R (2017) Vermiremoval of methylene blue using Eisenia fetida: a potential strategy for bioremediation of synthetic dye-containing effluents. Ecol Eng 106:200–208

Domínguez J (2004) State of the art and new perspectives on vermicomposting research. In: Edwards CA (ed) Earthworm ecology. CRC Press, Boca Raton, FL, pp 401–424

Elvira C, Sampedro L, Benitez E, Nogales R (1998) Vermicomposting of sludges from paper-mill and dairy industries with *Eisenia andrei*: a pilot-scale study. Bioresour Technol 63:205–211

Fracchia L, Dohrmann AB, Martinotti MG, Tebbe CC (2006) Bacterial diversity in a finished compost and vermicompost: differences revealed by cultivation independent analyses of PCR-amplified 16S rRNA genes. Appl Microbiol Biotechnol 71:942–952

Garg P, Gupta A, Satya S (2006) Vermicomposting of different types of waste using Eisenia fetida: a comparative study. Bioresour Technol 97:391–395

Kaushik P, Garg VK (2004) Dynamics of biological and chemical parameters during vermicomposting of solid textile mill sludges mixed with cow dung and agricultural residues. Bioresour Technol 4:203–209

Meena VS, Maurya BR, Verma JP (2014) Does a rhizospheric microorganism enhance K+ availability in agricultural soils? Microbiol Res 169:337–347

Meena VS, Bahadur I, Maurya BR, Kumar A, Meena RK, Meena SK, Prakash Verma J (2016) Potassium-solubilizing microorganism in evergreen agriculture: an overview, Chapter 1. In: Meena VS, Maurya BR, Prakash Verma J, Meena RS (eds) Potassium solubilizing microorganisms for sustainable agriculture, 1st edn. Springer, New Delhi, pp 1–20, ISBN: 978-81-322-2774-8

Ministry of Chemical and Fertilizers (2015) Department of Fertilizers, Government of India. http://fert.nic.in/node/3808

Ndegwa PM, Thompson SA (2000) Effects of stocking density and feeding rate on vermicomposting of biosolids. Bioresour Technol 71:5–12

Nishanth D, Biswas DR (2008) Kinetics of phosphorus and potassium release from rock phosphate and waste mica enriched compost and their effect on yield and nutrient uptake by wheat (Triticum aestivum). Bioresour Technol 99:3342–3353

Pramanik P, Chung YR (2011) Changes in fungal population of fly ash and vinasse mixture during vermicomposting by *Eudrilus eugeniae* and *Eisenia fetida*: documentation of cellulase isozymes in vermicompost. Waste Manag 31:1169–1175

Pramanik P, Kalita C (2019) Evaluated potential of mica waste to release sufficient potassium to support the growth of plants in acidic tea-growing soil in Northeast India. Commun Soil Sci Plant Anal 50:1258–1266

Pramanik P, Ghosh GK, Ghosal PK, Banik P (2007) Changes in organic C, N, P and K and enzyme activities in vermicompost of biodegradable organic wastes under liming and microbial inoculants. Bioresour Technol 98:2485–2494

Pramanik P, Ghosh GK, Banik P (2009) Effect of microbial inoculation during vermicomposting of different organic substrates on microbial status and quantification and documentation of acid phosphatase. Waste Manag 29:574–578

Pramanik P, Ghosh GK, Chung YR (2010) Changes in nutrient content, enzymatic activities and microbial properties of lateritic soil due to application of different vermicomposts: comparative study of ergosterol and chitin to determine fungal biomass in soil. Soil Use Manag 26:508–515

Pramanik P, Kim SY, Kim PJ (2012) Changes in fungal and bacterial diversity during vermicomposting of industrial sludge and poultry manure mixture: detecting the mechanism of plant growth promotion by vermicompost, Chapter 6. In: Matovic D (ed) Remote sensing of biomass: principles and applications (Book 4). InTech, Rijeka, pp 113–123, ISBN: 978-953-307-492-4

Pramanik P, Kalita C, Borah K (2019) Hastening potassium release from mica waste by treating humic substrate solution: an approach to adopt mica waste as potassium amendment in tea-growing soil. Commun Soil Sci Plant Anal 50:1854–1863

Reinecke AJ, Viljioen SA, Saayman RJ (1992) The suitability of *Eudrilus eugeniae*, *Perionyx excavatus* and *Eisenia fetida* (Oligochaete) for vermicomposting in southern Africa in terms of their temperature requirements. Soil Biol Biochem 24:1295–1307

Singh G, Biswas DR, Marwaha TS (2010) Mobilization of potassium from waste mica by plant growth promoting rhizobacteria and its assimilation by maize (*Zea mays*) and wheat (*Triticum aestivum* L.): a hydroponics study under phytotron growth chamber. J Plant Nutr 33:1236–1251

Sitienei K, Home PG, Kamau DM, Wanyoko JK (2013) Nitrogen and potassium dynamics in tea cultivation as influenced by fertilizer type and application rates. Am J Plant Sci 4:59–65

Suthar S, Singh S (2008) Vermicomposting of domestic waste by using two epigeic earthworms (*Perionyx excavatus* and *Perionyx sansibaricus*). Int J Environ Sci Technol 5:99–106

Türkmen Ö, Dursun A, Turan M, Erdinc C (2004) Calcium and humic acid affect seed germination, growth and nutrient content of tomato (Lycopersicon esculentum L.) seedlings under saline soil conditions. Acta Agric Scand B Soil Plant Sci 54:168–174

Xiao Y, Wang X, Chen W, Huang Q (2017) Isolation and identification of three potassium-solubilizing bacteria from rape rhizospheric soil and their effects on ryegrass. Geomicrobiol J 34:873–880

Manash Deep Dey, Subhasish Das, Rupesh Kumar, Robin Doley, Satya Sundar Bhattacharya, Rupak Mukhopadhyay, (2017) Vermiremoval of methylene blue using Eisenia fetida : A potential strategy for bioremediation of synthetic dye-containing effluents. Ecological Engineering 106:200-208

Chapter 14
PGPR and Earthworm-Assisted Phytoremediation of Heavy Metals

Pooja Sharma, Palak Bakshi, Jaspreet Kour, Arun Dev Singh, Shalini Dhiman, Pardeep Kumar, Ibrahim, Ashutosh Sharma, Bilal Ahmad Mir, and Renu Bhardwaj

Abstract The pronounced and major effects of contamination of the environment with heavy metals and other xenobiotic compounds have become a major problem worldwide. Soil contaminated with heavy metals poses serious threat to plants, animals as well as human health. Heavy metals due to their toxicity reduces the soil fertility, affects the plant photosynthetic efficiency, reduces yield of the crops, and causes nutrient imbalance. Phytoremediation an eco-friendly, clean, and green technology helps to remove contaminants from the polluted soils. The use of beneficial microorganisms along with plants is considered as an effective method for increasing the efficiency of remediation of contaminated soils. Earthworms also play an important role in remediation process. Interaction of plants with microflora plays a vital role in bioavailability of the metals and their bioaccumulation in plants.

Keywords Plant growth-promoting rhizobacteria · Earthworms · Fungi · Heavy metal · Plants

---

P. Sharma
Department of Botanical and Environmental Sciences, Guru Nanak Dev University, Amritsar, Punjab, India

Department of Microbiology, DAV University, Jalandhar, Punjab, India

P. Bakshi · J. Kour · A. D. Singh · S. Dhiman · P. Kumar · Ibrahim · R. Bhardwaj (✉)
Department of Botanical and Environmental Sciences, Guru Nanak Dev University, Amritsar, Punjab, India

A. Sharma
Department of Agriculture Sciences, DAV University, Jalandhar, Punjab, India

B. A. Mir
Department of Botany, School of Life Sciences, Satellite Campus Kargil, University of Kashmir, Srinagar, Jammu and Kashmir, India

S. A. Bhat et al. (eds.), *Earthworm Assisted Remediation of Effluents and Wastes*,
https://doi.org/10.1007/978-981-15-4522-1_14

14.1 Introduction

Ecological rehabilitation of polluted soils is one of the major concerns in recent decades due to human activities. Increase in pace of human activities are considered as unprecedented threats to biosphere. Soil is one of the life-supporting factors of this planet but continued industrialization, urbanization, intense agricultural practices, etc. are degrading the quality of the land. More than 30% of the land has been degraded by various anthropogenic activities (Abhilash et al. 2012; Ma et al. 2011a). Pollution, salinity stress, soil erosion, etc. are the processes which deteriorate the soil properties. Large numbers of chemicals like pesticides, heavy metals, persistent organic pollutants, chlorinated solvents, etc. have been detected from soil, water, and air. These discharged entities alter the ecosystem balance that would ultimately deteriorate the quality of food and affect human health (Sivarajasekar and Baskar 2015a, b; Muthusaravanan et al. 2018). They have been documented to cause serious health hazards which include cancer, kidney malfunctioning, impaired central nervous system, skin problems, weakening of bones, etc. (Jarup 2003). Emamverdian et al. (2015) in his study reported oxidative stress in rice seedlings grown in the presence of heavy metals. Similar reports have been reported by many researchers on the generation of oxidative stress under various heavy metals (Pb, Cd, Cr, etc.) (Kohli et al. 2018; Kaur et al. 2018). Plants itself have innate defense response to thrive in stressed environment which includes immobilization, chelation, vacuolar sequestration, expression of defense-related genes, and metal transporters (Sharma et al. 2019; Khanna et al. 2019; Bali et al. 2019). But due to overexploitation of resources, these contaminants are gaining momentum and are considered as challenging mission due to their persistent nature in the ecosystem. Contaminated lands show direct and indirect effect on social, economic, and environmental half of the society. Conventional cleanup methods like physicochemical methods are very costly and laborious.

Bioremediation is an ecofriendly process in which living organisms are used to get rid of various contaminants present in the soil (Abhilash et al. 2009). There is an increasing demand for utilization of plants and their microflora for remediation of contaminated soils which proves to be effective and cost-friendly. Several plant species have potential to remediate the contaminated soil like *Brassica*, castor, *Salix*, *Populus*, etc. (Kohli et al. 2018; Huang et al. 2011; Pulford and Watson 2003). The indirect benefits of growing plants in contaminated sites such as improvement in soil quality, biofuel production, carbon sequestration, biodiversity balance, and aesthetic value have been studied by Abhilash et al. (2019). Phytoremediation potential of the contaminated soil depends upon the extent of contaminants present in the soil, edaphic factors such as biotic diversity of the soil, organic matter, pH, plant-soil-water interaction, availability of nutrients, etc. (Weyens et al. 2009). Plant microbiota which includes plant growth-promoting rhizobacterias (PGPRs), earthworm, fungi, actinomycetes, etc. plays an effective role in the remediation of contaminated sites.

In the present chapter, an attempt has been made to review the recent advances on phytoremediation potential of heavy metal-contaminated soils with the help of plant growth-promoting rhizobacteria, earthworms, fungi, etc.

14.2 Toxicity of Heavy Metals and Other Xenobiotics on Plants

Heavy metals released from industrial effluents and the excessive use of pesticides in agricultural fields are reducing crop production by polluting environment and degrading the food quality (Shahzad et al. 2018). These heavy metals are taken up from the soil through the roots of plants and are translocated to other parts of plants or are accumulated in the roots of plants. The increase in the level of heavy metals in plant's rhizosphere has caused various negative effects on physiological and biochemical processes of plants (Shahzad et al. 2016). Heavy metal toxicity induces formation of free radicals which leads to generation of various reactive species and causes oxidative, nitrosative stress in plants (Dubey et al. 2018). Heavy metal stress disturbed the structure as well as permeability of the organelles and membranes; decreased photosynthetic rate, transpiration rate, and nutrient imbalance; etc. (Green et al. 2003). Increased amount of these pollutants in soil render the soil unfit for vegetation and hamper the biodiversity (Setia et al. 2008). Different heavy metals, pesticides, and the toxicity caused in plants due to these pollutants are given in Table 14.1.

14.3 Phytoremediation

Phytoremediation is an emerging, green, and eco-friendly technology for removing contaminants from the soils with the help of plants (Yang et al. 2019). Plants have the potential to cope up with various contaminants present in soil by changing soil properties, fertility, and efficacy. Phytoremediation approach includes phytoextraction, phytovolatilization, phytodegradation, phytotransformation, rhizodegradation, and phytostabilization (Fig. 14.1) (Marques et al. 2009; Bhargava et al. 2012).

Baker (1981) in his study categorized the plants into three main categories—excluders, accumulators, and indicators. Excluders play a role in the stabilization of soil by preventing soil erosion, reducing uptake of heavy metals to critical level (Lasat 2002). Accumulators have the potential to bioaccumulate the contaminants, whereas indicators show the level of contamination (McGrath et al. 2002). Hyperaccumulators are considered as subclass of accumulators which have very high potential to accumulate the contaminants in their aboveground tissues without

Table 14.1 Adverse effects of pollutants on plants

S. no.	Heavy metal	Heavy metal toxicity in plants	Reference
1	Zinc	Causes chlorosis, accumulation of other heavy metals like manganese (Mn) and copper (Cu) in roots and shoots, chromosomal aberrations and affects chlorophyll fluorescence	Nagajyoti et al. (2010) Huang et al. (2019) Chen et al. (2017) Yamaguchi et al. (2012)
2	Cadmium	Causes retarded growth of plants, chlorosis in leaves	Nagajyoti et al. (2010) Kaya et al. (2020)
3	Copper	Causes chlorosis and inhibited growth retardation in the growth of the roots	Jisha et al. (2017)
4	Mercury	Affects mitochondrial and chloroplast activities, causes chlorosis and necrosis in dose-dependent manner	Nagajyoti et al. (2010) Azevedo and Rodriguez (2012)
5	Chromium	Causes various phytotoxic effects in plants like chlorosis, growth retardation, alteration in the synthesis of essential pigments; inhibits the cell division and elongation of plant roots, thus shortening the overall length of roots	Ozturk et al. (2012) Boonyapookana et al. (2002) Chen et al. (2017)
6	Nickel	Causes necrosis and chlorosis, disturbs the functions of cell membrane, and disrupts water balance which led to poor nodulation and resulted in reduced yield	Yadav (2010) Turan et al. (2018)
7	Lead	Causes retardation and inhibition in germination Affects the Calvin cycle, and electron transport chain is also disturbed and causes carbon dioxide shortage by closing stomata pores and affects chlorophyll florescence	Sharma and Dubey (2005) Chen et al. (2017) Huang et al. (2019)
8	Aluminum	Inhibits root growth, influences reactive oxygen species (ROS) generation, causes signal transduction disorder, and lowers cell wall extensibility, thereby impairing plant growth	Von Uexküll and Mutert (1995) Guo et al. (2017) Dai et al. (2013) Liu et al. (2014)
9	Manganese	Generates oxidative stress, inhibiting the activities of several key enzymes and disturbing the absorption and translocation of other mineral elements, impairing leaf structure and chlorophyll biosynthesis, impeding photosynthesis and respiration	González et al. (1998) Führs et al. (2011) Millaleo et al. (2012) Shi and Zhu (2008)
10	Selenium	Decreases glutathione level and increases the lipid peroxidation in wheat seedlings	Hugouvieux et al. (2009) Łabanowska et al. (2012)
11	Organochlorine pesticides	Cause carbon and nitrogen imbalance and increase in oxidized fatty acids	Blondel et al. (2016)

(continued)

Table 14.1 (continued)

S. no.	Heavy metal	Heavy metal toxicity in plants	Reference
12	Fungicides	Causes changes in ascorbate, lipid, sugar, and nucleotide metabolisms and amino acids levels	Pereira et al. (2014)
13	Insecticide (diazinon)	Produces alterations in metabolites involved in the biosynthesis and metabolism of phenylpropanoids, sugars, amino acids, and organic acids; induces changes in antioxidant defense	Mahdavi et al. (2015)
14	Pesticides	Leads to upregulation of carbohydrates, phenols, and antioxidants in rice leaves and amino acids, and other carbohydrates were downregulated in rice seed	Zhao et al. (2015)
15	Nanoparticles	Induces metabolite changes in *Arabidopsis thaliana* and also affects metabolite pathways	Ke et al. (2018)

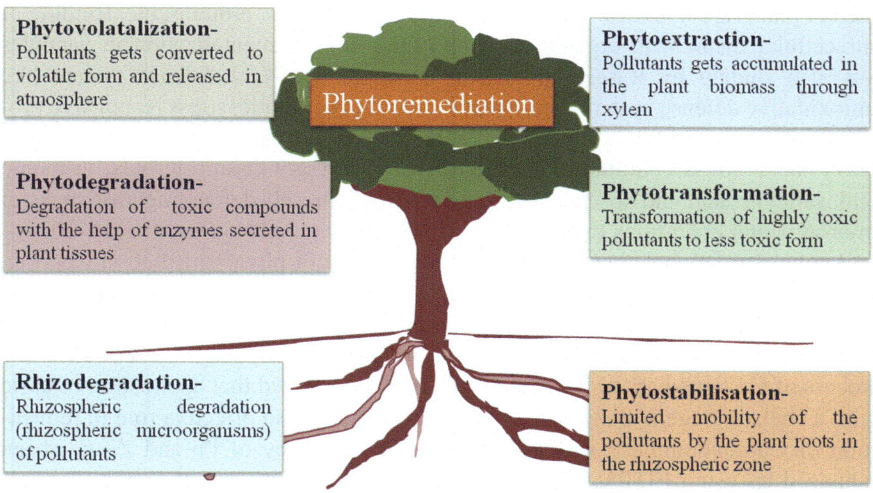

Fig. 14.1 Classification of phytoremediation

showing any toxic symptoms. McGrath and Zhao (2003) in his study mentioned the bioconcentration factor which is the ratio of heavy metal concentration above the ground to the soil. Hyperaccumulators have high bioaccumulation factor which makes it as a feasible option for phytoremediation technique.

14.3.1 Mechanism of Heavy Metal Detoxification in Plants

Various cellular and molecular mechanisms are involved in detoxification and tolerance and maintaining the homeostasis of metal-tolerant plant species (Hall 2002; Bhargava et al. 2012). Metal uptake by plants depends upon their bioavailability in soil which varies due to their low solubility in water and strong interaction with soil particles. Metal present in soil first binds with the cell wall of the root which contains various ion transport channels. High affinity to the binding sites present in the cell wall mediates the uptake of the metal ions through various transporters like P-type ATPase (Bali et al. 2019; Khanna et al. 2019), iron-regulated transporters (IRT), zinc-regulated transporters (ZRP) (Dubey et al. 2018), and channel proteins (Chaney et al. 2007). Absorbed metals from the roots are then transported to the aerial parts of the plant through the xylem by root pressure and transpiration pull (Robinson et al. 2003). Effective loading of the metal ions requires radial symplastic path along with xylem loading. Plant changes the chemistry of these ions by catalyzing and chelation to nontoxic forms and finally ends in their vacuolar compartmentalization for accumulating the heavy metals (McGrath and Zhao 2003). It was assumed by Verbruggen et al. (2009) in his study that metals accumulated in plants are bounded with amino acids, organic acids, proteins, etc. Binding of intracellular metal with organic acids and their compartmentalization in vacuoles was also studied by Rascio and Navari-Izzo (2011). Plant defense system like antioxidative defense system—enzymatic and non-enzymatic, stress responsive proteins, various transporters have been well identified in playing protective role in heavy metal stress conditions. Plant hormones such as brassinosteroids, jasmonic acid, salicylic acid and abscisic acid play a key role in self-defense against stressful environmental conditions (Sharma et al. 2017; Bali et al. 2019; Kohli et al. 2018).

Application of chelators in enhancing the process of phytoremediation has been documented by many researchers and is considered as an effective approach in increasing the metal uptake (Lasat 2002). In his study Wang et al. (2019a) mentioned the significant role of chelators along with *Amaranthus hypochondriacus* L. in the process of Cd remediation. Arsenov et al. (2020) suggested that application of citric acid, a chelating agent along with *Salix viminalis*, plays an effective role in remediation of Cd-contaminated soil. Phytoextraction efficiency of Pb and Zn has been enhanced by using EDTA along with *Raphanus sativus* L. and *Brassica oleracea* L. (Chaturvedi et al. 2019). Similar results were observed by Liphadzi and Kirkham (2006) with EDTA- on Pb-contaminated soil. Sharma et al. (2019) found that exogenous application of ascorbic acid on copper-stressed *Brassica juncea* seedlings enhanced various osmoprotectants like cysteine, proline, free amino acids, and metal chelators like thiols.

14.4 Assisted Phytoremediation

Assisted phytoremediation includes remediation with the help of plants and microflora which includes microbes, earthworms, fungi, and actinomycetes. Microflora plays a crucial role in shaping the soil ecosystem. There are two types of processes: one is direct in which solubilization of heavy metals and their accumulation takes place directly with plants, whereas indirect process involves prevention form phytopathogens, enhanced growth of plants, and heavy metal accumulation with the help of microflora.

14.4.1 Mechanism of Plant Growth-Promoting Microorganisms in Phytoremediation

Rhizosphere (i.e., an area around the plant roots) of plants is a natural habitat for microorganisms. Plants dwelling in the metal-polluted soil provide shelter to many soil microorganisms which are able to tolerate the effects of metals in the soil and play beneficial role in the bioremediation of heavy metals (Zarei et al. 2010). Plant roots release ample of compounds known as root exudates which include carbohydrates, amino acids, carboxylic acids, etc. that serve as nutrient media for microbial population. Microbial cells generate and sense the signal molecules and undergo the process of quorum sensing (Ullah et al. 2015). Some soil microbiota has metal-resistant bacterial population. They play effective role in providing nutrients and reducing the heavy metal toxicity. Microbial degradation of organic and inorganic pollutants has been well documented by many researchers (Gadd 2010; Khanna et al. 2019). Microbes secrete siderophores, organic acids, and plant growth regulators which help in mobilization, binding of the heavy metal ions present in the soil, and increase their bioavailability for plants (Fig. 14.2) (Visioli et al. 2014). A study conducted by Braud et al. (2009) documented that *P. aeruginosa* enhanced the bioavailability of Pb and Cr to the maize. In *Alyssum serpyllifolium*, increased Ni uptake has been observed on inoculation of *Pseudomonas* sp. A3R3 (Ma et al. 2011b). Sulfur-oxidizing bacteria increase the Cu mobilization and its uptake in *Hordeum vulgare* (Shi et al. 2012). In his study Glick (2012) documented that plant growth-promoting bacteria have potential to produce various plant hormones like auxin, cytokinins, and gibberellins which play effective role in stress amelioration.

Oladipo et al. (2018) documented that many filamentous fungi genera (*Trichoderma*, *Aspergillus*, *Mucor*, *Penicillium*) have the ability to tolerate metal toxicity. Various functional groups like amine, carboxylic, phosphate, etc. are present on the fungal cell wall which generates overall negative charge that helps in binding the metal (Ong et al. 2017). Arbuscular mycorrhizal fungi (AMF) are important entity of soil microflora which have potential to withstand heavy metal stress. They establish synergistic association with plants by increasing their root

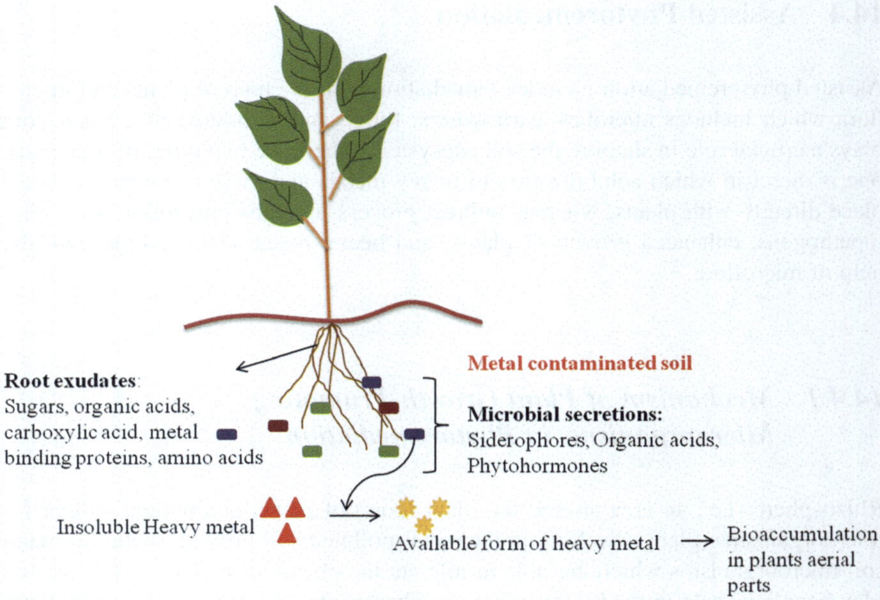

Fig. 14.2 Plant-microbe interaction

surface area and enhancing nutrient availability (Giasson et al. 2006; Saxena et al. 2017). AM fungi reported to increase arsenic (As) uptake in *Pteris vittata* (Al Agely et al. 2005). Jan and Parray (2016) hypothesized the mechanism AMF action of binding heavy metals to the fungal cell wall and transporting them to the vacuoles, chelating heavy metals by phytochelatins present in fungal as well as plant cells, and transporting metal ions by transporters to the tonoplast of fungi and plants. A study conducted by Ma et al. (2019) observed enhanced accumulation of Ni in *Helianthus annuus* when inoculated with *Pseudomonas libanensis* TR1 and *Claroideoglomus claroideum* BEG210.

14.5 Plant-Microbe Interaction in the Remediation of Heavy Metals

Plant-microbe interactions have been studied as a beneficial, co-friendly, and cost-effective method. The scavenging behavior of microbes for reducing the metal stress in soil and toward the plants is of utmost importance (Saha and Rao 2017). They affect the metal solubility in soil and thus their availability to plants. Their influence on roots such as changes in the root morphology and their growth through the release of chelating compounds and solubilizing metal phosphate complexes plays an important role in the metal detoxification by plant-microbe complex. There are a number of mechanisms followed by these microbes and may include elimination through permeability barrier, enzymatic reduction, and capturing and sequestering in the cell active efflux pumps (Nies 1999; Bruins et al. 2000). There are many cases in which different microbes are used in association with plants which enhance the uptake and availability of heavy metals in soil and to plants (shown in Table 14.2).

14.6 Earthworm-Assisted Remediation of Contaminated Soil

Earthworms are the prevailing groups of macroinvertebrates of soil in numerous terrestrial ecosystems comprising almost 80% of the biomass of soil. They are known as key ecosystem engineers as they help in modifying and maintaining the soil habitat. They play a key role in decomposition of organic matter, nutrient cycling, and improving soil condition (Lemtiri et al. 2016). Earthworms slightly decrease the pH of soil by the secreting fulvic and humic acids through their gut microflora which increase the nutrient and heavy metal availability to soils (Drake and Horn 2007). They promote the plant growth by enhancing the accessibility of nutrients and microbes in the rhizosphere. The increased uptake of nutrients results in increased biomass, shoot length, and root length (Varghese and Prabha 2014). Earthworms excrete most of the ingested materials as vermicasts which are mucus-coated granular aggregates. The vermicasts are rich in nutrients like nitrogen, phosphorus, potassium, magnesium, and calcium and have low carbon:nitrogen ratio and good moisture holding capacity (Sharma et al. 2017). The dual action of earthworms and improved soil properties result in enhanced metal uptake of heavy metals (Kaur et al. 2017). The applications of earthworm by-products are also reported to have positive impact on plant growth parameters and metal uptake. Various reports of using earthworms and vermicompost for assisted phytoremediation are mentioned in Table 14.3.

14.7 Conclusion

Heavy metal pollution is one of the leading causes of major environmental issues due to their persistence in the environment and their nondegradable character. Bioaccumulation of these metals in the food chain and their toxicity points a serious threat for living organisms. The pronounced effects of the heavy metals have attracted attention worldwide to get effective solutions for remediating contaminated soils. Many physiological, chemical, and biological approaches have been figured. Phytoremediation is considered as one of the safe, eco-friendly, innovative strategy. Assisted phytoremediation with plant growth-promoting microbes, earthworms, fungi, etc. enhances the efficiency of the remedial technology. Ample amount of work has been done on various plant growth regulators and earthworms that show

Table 14.2 Effect of association of microbes with plants on metal uptake

S. no.	Microbes	Plant	Metal/ abiotic stress	Response	References
1	*Bacillus pichinotyi*	*Triticum aestivum*	Se	Enhanced concentration of Se was observed in wheat kernels and stem Enhanced the growth of plant in both presence and absence of metal	Yasin et al. (2015)
2	*Phyllobacterium myrsinacearum*	*Sedum plumbizincicola*	Cd	Metal uptake was increased and also caused the enhancement of plant growth	Ma et al. (2013)
3	*Pseudomonas aeruginosa* *Bacillus subtilis* *Cupriavidus taiwanesis* *Beauveria bassiana*	*Oryza sativa*	Cd	Increases the plant growth and decreases the accumulation of metal in the plant as metal was accumulated in the microbial tissues	Suksabye et al. (2016)
4	*Brevibacterium frigoritolerans* *Bacillus paralicheniformis*	*Brassica juncea*	Pb	Enhanced metal uptake by *B. juncea* and improved the plant growth	Yahaghi et al. (2018)
5	*Bacillus* sp.	*Raphanus sativus*	Ni	Improved Ni extraction and enhanced the plant growth	Akhtar et al. (2018)
6	*Pseudomonas aeruginosa* *Burkholderia gladioli*	*Lycopersicum esculentum*	Cd	Metal was accumulated in the microbes and thus reduced the uptake by plants and helped in the growth of plants	Khanna et al. (2019)
7	*Pseudomonas libanensis* *Claroideoglomus claroideum*	*Helianthus annuus*	Ni Salinity	Increased phytoremediation by reducing metal uptake by plants and increasing plant growth	Ma et al. (2019)
8	*Alcaligenes faecalis* *Bacillus cereus*	*Sorghum bicolor*	Cu Pb Cd Zn	Help in the enhancement of plant growth and the translocation of metals from roots to shoots with the help of microbes	El-Meihy et al. (2019)
9	*Pseudomonas* sp.	*Pteris vittata*	As Pb Cd	Improves the process of bioremediation by accumulating the metal in microbial tissues and reducing the metal uptake in plants	Manzoor et al. (2019)
10	*Burkholdria* sp. *Sphingomonas* sp.	*Brassica juncea*	Cd	Helps in the growth of plant and increases the accumulation of metal in the root and shoot	Wang et al. (2019b)
11	*Pseudomonas plecoglossicida* *Staphylococci saprophyticus*	*Spinacia oleracea*	Cr	Promotes the plant growth and increase in the biomass was observed	Arshad et al. (2019)

(continued)

Table 14.2 (continued)

S. no.	Microbes	Plant	Metal/ abiotic stress	Response	References
12	*Paenibacillus mucilaginosus* *Sinorhizobium meliloti*	*Alfa alfa*	Cu	Metal uptake was enhanced and also increased the biochemical responses in plants and phytomanagement of soil contaminated with the metal	Ju et al. (2019)
13	*Bacillus cereus* *Enterobacter cloacae*	*Zea mays*	Cd Pb Salinity	Decreases the accumulation of metal in plant as metal accumulated in the microbial tissues and increases the growth of plant	Abedinzadeh et al. (2019)
14	*Enterobacter cloacae*	*Oryza sativa*	Cr	Uplifts the process of bioremediation by reducing the metal	Pattnaik et al. (2020)
15	*Rhizobium leguminosarum* *Sinorhizobium meliloti* *Pseudomonas fluorescens* *Luteibacter* sp. *Variovorax* sp. *Bacillus simplex* *Bacillus megaterium*	*Lathyrus sativus*	Cu Zn Pb Cd	Accumulation in the plant parts increased by removing metal from the soil	Abdelkrim et al. (2019)
16	*Microbacterium oxydans* *Pseudomonas thivervalensis* *Burkholderia cepacia*	*Brassica napus*	Cu	Increases the metal uptake and physiological processes in plant by reducing the metal stress	Ren et al. (2019)
17	*Bacillus anthracis*, *Bacillus cereus*	*Lens culinaris*	Cr	Increased the fresh weight of plants and also enhanced morphological and the biochemical parameters of the plant	Fatima and Ahmed (2018)
18	*Bacillus subtilis* *B. thuringiensis* *B. megaterium*	*Cicer arietinum*	Cd	Accumulation of metal was observed in the shoot	Khan and Bano (2019)
19	*Pseudomonas koreensis*	*Miscanthus sinensis*	Pb	Enhanced plant biomass, photosynthetic pigments, antioxidative defense in plants, and metal uptake	Babu et al. (2015)

Table 14.3 Earthworm assisted phytoremediation

S. no.	Metal stress	Plant species	Earthworm/vermicompost	References
1	Cr and Ni	*Acacia mangium*	*Pontoscolex corethrurus*	Bongoua-Devisme et al. (2019)
2	Cu	*Canavalia ensiformis*	*Rhizoglomus clarum*	Santana et al. (2019)
3	Cd	*Brassica juncea*	*Eisenia fetida*	Kaur et al. (2018)
4	Cu	*Canavalia ensiformis*	Bovine manure vermicompost	Santana et al. (2018)
5	Cd	*Avena sativa* *Helianthus annuus*	*Aporrectodea caliginosa* and *Eisenia fetida*	Elyamine et al. (2018)
6	Pb	*Ocimum basilicum*	Vermicompost	Golrizkhatami et al. (2018)
7	Cd	*Vicia faba*	*Eisenia fetida*	Lemtiri et al. (2016)
8	Pb	*Medicago polymorpha* *Medicago rigidula* *Onobrychis sativa*	Vermicompost	Akhzari et al. (2016)
9	Cd, Cr, and Pb	*Avena strigosa*	Vermicompost using *Einsenia Andrei*	Hoehne et al. (2016)
10	Cr, Pb, Cd, and Zn	*Croton sparsiflorus*	Vermicompost	Kumar et al. (2014)
11	Cr, Pb, Cd, and Zn	*Eclipta alba*	Vermicompost	Kumar et al. (2014)
12	Cd	*Zea mays*	*Eisenia fetida*	Du et al. (2014)
13	Pb	*Lantana camara*	*Pontoscolex corethrurus*	Jusselme et al. (2012)
14	Ni and Pb	*Mentha arvensis*	Vermicompost	Chand et al. (2012a)
15	Ni and Cd	*Matricaria chamomilla*	Vermicompost	Chand et al. (2012b)
16	Cd	*Sedum alfredii*	Pig manure vermicompost	Wang et al. (2012)
17	Pb and Zn	Maize and Barley	*Lumbricus terrestris*	Ruiz et al. (2011)
18	Cr	*Sorghum bicolor*	*Eudrilus eugeniae*	Revathi et al. (2011)
19	Zn and Pb	*Lolium multiflorum*	*Pheretima* sp.	Cheng and Wong (2008)
20	Cd, Cu, Ni, Pb, and Zn	*Helianthus annuus*	Vermicompost	Jadia and Fulekar (2008)
21	Cu	Ryegrass	*Metaphire guillelmi*	Dandan et al. (2007)
22.	Cu, Ni, Zn	*Lactuca sativa*	Vermicompost	Jordao et al. (2007)

the increased heavy metal accumulation in plants. Production of siderophores, phytohormones, organic acids, and chelators improved the plant growth as well as phytoremediation efficiency.

References

Abdelkrim S, Jebara SH, Saadani O, Chiboub M, Abid G, Mannai K, Jebara M (2019) Heavy metal accumulation in *Lathyrus sativus* growing in contaminated soils and identification of symbiotic resistant bacteria. Arch Microbiol 201(1):107–121

Abedinzadeh M, Etesami H, Alikhani HA (2019) Characterization of rhizosphere and endophytic bacteria from roots of maize (Zea mays L.) plant irrigated with wastewater with biotechnological potential in agriculture. Biotechnol Rep 21:e00305

Abhilash PC, Edrisi SA, Chotte JL (2019) Land restoration for achieving SDG 2: end. Sustain Dev Goals 33

Abhilash PC, Jamil S, Singh N (2009) Transgenic plants for enhanced biodegradation and phytoremediation of organic xenobiotics. Biotechnol Adv 27(4):474–488

Abhilash PC, Powell JR, Singh HB, Singh BK (2012) Plant–microbe interactions: novel applications for exploitation in multipurpose remediation technologies. Trends Biotechnol 30 (8):416–420

Akhtar MJ, Ullah S, Ahmad I, Rauf A, Nadeem SM, Khan MY et al (2018) Nickel phytoextraction through bacterial inoculation in *Raphanus sativus*. Chemosphere 190:234–242

Akhzari D, Khedmati M, Soleymani AR, Pessarakli M (2016) Growth, survival, protein content, and phytoremediation potency of various rangeland plant species (*Medicago polymorpha* L., *Medicago rigidula* L., and *Onobrychis sativa* L.) grown in vermicompost-containing potting media. Commun Soil Sci Plant Anal 47(20):2261–2270

Al Agely A, Sylvia DM, Ma LQ (2005) Mycorrhizae increase arsenic uptake by the hyperaccumulator Chinese brake fern (*Pteris vittata* L.). J Environ Qual 34(6):2181–2186

Arsenov D, Župunski M, Borišev M, Nikolić N, Pilipovic A, Orlovic S et al (2020) Citric acid as soil amendment in cadmium removal by Salix viminalis L., alterations on biometric attributes and photosynthesis. Int J Phytoremediation 22(1):29–39

Arshad M, Javaid A, Manzoor M, Hina K, Ali MA, Ahmed I (2019) Isolation and identification of chromium-tolerant bacterial strains and their potential to promote plant growth. In: E3S web of conferences, vol 96. EDP Sciences, Les Ulis, p 01005

Azevedo R, Rodriguez E (2012) Phytotoxicity of mercury in plants: a review. J Bot 2012:848614

Babu AG, Shea PJ, Sudhakar D, Jung IB, Oh BT (2015) Potential use of *Pseudomonas koreensis* AGB-1 in association with *Miscanthus sinensis* to remediate heavy metal(loid)-contaminated mining site soil. J Environ Manag 151:160–166

Baker AJ (1981) Accumulators and excluders-strategies in the response of plants to heavy metals. J Plant Nutr 3(1–4):643–654

Bali S, Jamwal VL, Kaur P, Kohli SK, Ohri P, Gandhi SG et al (2019) Role of P-type ATPase metal transporters and plant immunity induced by jasmonic acid against lead (Pb) toxicity in tomato. Ecotoxicol Environ Saf 174:283–294

Bhargava A, Carmona FF, Bhargava M, Srivastava S (2012) Approaches for enhanced phytoextraction of heavy metals. J Environ Manag 105:103–120

Blondel C, Khelalfa F, Reynaud S, Fauvelle F, Raveton M (2016) Effect of organochlorine pesticides exposure on the maize root metabolome assessed using high-resolution magic-angle spinning 1H NMR spectroscopy. Environ Pollut 214:539–548

Bongoua-Devisme AJ, Akotto OF, Guety T, Kouakou S, Edith AA, Ndoye F, Diouf D (2019) Enhancement of phytoremediation efficiency of Acacia mangium using earthworms in metal-contaminated soil in Bonoua, Ivory Coast. Afr J Biotechnol 18(27):622–631

Boonyapookana B, Upatham ES, Krutrachue M, Pokethitiyook P, Singhakaew S (2002) Phytoaccumulation and phytotoxicity of cadmium and chromium in duckweed Wolffia globosa. Int J Phytoremediation 4:87–100

Braud A, Jézéquel K, Bazot S, Lebeau T (2009) Enhanced phytoextraction of an agricultural Cr-and Pb-contaminated soil by bioaugmentation with siderophore-producing bacteria. Chemosphere 74(2):280–286

Bruins MR, Kapil S, Oehme FW (2000) Microbial resistance to metals in the environment. Ecotoxicol Environ Saf 45(3):198–207

Chand S, Pandey A, Patra DD (2012a) Influence of nickel and lead applied in combination with vermicompost on growth and accumulation of heavy metals by *Mentha arvensis* Linn. cv.'Kosi'. Indian J Nat Prod Resour 3:256–261

Chand S, Pandey A, Patra DD (2012b) Influence of vermicompost on dry matter yield and uptake of Ni and Cd by chamomile (Matricaria chamomilla) in Ni-and Cd-Polluted Soil. Water Air Soil Pollut 223(5):2257–2262

Chaney RL, Angle JS, Broadhurst CL, Peters CA, Tappero RV, Sparks DL (2007) Improved understanding of hyperaccumulation yields commercial phytoextraction and phytomining technologies. J Environ Qual 36(5):1429–1443

Chaturvedi R, Favas P, Pratas J, Varun M, Paul MS (2019) EDTA-assisted metal uptake in Raphanus sativus L. and Brassica oleracea L.: assessment of toxicity and food safety. Bull Environ Contam Toxicol 103(3):490–495

Chen Q, Zhang X, Liu Y, Wei J, Shen W, Shen Z, Cui J (2017) Hemin-mediated alleviation of zinc, lead and chromium toxicity is associated with elevated photosynthesis, antioxidative capacity; suppressed metal uptake and oxidative stress in rice seedlings. Plant Growth Regul 81 (2):253–264

Cheng J, Wong MH (2008) Effects of earthworm (Pheretima sp.) on three sequential ryegrass harvests for remediating lead/zinc mine tailings. Int J Phytoremediation 10(3):173–184

Dai H, Cao F, Chen X, Zhang M, Ahmed IM, Chen ZH et al (2013) Comparative proteomic analysis of aluminum tolerance in Tibetan wild and cultivated barleys. PLoS One 8(5):e63428

Dandan W, Huixin L, Feng H, Xia W (2007) Role of earthworm-straw interactions on phytoremediation of Cu contaminated soil by ryegrass. Acta Ecol Sin 27(4):1292–1298

Drake HL, Horn MA (2007) As the worm turns: the earthworm gut as a transient habitat for soil microbial biomes. Annu Rev Microbiol 61:169–189

Du YL, He MM, Xu M, Yan ZG, Zhou YY, Guo GL et al (2014) Interactive effects between earthworms and maize plants on the accumulation and toxicity of soil cadmium. Soil Biol Biochem 72:193–202

Dubey S, Shri M, Gupta A, Rani V, Chakrabarty D (2018) Toxicity and detoxification of heavy metals during plant growth and metabolism. Environ Chem Lett 16(4):1169–1192

El-Meihy RM, Abou-Aly HE, Youssef AM, Tewfike TA, El-Alkshar EA (2019) Efficiency of heavy metals-tolerant plant growth promoting bacteria for alleviating heavy metals toxicity on sorghum. Environ Exp Bot 162:295–301

Elyamine A, Moussa M, Ismael M, Wei J, Zhao Y, Wu Y, Hu C (2018) Earthworms, rice straw, and plant interactions change the organic connections in soil and promote the decontamination of cadmium in soil. Int J Environ Res Public Health 15(11):2398

Emamverdian A, Ding Y, Mokhberdoran F, Xie Y (2015) Heavy metal stress and some mechanisms of plant defense response. Sci World J 2015:756120

Fatima H, Ahmed A (2018) Micro-remediation of chromium contaminated soils. Peer J 6:e6076 0

Führs H, Specht A, Erban A, Kopka J, Horst WJ (2011) Functional associations between the metabolome and manganese tolerance in *Vigna unguiculata*. J Exp Bot 63(1):329–340

Gadd GM (2010) Metals, minerals and microbes: geomicrobiology and bioremediation. Microbiology 156(3):609–643

Giasson P, Jaouich A, Cayer P, Gagné S, Moutoglis P, Massicotte L (2006) Enhanced phytoremediation: A study of mycorrhizoremediation of heavy metal–contaminated soil. Remed J Environ Cleanup Costs Technol Techn 17(1):97–110

Glick BR (2012) Plant growth-promoting bacteria: mechanisms and applications. Scientifica 2012:963401

Golrizkhatami F, Farsad F, Rafati M (2018) The combined effect of EDTA and vermicompost on removal of lead from soil by *Ocimum basilicum*. Asian J Water Environ Pollut 15(4):41–45

González A, Steffen KL, Lynch JP (1998) Light and excess manganese: implications for oxidative stress in common bean. Plant Physiol 118(2):493–504

Green CE, Chaney RL, Bouwkamp J (2003) Interactions between cadmium uptake and phytotoxic levels of zinc in hard red spring wheat. J Plant Nutr 26(2):417–430

Guo P, Qi YP, Yang LT, Lai NW, Ye X, Yang Y, Chen LS (2017) Root adaptive responses to aluminum-treatment revealed by RNA-Seq in two Citrus species with different aluminum-tolerance. Front Plant Sci 8:330

Hall JL (2002) Cellular mechanisms for heavy metal detoxification and tolerance. J Exp Bot 53 (366):1–11

Hoehne L, de Lima CV, Martini MC, Altmayer T, Brietzke DT, Finatto J et al (2016) Addition of vermicompost to heavy metal-contaminated soil increases the ability of black oat (*Avena strigosa* Schreb) plants to remove Cd, Cr, and Pb. Water Air Soil Pollut 227(12):443

Huang H, Yu N, Wang L, Gupta DK, He Z, Wang K et al (2011) The phytoremediation potential of bioenergy crop Ricinus communis for DDTs and cadmium co-contaminated soil. Bioresour Technol 102(23):11034–11038

Huang XH, Zhu F, Yan WD, Chen XY, Wang GJ, Wang RJ (2019) Effects of Pb and Zn toxicity on chlorophyll fluorescence and biomass production of *Koelreuteria paniculata* and *Zelkova schneideriana* young plants. Photosynthetica 57(2):688–697

Hugouvieux V, Dutilleul C, Jourdain A, Reynaud F, Lopez V, Bourguignon J (2009) Arabidopsis putative selenium-binding protein1 expression is tightly linked to cellular sulfur demand and can reduce sensitivity to stresses requiring glutathione for tolerance. Plant Physiol 151 (2):768–781

Jadia CD, Fulekar MH (2008) Phytoremediation: the application of vermicompost to remove zinc, cadmium, copper, nickel and lead by sunflower plant. Environ Eng Manag J 7(5):547–558

Jan S, Parray JA (2016) Approaches to heavy metal tolerance in plants. Springer, Singapore, pp 1–18

Jarup L (2003) Hazards of heavy metal contamination. Br Med Bull 68(1):167–182

Jisha CK, Bauddh K, Shukla SK (2017) Phytoremediation and bioenergy production efficiency of medicinal and aromatic plants. In: Phytoremediation potential of bioenergy plants. Springer, Singapore, pp 287–304

Jordao CP, Fialho LL, Neves JCL, Cecon PR, Mendonça ES, Fontes RLF (2007) Reduction of heavy metal contents in liquid effluents by vermicomposts and the use of the metal-enriched vermicomposts in lettuce cultivation. Bioresour Technol 98(15):2800–2813

Ju W, Liu L, Fang L, Cui Y, Duan C, Wu H (2019) Impact of co-inoculation with plant-growth-promoting rhizobacteria and rhizobium on the biochemical responses of alfalfa-soil system in copper contaminated soil. Ecotoxicol Environ Saf 167:218–226

Jusselme MD, Poly F, Miambi E, Mora P, Blouin M, Pando A, Rouland-Lefèvre C (2012) Effect of earthworms on plant Lantana camara Pb-uptake and on bacterial communities in root-adhering soil. Sci Total Environ 416:200–207

Kaur P, Bali S, Sharma A, Vig AP, Bhardwaj R (2017) Effect of earthworms on growth, photosynthetic efficiency and metal uptake in *Brassica juncea* L. plants grown in cadmium-polluted soils. Environ Sci Pollut Res 24(15):13452–13465

Kaur P, Bali S, Sharma A, Vig AP, Bhardwaj R (2018) Role of earthworms in phytoremediation of cadmium (Cd) by modulating the antioxidative potential of Brassica juncea L. Appl Soil Ecol 124:306–316

Kaya C, Ashraf M, Alyemeni MN, Ahmad P (2020) Responses of nitric oxide and hydrogen sulfide in regulating oxidative defence system in wheat plants grown under cadmium stress. Physiol Plant 168:345–360

Ke M, Qu Q, Peijnenburg WJGM, Li X, Zhang M, Zhang Z et al (2018) Phytotoxic effects of silver nanoparticles and silver ions to *Arabidopsis thaliana* as revealed by analysis of molecular responses and of metabolic pathways. Sci Total Environ 644:1070–1079

Khan N, Bano A (2019) Rhizobacteria and abiotic stress management. In: Plant growth promoting rhizobacteria for sustainable stress management. Springer, Singapore, pp 65–80

Khanna K, Jamwal VL, Gandhi SG, Ohri P, Bhardwaj R (2019) Metal resistant PGPR lowered Cd uptake and expression of metal transporter genes with improved growth and photosynthetic pigments in Lycopersicon esculentum under metal toxicity. Sci Rep 9(1):5855

Kohli SK, Handa N, Sharma A, Gautam V, Arora S, Bhardwaj R et al (2018) Combined effect of 24-epibrassinolide and salicylic acid mitigates lead (Pb) toxicity by modulating various metabolites in Brassica juncea L. seedlings. Protoplasma 255(1):11–24

Kumar BA, Jothiramalingam S, Thiyagarajan SK, Hidhayathullakhan T, Nalini R (2014) Phytoremediation of heavy metals from paper mill effluent soil using Croton sparsiflorus. Int Lett Chem Phys Astron 17(1):1–9

Łabanowska M, Filek M, Kościelniak J, Kurdziel M, Kuliś E, Hartikainen H (2012) The effects of short-term selenium stress on Polish and Finnish wheat seedlings—EPR, enzymatic and fluorescence studies. J Plant Physiol 169(3):275–284

Lasat MM (2002) Phytoextraction of toxic metals. J Environ Qual 31(1):109–120

Lemtiri A, Liénard A, Alabi T, Brostaux Y, Cluzeau D, Francis F, Colinet G (2016) Earthworms Eisenia fetida affect the uptake of heavy metals by plants *Vicia faba* and *Zea mays* in metal-contaminated soils. Appl Soil Ecol 104:67–78

Liphadzi MS, Kirkham MB (2006) Availability and plant uptake of heavy metals in EDTA-assisted phytoremediation of soil and composted biosolids. S Afr J Bot 72(3):391–397

Liu J, Piñeros MA, Kochian LV (2014) The role of aluminum sensing and signaling in plant aluminum resistance. J Integr Plant Biol 56(3):221–230

Ma Y, Prasad MNV, Rajkumar M, Freitas H (2011a) Plant growth promoting rhizobacteria and endophytes accelerate phytoremediation of metalliferous soils. Biotechnol Adv 29(2):248–258

Ma Y, Rajkumar M, Luo Y, Freitas H (2011b) Inoculation of endophytic bacteria on host and non-host plants—effects on plant growth and Ni uptake. J Hazard Mater 195:230–237

Ma Y, Rajkumar M, Luo Y, Freitas H (2013) Phytoextraction of heavy metal polluted soils using *Sedum plumbizincicola* inoculated with metal mobilizing *Phyllobacterium myrsinacearum* RC6b. Chemosphere 93(7):1386–1392

Ma Y, Rajkumar M, Oliveira RS, Zhang C, Freitas H (2019) Potential of plant beneficial bacteria and arbuscular mycorrhizal fungi in phytoremediation of metal-contaminated saline soils. J Hazard Mater 379:120813

Mahdavi V, Farimani MM, Fathi F, Ghassempour A (2015) A targeted metabolomics approach toward understanding metabolic variations in rice under pesticide stress. Anal Biochem 478:65–72

Manzoor M, Abid R, Rathinasabapathi B, De Oliveira LM, da Silva E, Deng F et al (2019) Metal tolerance of arsenic-resistant bacteria and their ability to promote plant growth of Pteris vittata in Pb-contaminated soil. Sci Total Environ 660:18–24

Marques AP, Rangel AO, Castro PM (2009) Remediation of heavy metal contaminated soils: phytoremediation as a potentially promising clean-up technology. Crit Rev Environ Sci Technol 39(8):622–654

McGrath SP, Zhao FJ (2003) Phytoextraction of metals and metalloids from contaminated soils. Curr Opin Biotechnol 14(3):277–282

McGrath SP, Zhao FJ, Lombi E (2002) Phytoremediation of metals, metalloids, and radionuclides. Adv Agron 75:1–56

Millaleo R, Reyes-Díaz M, Alberdi M, Ivanov AG, Krol M, Hüner NPA (2012) Excess manganese differentially inhibits photosystem I versus II in Arabidopsis thaliana. J Exp Bot 64(1):343–354

Muthusaravanan S, Sivarajasekar N, Vivek JS, Paramasivan T, Naushad M, Prakashmaran J et al (2018) Phytoremediation of heavy metals: mechanisms, methods and enhancements. Environ Chem Lett 16(4):1339–1359

Nagajyoti PC, Lee KD, Sreekanth TVM (2010) Heavy metals, occurrence and toxicity for plants: a review. Environ Chem Lett 8(3):199–216

Nies DH (1999) Microbial heavy-metal resistance. Appl Microbiol Biotechnol 51(6):730–750

Oladipo OG, Awotoye OO, Olayinka A, Bezuidenhout CC, Maboeta MS (2018) Heavy metal tolerance traits of filamentous fungi isolated from gold and gemstone mining sites. Braz J Microbiol 49(1):29–37

Ong GH, Ho XH, Shamkeeva S, Manasha Savithri Fernando AS, Wong LS (2017) Biosorption study of potential fungi for copper remediation from Peninsular Malaysia. Remediat J 27 (4):59–63

Ozturk S, Kaya T, Aslim B, Tan S (2012) Removal and reduction of chromium by Pseudomonas spp. and their correlation to rhamnolipid production. J Hazard Mater 231:64–69

Pattnaik S, Dash D, Mohapatra S, Pattnaik M, Marandi AK, Das S, Samantaray DP (2020) Improvement of rice plant productivity by native Cr (VI) reducing and plant growth promoting soil bacteria Enterobacter cloacae. Chemosphere 240:124895

Pereira SI, Figueiredo PI, Barros AS, Dias MC, Santos C, Duarte IF, Gil AM (2014) Changes in the metabolome of lettuce leaves due to exposure to mancozeb pesticide. Food Chem 154:291–298

Pulford ID, Watson C (2003) Phytoremediation of heavy metal-contaminated land by trees—a review. Environ Int 29(4):529–540

Rascio N, Navari-Izzo F (2011) Heavy metal hyperaccumulating plants: how and why do they do it? And what makes them so interesting? Plant Sci 180(2):169–181

Ren X, Guo S, Tian W, Chen Y, Han H, Chen E et al (2019) Effects of plant growth-promoting bacteria (PGPB) inoculation on the growth, antioxidant activity, Cu uptake, and bacterial community structure of rape (*Brassica napus* L.) grown in Cu-contaminated agricultural soil. Front Microbiol 10:1455

Revathi K, Haribabu TE, Sudha PN (2011) Phytoremediation of chromium contaminated soil using sorghum plant. Int J Environ Sci 2(2):418

Robinson B, Fernández JE, Madejón P, Marañón T, Murillo JM, Green S, Clothier B (2003) Phytoextraction: an assessment of biogeochemical and economic viability. Plant Soil 249 (1):117–125

Ruiz E, Alonso-Azcárate J, Rodríguez L (2011) Lumbricus terrestris L. activity increases the availability of metals and their accumulation in maize and barley. Environ Pollut 159 (3):722–728

Saha P, Rao KV (2017) Biosurfactants—a current perspective on production and applications. Nat Environ Pollut Technol 16(1):181–188

Santana NA, Rabuscke CM, Soares VB, Soriani HH, Nicoloso FT, Jacques RJS (2018) Vermicompost dose and mycorrhization determine the efficiency of copper phytoremediation by *Canavalia ensiformis*. Environ Sci Pollut Res 25(13):12663–12677

Santana NA, Ferreira PAA, Tarouco CP, Schardong IS, Antoniolli ZI, Nicoloso FT, Jacques RJS (2019) Earthworms and mycorrhization increase copper phytoextraction by Canavalia ensiformis in sandy soil. Ecotoxicol Environ Saf 182:109383

Saxena B, Shukla K, Giri B (2017) Arbuscular mycorrhizal fungi and tolerance of salt stress in plants. In: Arbuscular mycorrhizas and stress tolerance of plants. Springer, Singapore, pp 67–97

Setia RC, Kaur N, Setia N, Nayyar H (2008) Heavy metal toxicity in plants and phytoremediation. In: Crop improvement: strategies and applications. IK International Publishing House Pvt. Ltd., New Delhi, pp 206–218

Shahzad B, Tanveer M, Che Z, Rehman A, Cheema SA, Sharma A et al (2018) Role of 24-epibrassinolide (EBL) in mediating heavy metal and pesticide induced oxidative stress in plants: a review. Ecotoxicol Environ Saf 147:935–944

Shahzad B, Tanveer M, Hassan W, Shah AN, Anjum SA, Cheema SA, Ali I (2016) Lithium toxicity in plants: reasons, mechanisms and remediation possibilities–A review. Plant Physiol Biochem 107:104–115

Sharma P, Dubey RS (2005) Lead toxicity in plants. Braz J Plant Physiol 17(1):35–52

Sharma DK, Tomar S, Chakraborty D (2017) Role of earthworm in improving soil structure and functioning. Curr Sci 113:1064–1071

Sharma R, Bhardwaj R, Thukral AK, Al-Huqail AA, Siddiqui MH, Ahmad P (2019) Oxidative stress mitigation and initiation of antioxidant and osmoprotectant responses mediated by ascorbic acid in Brassica juncea L. subjected to copper (II) stress. Ecotoxicol Environ Saf 182:109436

Shi Q, Zhu Z (2008) Effects of exogenous salicylic acid on manganese toxicity, element contents and antioxidative system in cucumber. Environ Exp Bot 63(1–3):317–326

Shi R, Weber G, Köster J, Reza-Hajirezaei M, Zou C, Zhang F, von Wirén N (2012) Senescence-induced iron mobilization in source leaves of barley (Hordeum vulgare) plants. New Phytol 195 (2):372–383

Sivarajasekar N, Baskar R (2015a) Biosorption of basic violet 10 onto activated Gossypium hirsutum seeds: Batch and fixed-bed column studies. Chin J Chem Eng 23(10):1610–1619

Sivarajasekar N, Baskar R (2015b) Agriculture waste biomass valorisation for cationic dyes sequestration: a concise review. J Chem Pharm Res 7:737–748

Suksabye P, Pimthong A, Dhurakit P, Mekvichitsaeng P, Thiravetyan P (2016) Effect of biochars and microorganisms on cadmium accumulation in rice grains grown in Cd-contaminated soil. Environ Sci Pollut Res 23(2):962–973

Turan V, Ramzani PMA, Ali Q, Abbas F, Iqbal M, Irum A, Khan WUD (2018) Alleviation of nickel toxicity and an improvement in zinc bioavailability in sunflower seed with chitosan and biochar application in pH adjusted nickel contaminated soil. Arch Agron Soil Sci 64(8):1053–1067

Ullah A, Heng S, Munis MFH, Fahad S, Yang X (2015) Phytoremediation of heavy metals assisted by plant growth promoting (PGP) bacteria: a review. Environ Exp Bot 117:28–40

Varghese SM, Prabha ML (2014) Biochemical characterization of vermiwash and its effect on growth of *Capsicum frutescens*. Malaya J Biosci 1(2):86–91

Verbruggen N, Hermans C, Schat H (2009) Molecular mechanisms of metal hyperaccumulation in plants. New Phytol 181(4):759–776

Visioli G, D'Egidio S, Vamerali T, Mattarozzi M, Sanangelantoni AM (2014) Culturable endophytic bacteria enhance Ni translocation in the hyperaccumulator Noccaea caerulescens. Chemosphere 117:538–544

Von Uexküll HR, Mutert E (1995) Global extent, development and economic impact of acid soils. Plant Soil 171(1):1–15

Wang K, Zhang J, Zhu Z, Huang H, Li T, He Z et al (2012) Pig manure vermicompost (PMVC) can improve phytoremediation of Cd and PAHs co-contaminated soil by *Sedum alfredii*. J Soils Sediments 12(7):1089–1099

Wang K, Liu Y, Song Z, Wang D, Qiu W (2019a) Chelator complexes enhanced *Amaranthus hypochondriacus* L phytoremediation efficiency in Cd-contaminated soils. Chemosphere 237:124480

Wang Q, Ma L, Zhou Q, Chen B, Zhang X, Wu Y et al (2019b) Inoculation of plant growth promoting bacteria from hyperaccumulator facilitated non-host root development and provided promising agents for elevated phytoremediation efficiency. Chemosphere 234:769–776

Weyens N, van der Lelie D, Taghavi S, Newman L, Vangronsveld J (2009) Exploiting plant–microbe partnerships to improve biomass production and remediation. Trends Biotechnol 27 (10):591–598

Yadav SK (2010) Heavy metals toxicity in plants: an overview on the role of glutathione and phytochelatins in heavy metal stress tolerance of plants. S Afr J Bot 76(2):167–179

Yahaghi Z, Shirvani M, Nourbakhsh F, De La Pena TC, Pueyo JJ, Talebi M (2018) Isolation and characterization of Pb-solubilizing bacteria and their effects on Pb uptake by *Brassica juncea*: implications for microbe-assisted phytoremediation. J Microbiol Biotechnol 28(7):1156–1167

Yamaguchi N, Ishikawa S, Abe T, Baba K, Arao T, Terada Y (2012) Role of the node in controlling traffic of cadmium, zinc, and manganese in rice. J Exp Bot 63(7):2729–2737

Yang W, Wang Y, Liu D, Hussain B, Ding Z, Zhao F, Yang X (2019) Interactions between cadmium and zinc in uptake, accumulation and bioavailability for Salix integra with respect to phytoremediation. Int J Phytoremediation 22(6):628–637

Yasin M, El-Mehdawi AF, Pilon-Smits EA, Faisal M (2015) Selenium-fortified wheat: potential of microbes for biofortification of selenium and other essential nutrients. Int J Phytoremediation 17 (8):777–786

Zarei M, Hempel S, Wubet T, Schäfer T, Savaghebi G, Jouzani GS et al (2010) Molecular diversity of arbuscular mycorrhizal fungi in relation to soil chemical properties and heavy metal contamination. Environ Pollut 158(8):2757–2765

Zhao Y, Zhang L, Zhao C, Hu C, Li Y, Zhao J et al (2015) Metabolic responses of rice leaves and seeds under transgenic backcross breeding and pesticide stress by pseudotargeted metabolomics. Metabolomics 11(6):1802–1814

Chapter 15
Waste Management Practices and Their Impact on Earthworms

Harsimran Kaur, Puttaganti Vijaya, and Suman Sharma

Abstract **Earthworms in vermicomposting**: Earthworms play an important role in the fragmentation process and provide substrate so as to increase the surface area for growing microorganisms. The vermicomposting process involves bio-oxidation and stabilization of organic wastes through the interaction between earthworms and microorganisms. Epigeic earthworms, that is, *Eisenia fetida*, are small in size, are very active, and have regenerative capacity within the short period of time. They are richly pigmented worms and are phytophagous. The role of earthworm in waste management has been well documented, and this technology has been implemented to process the waste to produce an efficient bio-product vermicompost. Earthworms have an ability to decompose various toxic wastes, such as agricultural wastes, food wastes, weed biomass, leaf litters, and municipal wastes. *Parthenium hysterophorus* is one of the toxic weeds which is dangerous, allergic, poisonous, pernicious, and aggressive, and it is known to pose a serious threat to livestock and human beings. During the process of vermicomposting of *Parthenium hysterophorus*-rich substrate, the worm's fundamental health and physiology are known to be affected, as it causes oxidative damage and alterations in its intestinal tissue. Therefore, appropriate concentration of *Parthenium hysterophorus* is to be mixed with cow dung to convert it into useful manure. This study illustrates the role of earthworm in converting *Parthenium hysterophorus*-amended cow dung into a useful manure and the deleterious consequences of parthenium weed on the health of *Eisenia fetida*.

Keywords Vermicomposting · Organic waste · *Eisenia fetida* · *Parthenium hysterophorus* · Oxidative damage

---

H. Kaur (✉)
Department of Agriculture, Sri Guru Granth Sahib World University, Fatehgarh Sahib, Punjab, India

P. Vijaya
Department of Zoology, Bangalore University, Bangalore, Karnataka, India

S. Sharma
Department of Zoology and Environmental Sciences, Punjabi University, Patiala, Punjab, India

S. A. Bhat et al. (eds.), *Earthworm Assisted Remediation of Effluents and Wastes*,
https://doi.org/10.1007/978-981-15-4522-1_15

15.1 Introduction

Solid-waste management is a vital issue in many cities of developing countries including India. The present day environmental deterioration is mainly due to unscientific disposal of biowastes, which mainly comprises agricultural wastes, food wastes, both aquatic and terrestrial weed biomass, leaf litters, municipal wastes, industrial wastes, and other organic wastes (Dhande et al. 2005). Generation of huge amounts of organic waste around the world causes serious environmental and disposal problems, such as contamination of ground water as well as soil and offensive odors (Edwards and Bater 1992). Large-scale industrialization, urbanization, and population expansion have affected the strong relationship between man and nature (Elvira et al. 1995).

Weeds are one of the undesirable plants that compete with other plants for various natural resources, such as water, nutrients, and sunlight. There are many reasons for the establishment of weed in any ecosystem, which includes adaptive nature, high growth rate, and high reproductive potential, and the most significant reason is allelopathy (Kohil and Rani 1994). In the past years, many weed species have been introduced in the country; many of them are the major source of contamination in grains and food. Some of these weeds have found the suitable conditions due to which they spread far with many serious consequences (Mooney et al. 2005).

Parthenium hysterophorus L. family Asteraceae is an exotic species which originated in Mexico and Argentina and then accidently introduced in India in 1955 through the imported food grains. It is commonly known as "congress grass," which becomes neutralized and is spreading at an alarming rate all over India due to its huge biotic potential and ability to enormously proliferate, remarkable power of generation, and capability to overcome environmental resistance (Sivakumar et al. 2009). The plant contains various important chemical constituents, viz. histamine, glucosides, saponin, triterpenes, toxins, etc., and also has the potential to decrease the crop productivity, biodiversity depletion, and fodder scarcity. Due to its allelopathic effect, it inhibits the germination and growth of neighboring plant species by releasing various allelochemicals, such as water-soluble phenolics and sesquiterpene lactones, including coronopilin and parthenin. At present, it is one of the most troublesome and obnoxious weeds of forest, wasteland, pasture, and agricultural lands in India (Bakthavathsalam and Geetha 2004). Several attempts have been made for its prevention, eradication, and control, but till date, there is no success (Kavita and Nagendra 2000).

Different physical, chemical, and microbiological methods are currently in practice for the disposal of organic waste, but these methods are time consuming and involve very high costs. Therefore, vermitechnology is a cost-effective alternative method of shorter duration, which is particularly suitable to the Indian conditions (Logsdson 1994). This method is about a century old, but it is now being received worldwide attention with several ecological objectives, including waste management, regeneration, and soil detoxification. Recycling of waste material through the method of vermicomposting minimizes the problem of nonutilization of agro waste (Kale 2000).

Earthworm is a ubiquitous animal species that represents the maximum proportion of soil faunal biomass. They are known to maintain the ecological functions of soil (Andriuzzi et al. 2016) and also enhance the soil fertility and nutrient cycling (Hoang et al. 2016). Earthworms are universally designated as an indicator species of ecosystem during several ecotoxicological studies (Xu et al. 2010). Earthworms can be used for "early warning" or "diagnosis" of soil pollution (Velki and Hackenberger 2013). Epigeic earthworm species, *Eisenia fetida* (Savigny 1826), are used in many ecotoxicological studies as they have a high reproductive rate, short life cycle, and fine sensitivity to variety of toxicants. According to Nahmani et al. (2007), *Eisenia fetida* is the best model organism for analyzing the soil toxicity and is also used in various standardized toxicity tests (ISO 1998).

Earthworms have been used to analyze biochemical responses of pesticides, heavy metals, polychlorinated biphenyls, and polycyclic hydrocarbons (Łaszczyca et al. 2004) and have an ability to accumulate adequate quantities of inorganic and organic pollutants in their bodies (Hobbelen et al. 2006). Different earthworm species are also used to analyze the oxidative stress caused by numerous toxic pollutants (Chen et al. 2011) and to estimate the biological biomarkers such as lipid peroxidation, activity of antioxidant enzymes, and DNA damage (Liu et al. 2012). An imbalance between generation of reactive oxygen species (ROS) and their neutralization results in the elevated levels of reactive oxygen species, which results to oxidative stress. The toxic levels of ROS can cause serious damage to macromolecules (Kelly et al. 1998). These reactive oxygen species are responsible in normal physiology of every cell. The reactive oxygen species react with the macromolecules, and the cells have a capability to counteract them through the production of antioxidant molecules such as glutathione or through the induction of various antioxidant enzymes, including catalase, superoxide dismutase, glutathione reductase, and glutathione peroxidase (Saint-Denis et al. 1998). The antioxidant enzymes protect the cells from oxidative damage caused by ROS and are considered the important biomarkers for evaluating the environmental impact of the toxic substances (Ferreira-Cravo et al. 2009).

The main objective of this chapter is to examine the capability of earthworm, *Eisenia fetida*, to convert *Parthenium hysterophorus*-mixed cow dung into a useful manure by analyzing various physicochemical parameters of final product, to analyze oxidative damage caused by parthenium weed in earthworms, and furthermore, to study the histomorphological alterations induced by *P. hysterophorus* on the intestine of *Eisenia fetida*.

15.2 Effects of *Parthenium hysterophorus*

15.2.1 Physicochemical Parameters

The physicochemical parameters confirm the rapid decomposition and mineralization of the substrate by earthworms (Bhat et al. 2017a). *Parthenium hysterophorus* disturbed the pH and total organic carbon (TOC) and showed gradual variations in the electrical conductivity (EC), nitrogen, phosphorus, and potassium (NPK) content of the compost that pose impact on the health of earthworm. During vermicomposting, earthworms were found to facilitate maturation of the substrate material as a result vermicompost was found to be converted into a much more homogenous product (Atiyeh et al. 2000), probably due to their interaction with the decomposers.

The pH of organic matter has a significant impact on the potency of vermicomposting process, and it plays an important role in survival and growth of earthworms. The low pH may be due to the production of CO_2 and other organic acids, which were produced by the combined action of earthworms and microbial decomposition (Garg and Gupta 2011). The change in pH was probably related to high mineral nitrogen content of the substrate and alterations in the ammonium-nitrate equilibrium and may be due to the production of humic and fulvic acids during vermicomposting process (Dominguez and Edwards 2004). The lower pH during vermicomposting may be due to conversion of nitrogen and phosphorus into nitrates and orthophosphates as investigated by other researchers (Sharma and Garg 2018). The main characteristic of good-quality compost was the alteration in the pH from alkaline to slightly acidic (Nath et al. 2009). Das et al. (2016) also reported a shift in pH from alkaline to neutral during vermiremediation of organic wastes. Earthworms do maintain the neutral range of pH by intestinal calcium and ammonia secretion that neutralizes phenolic and carboxylic groups of humic acids produced during the vermicomposting process, so the joint effect of these two oppositely charged groups actually modulates the pH of substrate material, which further results in a shift of pH toward acidity or neutrality (Pramanik et al. 2007). The low pH was also observed by Gunadi and Edwards (2003) on vermicomposting of fruit and vegetable wastes.

Electrical conductivity (EC) indicates the salinity of the material, and it is a good indicator of the utility and applicability of a substrate used for agricultural purposes. *P. hysterophorus* showed increment in electrical conductivity, which was due to the elevation in soluble salt level present in organic substances or through the mineralization action of microbes present in earthworm's gut (Karmegam and Daniel 2009). The decomposition of organic matter that releases mineral ions, such as potassium, ammonium, and phosphate, may affect the electrical conductivity (Huang et al. 2004). Garg et al. (2006) stressed that the change in EC was due to the generation of free available minerals and ions during ingestion as well as excretion of the earthworms. It was also documented that these earthworms produce organic-mineral compound after the digestion of organic matter as a food product, and these minerals get accumulated in the final substrate (Deka et al. 2011).

The carbon mineralization or total organic carbon (TOC) is an important indicator of waste stabilization (Suthar 2012). Various kinds of biodegradable wastes have lower TOC, but there is 20–45% reduction of TOC in the form of CO_2 during vermiconversion of industrial or municipal wastes (Kaviraj and Sharma 2003) because earthworms promote biochemical degradation and their activity led to colonization of communities of decomposers in the waste. Dominguez and Edwards (2004) asserted that the material ingested by earthworms was fragmented and homogenized by the muscular action of their foregut that added mucus and enzymes to the material, which was ingested by increasing the surface area for the action of microbes, while the microorganisms brought about the degradation of wastes in the worm's gut by providing some extra cellular enzymes. During this process, the substrate material was modified by the earthworms by losing carbon through microbial respiration in the form of carbon dioxide (Aira et al. 2007). Suthar (2010) described that the inoculated earthworms in substrate material digest the carbohydrates and some other polysaccharides, which causes reduction in carbon which may be due to conversion of organic carbon to worm biomass through assimilation process that further led to reduction of carbon budget of substrate material during vermicomposting of organic waste.

Earthworms play significant role in improving and enhancing the nitrogen content (N) of the substrate (Muthukumaravel et al. 2008). The elevation in nitrogen content is due to the action of N-fixing bacteria and mineralization of C-rich materials as well as low pH that helps in nitrogen retention, as at higher pH it may be lost as ammonia (Plaza et al. 2007). The nitrogen in vermicompost is dependent on two factors:

1. Nitrogen content present in the initial substrate material
2. Rate of decomposition process

The N content was enhanced due to the addition of nitrogenous substances, mucus, growth-stimulating hormones, and enzymes during digestion and fragmentation of organic matter by the earthworms (Varma et al. 2015). Even the loss of organic carbon content, conversion of ammonium nitrogen into nitrate, and mineralization of the organic matter containing proteins may also play an important role in the increase of nitrogen content in vermicompost (Yadav and Garg 2011b). Earthworms also promote the decomposition of organic materials that accelerates the mineralization process of nitrogen and leads to change in nitrogen profile of the substrate (Benitez et al. 1999).

Phosphorus (P) plays a significant role in plant physiology as it is essentially required for the process of photosynthesis and other metabolic pathways of plant, so it is formed by the P-mineralization process in vermibeds during vermicomposting (Suthar 2012). The small amount of phosphorus is converted to more available forms by the action of gut enzymes of earthworm, that is, alkaline phosphatases and acid phosphatases, which makes it more available to plants during vermiconversion process (Swarnam et al. 2016). Even the actions of phosphorus-solubilizing microorganisms present in the casts of earthworms were responsible for the release of phosphorus (Prakash and Karmegam 2010) along with the mobilization, and mineralization of phosphorus due to combined activity of earthworms and microbes might be responsible for the increase of total available phosphorus (Goswami et al. 2013). Microorganisms also produce acids during the decomposition of organic matter, which converts insoluble phosphorus into soluble forms which lead to increase in total phosphorus content (Suthar 2010). The higher potassium content in vermicompost indicates its potential in agronomy and its suitability for plant growth (Suthar 2012).

The potassium content (K) of the compost was significantly affected by *P. hysterophorus* treatment. The microflora present in the earthworm's gut might be responsible for the variation in potassium content during vermicomposting, and these microorganisms played a significant role in vermicomposting process (Suthar 2007). They further stressed that the various acids produced by the worms during vermicomposting may convert insoluble potassium to soluble potassium by increasing its concentration in the compost.

P. hysterophorus significantly affected the physicochemical properties of the compost, but it enhances the nutrient content when mixed with the organic waste, such as cow dung, and can be used as valuable manure. It has been proved that vermicomposting could be introduced as an effective technology to convert the *P. hysterophorus* into a nutrient-rich product for sustainable agriculture.

15.2.2 Concentration of Heavy Metals

Heavy metals are present in the form of complexes of humic or fulvic acids in the soil. The metal toxicity is not caused by the presence of metals, but it depends on many factors:

- Concentration of metal
- Metal toxicity and its mobility
- The route of uptake mechanism
- The bioavailability for the intake of plants

It is important to analyze the concentration of metals in the final product before being applied to soil due to their high toxicity and risk (Jamali et al. 2009).

The heavy metals were present in both bedding materials and the intestine of earthworms.

(a) **Heavy metal concentration in bedding material**: Heavy metals (cadmium, chromium, nickel, and iron) were in higher concentration in the final vermicompost than in the initial substrate. The increase in the metal content may be due to the degradation of organic matter and mineralization followed by reduction in volume of substrate material that led to the congregation of the material and elevation in heavy metal content (Sharma and Garg 2017) followed by the reduction in volume and mass of initial feed mixture. Many reports also stated that the heavy metal concentrations were enhanced in final mixture, but the levels were below the internationally permissible levels of metals for compost. They further analyzed that the variation in heavy metal concentration in the vermicompost considerably depends on various factors, including environmental conditions, nature of raw feed mixture, species of earthworms used, etc. Heavy metals were reported high in sewage soils, cultivated lands, winery wastes, paper-pulp mill sludge, and even in agricultural wastes because of the carbon loss by mineralization, and the overall biomass reduction due to the breakdown of organic matter during vermiconversion may be the ultimate reason for the enrichment of heavy metals in the composts (Yadav and Garg 2011a). Even the increase in the levels of humic fractions by earthworm mediated biodegradation process can contribute to the heavy metals content, which strongly immobilizes metals by forming stable metal–humus complexes (Kang et al. 2011). During vermicomposting process, there is an improvement in degree of humification that indicates the formation of highly stable aromatic compounds having great potential for sequestration of metals (Romero et al. 2007).

(b) **Heavy metal concentration in the intestine of *Eisenia fetida***: Many species of earthworms are known to accumulate metals in their intestine by synthesizing cysteine-rich metal-binding metallothioneins (Stürzenbaum et al. 2012). Metallothionein is a metal-inducible protein with low molecular weight that regulates the bioavailability-detoxification dynamics of nonessential and essential metals in the gut of earthworms (Maity et al. 2011). The physiological tolerance of heavy metals in many organisms is attributed to induction of metal-chelating proteins known as metallothioneins (Stürzenbaum et al. 1998). Heavy metal content in the intestine of earthworms was increased by *P. hysterophorus* treatment. The heavy metal content accumulated in the tissues of earthworm is directly related to the amount of metals present in organic materials consumed by them. The earthworms inhabiting metal soils of industrial areas contain more metals as their enzymes help to mineralize the organic waste to form simpler or short-chain organic acids, and these simpler acids bind to metals to form stable metal complexes during vermicomposting (Wang et al. 2013). Previous studies revealed two different metal-binding mechanisms in earthworms.

- First, the metals are retained in chloragosomes or insoluble calcium phosphate granules (Morgan et al. 2012), and these insoluble forms of metals cannot affect the normal biochemical processes in the cytoplasm.
- The sulfur-donating ligands of metallothionein chelate these insoluble metals, and the metals are then transported to chloragogenous tissues of the intestine of earthworms, where they get neutralized (Asensio et al. 2007). The chloragogenous tissue is a sheath of modified peritoneal cells that surround the outer wall of the gut (Fischer and Molnar 1993).

Earthworms consume organic matter to achieve sufficient nutrition, and during this process due to enzymatic actions in the gut, metals are liberated in free forms.

These available forms of metals are then further absorbed by the epithelial layer of gut during the waste transition (Suthar 2007). The chloragosomes in earthworms function as the cation-exchange system having a capability to take up and retain the heavy metals (Morgan and Morgan 1998), which is further excreted as chloragocyte fraction (Fischer 1976). Richardson et al. (2015) stated that the bioaccumulation process of heavy metals in earthworms appears to be site and species dependent, but it also depends on body mass and abundance of metals present in the substrate. Tang et al. (2017) revealed that the gut condition of earthworm may decrease the number of ligands or metal-binding sites due to excretion of digestive enzymes (Wallwork 1983) and chemical composition of gut liquid, which may alter the bioavailability of metals in the soil.

P. hysterophorus effected the heavy metal concentrations in both final substrates and the intestine of earthworm. These heavy metals are responsible for the toxic effects on the health of earthworms. The economic utilization of *P. hysterophorus* can be introduced in the vermicomposting technology and also a model for ecological engineering and for the current agricultural needs. These weed amendments up to 25% along with cow dung may help in its eradication and utilization.

15.2.3 Growth and Reproduction of Earthworms

The growth rate, survival, and reproduction capability of earthworms are highly affected by type, quality, and palatability of the food (Gajalakshmi et al. 2005). The rate of productivity of inoculated earthworms is a significant aspect in the waste decomposing system of the vermiomposting process. Population buildup in the form biomass and growth rate is a significant indicator of earthworm productivity in different feed mixtures, as the biomass of earthworms was higher on the final day than on the initial day of the study because the feedstock availability per unit of earthworm body mass in waste substrate also affected the rate of biomass gain in vermicomposting earthworms (Suthar 2011). It has been suggested that during vermicomposting process, the presence of fungi in substrate becomes additional food for the worms, which further contributes to the weight gain of the earthworms (Pramanik and Chung 2011). Suthar et al. (2017) documented that the development and growth of earthworms during vermicomposting process depend on:

1. Environmental conditions and quality
2. Growth-supporting nutrients present in feed mixture

The growth and development of earthworms were reported during vermicomposting of almost all the wastes. Lesser weight gain in earthworm appears to be a valuable indicator of physiological stress, which is related to the degree of intoxication and exposure time (Frampton et al. 2006). Lesser gain in earthworm biomass may also be due to the food deficiency and toxic nature of the waste mixture. Loh et al. (2005) reported that biomass gain was more in cattle dung as compared to goat dung as cattle dung provided more nutrition and suitable environment to *Eisenia fetida*. The essential factor that contributes to attain maximum growth of organism is nutrition. Nutrition resources for earthworms are organic material, live bacteria, algae, fungi, nematodes, and protozoa (Sarojini et al. 2009). Growth rate provides a comparative analysis, and it is an indicator of growth of earthworm in different feed mixtures (Vodounnou et al. 2016).

The maximum growth rate in cow dung treatment indicates that it could supply sufficient nutrients for the earthworms. The decrement in growth rate in the later phase of the experiment in earthworm may be due to the exhaustion of feed mixture, but they also stated that vegetable and food mixed with buffalo dung could be a good feedstock for vermicomposting of organic wastes (Sangwan et al. 2008). Suthar (2008b) reported that the growth of the earthworms depends on microbial communities and nutrient availability in feed mixture. The stocking density of earthworms may affect the growth patterns during vermicomposting (Suthar 2011). Various pollutants may also affect the growth of earthworms, and it can be altered by direct toxic effects on the physiology of earthworms or by variations in the energy budget to prevent such toxic effects (Spurgeon and Hopkin 1996). The energy required for combating the poisonous effects ultimately results in the reduction in energy required for growth and development (Donker et al. 1993). Vodounnou et al. (2016) reported that the variation in growth performance in different feed mixtures may be due to the quality of feedstock and the percentage of mixing substrate. The higher amount of crop residue in feed mixture also affects the growth and reproduction due to variation in nutrient content and palatability or may be due to some growth retarding substances (Suthar 2008a).

Cocoon is small, translucent, spherical capsule in which earthworms lay their eggs (Bhat et al. 2016b). There was lesser number of cocoons due to *P. hysterophorus* treatment. The onset as well as rate of cocoon formation depends on quality of feed mixtures (Bhat et al. 2015) as food source plays a significant role in cocoon formation. Rate of cocoon production depends on the biochemical differences between the feed qualities, which plays an important role in determining the time to reach sexual maturity (Edwards et al. 1998). During vermicomposting process, shortage of food and conversion of whole food into vermicompost are major factors responsible for low cocoon production (Suthar 2012). The variations in cocoon production of worms in different waste mixtures are related to the quality and palatability of food source (Bhat et al. 2017b). The feed which has sufficient amount of organic matter and nonassimilated carbohydrates promotes better growth and reproduction in earthworms (Edwards 1988). Suthar (2006) suggested that during vermicomposting, decomposition activities and microbial biomass are also important in determining the biomass of earthworm and cocoon production.

The hatchability was reduced by *P. hysterophorus* treatment. It can be correlated with the low number of cocoons production. During vermicomposting process, the hatchability directly depends on the rate of cocoon formation (Singh et al. 2016). Number of hatchlings may also be considered as an indicator of evaluating the effect of heavy metals on earthworms (Bhat et al. 2016a). Reinecke and Reinecke (1997) reported that the exposure to heavy metals leads to the ultrastructural damage in the sperms of earthworms, which resulted in the lower fertility and viability of cocoons.

The vermicompost is rich in nutrients, odor free, more mature, and stabilized than the initial waste mixture. Though P. hysterophorus enhanced the average body weight of earthworms, it reduced the growth rate of earthworms, cocoon formation, and number of hatchings during vermicomposting. Thus, it affected the growth and reproduction efficacy of earthworm. If P. hysterophorus was used in higher concentrations, it may inhibit the vermicomposting efficiency of earthworms.

15.2.4 Biochemical Parameters

Biological markers are regarded as fast, diagnostic, and prognostic early warning of toxicity. These markers are used to detect and assess the impact of wide range of contaminants present in the environment (Lourenço et al. 2011a). Living organisms exposed to contaminated environments showed significant deleterious effects (Ajani 2014). Stress caused by environmental conditions activates the endogenous production of reactive oxygen species (ROS), such as superoxide radicals (O^{2-}), hydrogen peroxide (H_2O_2), and hydroxyl radical (OH^-), most of which are side products of tissue respiration (Valavanidis et al. 2009). Due to their unstable nature, ROS affects biomolecules by interacting aggressively with polysaturated fatty acids, proteins, and DNA (Damien et al. 2004), and this further results in oxidative stress. Oxidative stress is an imbalance between the antioxidant defense systems and production of free radicals. The potential of oxidative damage is mainly affected by the level of target molecules, that is, polysaturated fatty acids (PUFA), which are easily oxidized by reactive oxygen species to lipid peroxides (Di Giulio et al. 1995). The antioxidant system constitutes various enzymes, such as superoxide dismutase (SOD), catalase (CAT), and glutathione peroxidase (GPx), which counteract this oxidative stress.

Malondialdehyde (MDA) is the oxidized product of lipid membrane, and it could be used as a sensitive biomarker of injury in cells (Xue et al. 2009). It is an important indicator of ROS peroxidation (Stohs and Bagchi 1995). The degree of lipid peroxidation may depend on the composition of fatty acids and levels of antioxidant enzymes in the organism (Kiran and Aruna 2010). The MDA level increases with the effect of toxicity. The elevated lipid peroxide level suggests hydroxyl radical activity and indicates oxidative damage (Kenneth 2000); as a result, it enhances the MDA level that is due to the formation of ROS. Luo et al. (2005) asserted that the peroxidation of membrane lipids is one of the major damaging effects of reactive oxygen species and their products in cells, which can be indicated by MDA detection.

Superoxide dismutase (SOD) is the first line of defense that catalyzes dismutation of superoxide anion into O_2 and H_2O_2. SOD shows increment due to the treatment with toxicity, which indicates that the SOD is biosynthesized to react with O^{2-} to avoid the oxidant damage of cells (Singh et al. 2006). The enhanced activity of SOD resulted in increased H_2O_2. Thus, the increment is due to increased activity of enzymes to remove H_2O_2 to reduce oxidative damage in the exposed earthworms (Zhang et al. 2015). Some toxic substances decrease the SOD activity. It may be due to inactivation of superoxide dismutase by singlet oxygen, hydrogen peroxide, and peroxyl radicals or elimination of highly reactive superoxide (Sun et al. 2007). The decline in SOD activity of earthworms might also be due to adaptation of the organisms to its environment (Sandrini et al. 2008).

Catalase (CAT) is a peroxisomal hydroperoxidase found in peroxisomes, cytosol, and mitochondria that converts H_2O_2 to water and oxygen. Initially, this antioxidant may show enhancement, but with the severity of the toxicity, CAT level may show decrement. Lin et al. (2010) suggested that the earthworm has a capacity to activate the antioxidant system to tolerate oxidative stress. The increasing CAT activity might be due to H_2O_2 produced from SOD mechanisms (Li et al. 2008), and to maintain the level of H_2O_2, the catalase activity gets enhanced (Liu et al. 2011). El-Shenawy et al. (2012) documented that the increased CAT activity was due to upregulation of catalase activity, which could be an adaptive mechanism of earthworms to prevent the ROS accumulation. The reduction in CAT activity is due to inhibition of enzyme by high cellular stress or due to high levels of ROS (Markad et al. 2012), or it might be due to inactivation by peroxyl radicals, singlet oxygen, and superoxide radicals (Kono and Fridovich 1982). When the ROS generation exceeds the ROS scavenging capacity of SOD and CAT, they become CAT inhibitors (Schreck et al. 2008). They further suggested that the decreased CAT activity may also be caused by damage of the antioxidant defense system, and the defensive effect of antioxidant enzymes would be suppressed by an increase in stress condition by toxicant.

Glutathione peroxidase (GPx) removes hydrogen peroxide by using reduced glutathione as a hydrogen donor (Markad et al. 2012). GSH is used as a hydrogen donor by GPx to eliminate H_2O_2. It also helps to convert the organic hydroperoxides into alcohols. The product of this reaction is GSSH, which is further reduced by GR in the presence of NADPH to restore the GSH level in the cell (Halliwell and Gutteridge 1999). The variation in the GPx activity may be regarded as a sign of hormesis (Calabrese and Baldwin 1998). The hormetic mechanism was interpreted as the result of instantaneous balance between the formation and degradation of specific protein including other processes also such as synthesis, regeneration, activation vs. inhibition, inactivation, and degradation, and this mechanism may signify temporary over adaptation. Maity et al. (2008) suggested that the reduction in concentration of glutathione and antioxidant enzymes may be due to their capability to remove the metal in biologically inactive form by the induction of metal-binding proteins (Homa et al. 2005).

The reduction of antioxidant enzymes SOD, CAT, and GPx may be due to the inactivation of SOD by singlet oxygen, peroxyl radicals, and hydrogen peroxide or removal of highly reactive superoxide (Sandrini et al. 2008). The biphasic response of antioxidant enzymes in treated earthworms was also reported by Łaszczyca et al. (2004). They further suggested that the hormetic-like phenomenon plays a significant role in assessing the degree of environment stress. Overall, the reduction of CAT and GPx activities in earthworms indicated severe damage to their antioxidant defense system (Wen et al. 2017). The variations in the activities of antioxidant enzymes may be due to the phenomenon that initially worms had an ability to

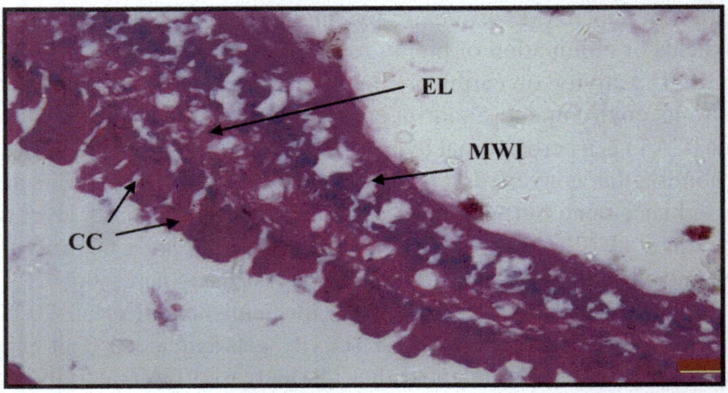

Fig. 15.1 Photomicrograph of normal intestine of earthworm showing chloragogenous cells (CC), epithelial layer (EL), and muscular wall of intestine (MWI). ×400

combat oxidative stress. The increased exposure time to toxicants may result in the decline of enzyme activities in worms which may be due to exceedance of earthworm's normal antioxidative capability, leading to a need for higher antioxidative enzyme production (Yang et al. 2012). Ukpabi et al. (2013) determined that the heavy metal toxicity also promoted the formation of reactive oxygen species, which may cause severe damage and injury to the biological system.

During oxidative stress, an organism needs sufficient amount of energy which is dispensed from reserve materials, such as glycogen, proteins, and lipids. In mild stress conditions, glycogen is used as an energy source, while in high stress, the energy stored in proteins and lipids may also be used by the organism (Mark 1961).

Histological changes: Histopathology signifies the damaging effects in organisms resulting from the exposure to toxic substances. Histology is a significant tool for analyzing the effect of organic wastes, agricultural pesticides, and industrial pollutants on the organism at the tissue level by providing information regarding the disorganization and damage of tissues (Kavitha and Anandhan 2018). These changes were reported in the intestine of earthworm, when exposed to any pollutant, organic and inorganic waste, as well as any other stress conditions.

Normal earthworms: The normal intestine comprises four different layers; the outermost layer known as visceral peritoneum is the covering layer of intestine. This layer consists of chloragogen cells (CC). These cells play a vital role in metabolic functions and are comparable to liver in vertebrates. These cells play an important role in excretion and are prime center for synthesis. These are the storage house for glycogen and fats. Following chloragogen cells (CC), two muscle layers are present, that is, longitudinal and circular muscle fibers. The fourth layer is the epithelial layer (EL), which is glandular and ciliated to form villi (Fig. 15.1).

***P. hysterophorus* treatment affects the earthworms**: *P. hysterophorus* showed deleterious effects on the health of earthworms, which was further confirmed with the antioxidant profile and heavy metal accumulation. *P. hysterophorus* treatment

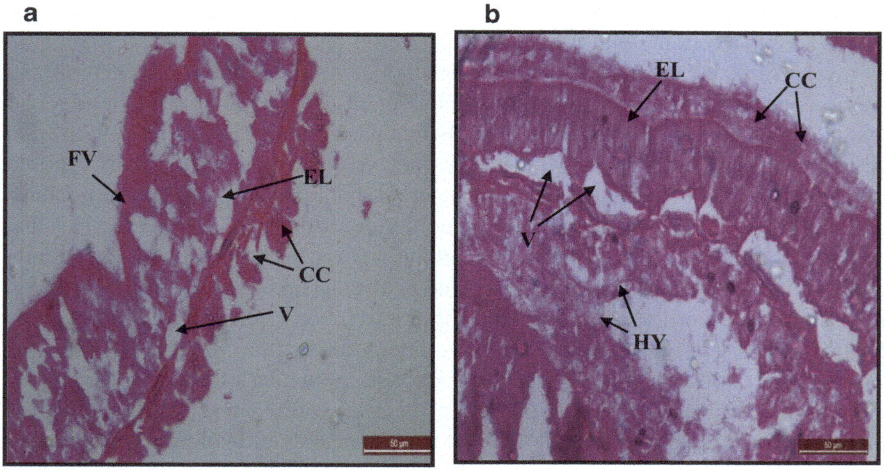

Fig. 15.2 (**a**, **b**) Photomicrographs of *P. hysterophorus* treated intestine of earthworm showing disintegration in chloragogenous cells (CC), discontinuous epithelial layer (EL) along with vacuolated (V) tissue showing hyperemia (HY). ×400

showed loss of structural integrity of intestinal tissue with intermingling of different layers. Alterations were observed in the form of Disintegration of chloragogenous cells (CC) along with mild vacuolation (V), pyknotic nuclei (PN), binucleated cell (BC), giant cell (GC) formation, and hyperemia (HY) in comparison to normal intestine (Fig. 15.2a, b). The toxicity also resulted in shedding of chloragogenous cells (CC), deformations in epithelial lining, and damage in the structure of villi and intervillous spaces along with fused villi (FV) of the intestine. The intensity of damage was enhanced with the increase in the time of exposure of toxicant.

Imbalance in the oxidant/antioxidant equilibrium results in oxidative stress, which is known to be a component of cellular tissue damage in a wide spectrum of human diseases (Klaunig and Kamendulis 2004). Excessive oxidative stress may result in various pathological alterations such as apoptotic and necrotic cell death that has been considered as a passive process (Samali et al. 1999). Earthworms are known to accumulate pesticides, organic chemicals, and heavy metals. Therefore, they are considered as model organisms for analyzing the significant impact of contamination in the environment (Callahan 1984). The histopathological alterations in earthworms have been considered as useful marker of toxicity (Lourenço et al. 2011b).

The chloragogenous tissue is the sheath of modified peritoneal cells that are located in the gastrointestinal canal. This tissue separates the absorptive epithelium from the coelom, and it is the main site for metal accumulation (Giovanetti et al. 2010). The chloragogenous cells play a vital role in eliminating excretory materials from the blood and also help in transporting nutrients from tissue to organs (Edwards and Lofty 1972). The elimination of the contaminants may be done by the extrusion of whole chloragocytes, which helps earthworms to tolerate high metal concentration at least to a certain point (Langdon et al. 2001). The histopathological changes in earthworm tissues are the major characteristics of heavy metal contamination and were reported by various toxicities. Babić et al. (2016) showed blunting and feathering of intestinal epithelium and disorganization of chloragocytes with mild granulation of chloragocytes followed with severe necrosis in earthworms exposed to sewage sludge. Popović et al. (2015) reported that industrial wastewater contains abundance of inorganic compounds, xenobiotics, and organic contaminants that produce significant histopathological alterations in earthworms. Bangarusamy et al. (2014) reported various histopathological changes such as villous atrophy, necrosis, and enteritis with cellular infiltration in the intestine of earthworms. Esaivani et al. (2015) observed extensive alterations in the intestinal tract of the earthworms exposed to inorganic fertilizer and pesticides. The chloragogenous tissue showed clear changes as cells lose their structural integrity. The epithelial lining showed fusion of intestinal villi that lead to the reduction of absorption area of nutrients from food. The chemical fertilizer-exposed earthworms showed disintegration of nuclei and enlargement of epithelial lining.

P. hysterophorus showed detrimental effects on the intestine of earthworm. The earthworm's growth, cocoon production, and antioxidant enzymes were in low level with *P. hysterophorus* treatment. During the management and risk assessment of parthenium, it was reported that this weed induces oxidative stress, and due to this, the earthworms adapt the morphometric characteristics of tissue including the epidermis as well as intestine of earthworm.

15.3 Conclusion

Various environmental wastes showed bioaccumulation of heavy metals in the tissues of earthworms, induced oxidative stress, and caused alterations in histomorphology of intestine as well as in antioxidant status of the organism along with the growth and fecundity of earthworms. Still the wonderworm *Eisenia fetida* was able to convert various toxic waste, that is, *P. hysterophorus*, into a useful manure. The appropriate concentration of *P. hysterophorus* in cow dung is an essential factor for the survival of earthworms without causing any harm so that it should be converted into effective manure. Thus, these organisms can be used on large scale to convert various environmental wastes into useful product. But some measures and precautions should be taken to protect these organisms from the ill effects of toxic environmental pollutants. It is also recommended that field trials should be conducted on different crops before using *P. hysterophorus*-mixed vermicompost in agricultural fields.

References

Aira M, Monroy F, Dominguez J (2007) Earthworms strongly modify microbial biomass and activity triggering enzymatic activities during vermicomposting independently of the application rates of pig slurry. Sci Total Environ 385:252–261

Ajani GE (2014) Oxidative stress in fish living in coastal water polluted with sawdust and wood waste along Lagos Lagoon, Nigeria. Researcher 6(3):1–5

Andriuzzi WS, Ngo P, Geisen S, Keith AM, Dumack K, Bolger T, Bonkowski M, Brussaard L, Faber JH, Chabbi A, Rumpel C, Schmidt O (2016) Organic matter composition and the protist and nematode communities around anecic earthworm burrows. Biol Fertil Soils 52:91–100

Asensio V, Kille P, Morgan AJ, Soto M, Marigomez I (2007) Metallothionein expression and Neutral Red uptake as biomarkers of metal exposure and effect in *Eisenia fetida* and *Lumbricus terrestris* exposed to Cd. Eur J Soil Biol 43:S233–S238

Atiyeh RM, Dominguez J, Subler S, Edwards CA (2000) Changes in biochemical properties of cow manure during processing by earthworms (*Eisenia Andrei Bouché*) and the effects on seedling growth. Pedobiologia 44:709–724

Babić S, Barišić J, Malev O, Klobučar G, Topić Popović N, Strunjak-Perović I, Krasnići N, Čož-Rakovac R, Sauerborn Klobučar R (2016) Sewage sludge toxicity assessment using earthworm *Eisenia fetida*: can biochemical and histopathological analysis provide fast and accurate insight. Environ Sci Pollut Res 23:12150–12163

Bakthavathsalam R, Geetha T (2004) Reproductive influence of the earthworm, *Lampito mauritii* cultured in different media of paddy chaff and weed plants materials. Environ Ecol 22 (3):574–578

Bangarusamy V, Karpagam S, Martin P (2014) Toxicity and histopathological effect of different organic waste on the earthworm (*Eudrillus eugeniae* and *Eisenia fetida*) under laboratory conditions. Int J Ethnomed Pharmacol Res 2(1):18–22

Benitez E, Nogales R, Elvira C, Masciandaro G, Ceccanti B (1999) Enzyme activities as indicators of the stabilization of sewage sludge composting with *Eisenia foetida*. Bioresour Technol 67:297–303

Bhat SA, Singh B, Vig AP (2015) Potential utilization of bagasse as feed material for earthworm *Eisenia fetida* and production of vermicompost. SpringerPlus 4(11):1–9

Bhat SA, Singh J, Vig AP (2016a) Recycling of sugar industrial wastes through earthworm technology. Int Lett Nat Sci 55:35–43

Bhat SA, Singh J, Vig AP (2016b) Effect on growth of earthworm and chemical parameters during vermicomposting of pressmud sludge mixed with cattle dung mixture. Proc Environ Sci 35:425–434

Bhat S, Singh J, Vig AP (2017a) Instrumental characterization of organic wastes for evaluation of vermicompost maturity. J Anal Sci Technol 8(2):9–12

Bhat SA, Singh J, Vig AP (2017b) Amelioration and degradation of pressmud and bagasse wastes using vermitechnology. Bioresour Technol 243:1097–1104

Calabrese EJ, Baldwin LA (1998) Hormesis as a biological hypothesis. Environ Health Perspect 106(1):357–362

Callahan CA (1984) Earthworms as ecotoxicological assessment tools. United States Environmental Protection Agency, EPA-600/D-48-272

Chen C, Zhou Q, Liu S, Xiu Z (2011) Acute toxicity, biochemical and gene expression responses of the earthworm *Eisenia fetida* exposed to polycyclic musks. Chemosphere 83:1147–1154

Damien C, Chantal VH, Pirouz S, Zerimech FH, Laurence J, Jean MH (2004) Cellular impact of metal trace elements in terricolous lichen *Diploschistes muscorum* (Scop.) R. Sant.—identification of oxidative stress biomarkers. Water Air Soil Pollut 152:55–69

Das D, Bhattacharyya P, Ghosh BC, Banika P (2016) Bioconversion and biodynamics of *Eisenia foetida* in different organic wastes through microbially enriched vermiconversion technologies. Ecol Eng 86:154–161

Deka H, Deka S, Baruah CK, Das J, Hoque S, Sarma H, Sarma NS (2011) Vermicomposting potentiality of *Perionyx excavatus* for recycling of waste biomass of *java citronella*—an aromatic oil yielding plant. Bioresour Technol 102:11212–11217

Dhande AD, Ingle ST, Attarde SB, Wagh ND (2005) Eco friendly approach of urban solid waste management—a case study of Jalgaon city. Maharashtra J Environ Biol 26(4):747–752

Di Giulio RT, Benson WH, Sanders BM, Van Veld PA (1995) Biochemical mechanisms: metabolism, adaptation and toxicity. In: Rand G (ed) Fundamentals of aquatic toxicology, effects, environmental fate and risk assessment, 1st edn. Taylor & Francis, London, pp 523–561

Dominguez J, Edwards CA (2004) Vermicomposting organic wastes: a review. In: Hanna SHS, Mikhail WZA (eds) Soil zoology for sustainable development in the 21st century. Palm Press, Cairo, pp 369–395

Donker MH, Zonneveld C, Van Straalen NM (1993) Early reproduction and reproductive allocation in metal adapted populations of the terrestrial isopod *Porcellio scaber*. Oecologia 96:316–323

Edwards CA (1988) Breakdown of animal, vegetable and industrial organic wastes by earthworms. In: Edwards CA, Neuhauser EP (eds) Earthworms in waste and environmental management. SPB Academic Publishing, The Hague, pp 21–31

Edwards CA, Bater JE (1992) The use of earthworm in environmental management. Soil Biol Biochem 24:1683–1689

Edwards CA, Lofty JR (1972) Biology of earthworms. Chapman & Hall, London, p 283

Edwards CA, Dominguez J, Neuhauser EF (1998) Growth and reproduction of *Perionyx excavatus* (Perr.) (Megascolecidae) as factors in organic waste management. Biol Fertil Soils 27:155–161

El-Shenawy NS, Mohammadden A, Al Fahmie ZH (2012) Using the enzymatic and non-enzymatic antioxidant defense system of the land snail *Eobania vermiculata* as biomarkers of terrestrial heavy metal pollution. Ecotoxicol Environ Saf 84:347–354

Elvira C, Dominguez J, Sampedro L, Mato S (1995) Vermicomposting for the paper-pulp industry. Biocycle 36(6):62–63

Esaivani C, Esakiammal B, Vasanthi K, Nithya R, Chairman K (2015) Vermiconversion of leaf wastes (*Ficus benghalensis* and *Ficus racemosa*) by employing *Eudrilus eugeniae*. Int J Adv Res 3(8):798–806

Ferreira-Cravo M, Ventura-Lima J, Sandrini JZ, Amado LL, Geracitano LA, Rebelo M, Bianchini A, Monserrat JM (2009) Antioxidant responses in different body regions of the polychaeta *Laeonereis acuta* (Nereididae) exposed to copper. Ecotoxicol Environ Saf 72:388–393

Fischer E (1976) Chloragogenzelle-Eleocyte-Transformation, induziert mit Benomyl-und Carbofuran-Vergiftung der Lumbriciden (Oligochaeta). Zool Anz 197:225–233

Fischer E, Molnar L (1993) Environmental aspects of the chloragogenous tissue of earthworms. Soil Biol Biochem 24:1723–1727

Frampton GK, Jänsch S, Scott-Fordsmand JJ, Römbke J, Brink PJ (2006) Effects of pesticides on soil invertebrates in laboratory studies: a review and analysis using species sensitivity distributions. Environ Toxicol Chem 25(9):2480–2489

Gajalakshmi S, Ramasamy EV, Abbasi SA (2005) Composting–vermicomposting of leaf litter ensuing from the trees of mango (*Mangifera indica*). Bioresour Technol 96:1057–1061

Garg VK, Gupta R (2011) Optimization of cow dung spiked pre-consumer processing vegetable waste for vermicomposting using *Eisenia fetida*. Ecotoxicol Environ Saf 74:19–24

Garg P, Gupta A, Satya S (2006) Vermicomposting of different waste using *Eisenia foetida*: a comparative study. Bioresour Technol 97:391–395

Giovanetti A, Fesenko S, Cozzella ML, Asencio LD, Sansone U (2010) Bioaccumulation and biological effects in the earthworm *Eisenia fetida* exposed to natural and depleted uranium. J Environ Radioact 101:509–516

Goswami L, Patel AK, Dutta G, Bhattacharyya P, Gogoi N, Bhattacharya SS (2013) Hazard remediation and recycling of tea industry and paper mill bottom ash through vermiconversion. Chemosphere 92(6):708–713

Gunadi B, Edwards CA (2003) The effect of multiple applications of different organic wastes on the growth, fecundity and survival of *Eisenia foetida* (Savigny) (Lumbricidae). Pedobiologia 47 (4):321–330

Halliwell B, Gutteridge JMC (1999) Free radicals in biology and medicine, 3rd edn. Oxford University Press, New York

Hoang DT, Pausch J, Razavi BS, Kuzyakova I, Banfield CC, Kuzyakov Y (2016) Hotspots of microbial activity induced by earthworm burrows, old root channels, and their combination in subsoil. Biol Fertil Soils 52:1105–1119

Hobbelen PH, Koolhaas JE, Van GCA (2006) Bioaccumulation of heavy metals in the earthworms *Lumbricus rubellus* and *Aporrectodea caliginosa* in relation to total and available metal concentrations in field soils. Environ Pollut 144(2):639–646

Homa J, Olchawa E, Stürzenbaum SR, Morgan AJ, Plytycz B (2005) Early phase immunodetection of metallothionein and heat shock proteins in extruded earthworm coelomocytes after dermal exposure to metal ions. Environ Pollut 135:275–280

Huang G, Wong J, Wu Q, Nagar B (2004) Effect of C/N on composting of pig manure with sawdust. Waste Manag 24:805–813

ISO (1998) Soil quality-effects of pollutants on earthworms (*Eisenia fetida*). 11268-2 determination of effects on reproduction. International Organization for Standardization, Geneva

Jamali MK, Kazi TG, Arain MB, Afridi HI, Jalbani N, Kandhro GA, Shah AQ, Baig JA (2009) Heavy metal accumulation in different varieties of wheat (*Triticum aestivum*) grown in soil amended with domestic sewage sludge. J Hazard Mater 164(2–3):1386–1391

Kale RD (2000) An evaluation of the vermitechnology process for the treatment of agro, sugar and food processing wastes. In: Technology appreciation programme on evaluation of biotechnological approaches to waste management held on 26 Oct 2000. Industrial Associationship of IIT, Madras, pp 15–17

Kang J, Zhang Z, Wang JJ (2011) Influence of humic substances on bioavailability of Cu and Zn during sewage sludge composting. Bioresour Technol 102:8022–8026

Karmegam N, Daniel T (2009) Investigating efficiency of *Lampito mauritii* (Kinberg) and *Perionyx ceylanensis* Michaelsen for vermicomposting of different types of organic substrates. Environmentalist 29:287–300

Kaviraj S, Sharma S (2003) Municipal solid waste management through vermicomposting employing exotic and local species of earthworms. Bioresour Technol 90:169–173

Kavita G, Nagendra B (2000) Effect of vermicompost of Parthenium on two cultivars of wheat. Indian J Ecol 27:177–180

Kavitha V, Anandhan R (2018) Sublethal effect of atrazine on the intestine of an Indian earthworm *Lampito mauritti* (Kinberg) (Annelida: Oligochaeta). Int J Zool Appl Biosci 3(4):322–326

Kelly SA, Havrilla CM, Brady TC, Abramo KH, Levin ED (1998) Oxidative stress in toxicology: established mammalian and emerging piscine model systems. Environ Health Perspect 106:375–384

Kenneth AC (2000) Dietary antioxidants during cancer chemotherapy: impact on chemotherapeutic effectiveness and development of side effects. Nutr Cancer 37:1–18

Kiran TR, Aruna HK (2010) Antioxidant enzyme activities and markers of oxidative stress in the life cycle of earthworm, *Eudrilus eugeniae*. Ital J Zool 77(2):144–148

Klaunig JE, Kamendulis LM (2004) The role of oxidative stress in carcinogenesis. Annu Rev Pharmacol Toxicol 44:239–267

Kohil KK, Rani D (1994) *Parthenium hysterophorus*—a review. Res Bull (Sci) Punjab Univ 44:105–149

Kono Y, Fridovich I (1982) Superoxide radical inhibits catalase. J Biol Chem 257:5751–5754

Langdon CJ, Piearce TG, Meharg AA, Semple KT (2001) Resistance to copper toxicity in populations of the earthworms *Lumbricus rubellus* and *Dendrodrilus rubidus* from contaminated mine wastes. Environ Toxicol Chem 20:2336–2341

Łaszczyca P, Augustyniak M, Babczyńska A, Bednarska K, Kafel A, Migula P, Wilczek G, Witas I (2004) Profiles of enzymatic activity in earthworms from zinc lead and cadmium polluted areas near Olkusz (Poland). Environ Int 30(7):901–910

Li YN, Zhou QX, Li FX, Liu XL, Luo Y (2008) Effects of tetrabromobisphenol A as emerging pollutant on wheat (*Triticum aestivum*) at biochemical levels. Chemosphere 74:119–124

Lin D, Zhou Q, Xie X, Liu Y (2010) Potential biochemical and genetic toxicity of triclosan as an emerging pollutant on earthworms (*Eisenia fetida*). Chemosphere 81:1328–1333

Liu X, Lee J, Ji K, Takeda S, Choi K (2012) Potentials and mechanisms of genotoxicity of six pharmaceuticals frequently detected in freshwater environment. Toxicol Lett 211:70–76

Liu S, Zhou Q, Wang Y (2011) Ecotoxicological responses of the earthworm *Eisenia fetida* exposed to soil contaminated with HHCB. Chemosphere 83(8):1080–1086

Logsdson G (1994) Worldwide progress in vermicomposting. Biocycle 35:63–65

Loh TC, Lee YC, Liang JB, Tan D (2005) Vermicomposting of cattle and goat manures by *Eisenia foetida* and their growth and reproduction preference. Bioresour Technol 96:111–114

Lourenço JI, Pereira RO, Silva AC, Morgado JM, Carvalho FP, Oliveira JM, Malta MP, Paiva AA, Mendo SA, Goncalves FJ (2011a) Genotoxic endpoints in the earthworms/sub-lethal assay to evaluate natural soils contaminated by metals and radionuclides. J Hazard Mater 186:788–795

Lourenço J, Silva A, Carvalho F, Oliveira J, Malta M, Mendo S, Gonçalves F, Pereira R (2011b) Histopathological changes in the earthworm *Eisenia andrei* associated with the exposure to metals and radionuclides. Chemosphere 85:1630–1634

Luo Y, Wang XR, Shi HH (2005) Electron paramagnetic resonance investigation of in vivo free radical formation and oxidative stress induced by 2,4-dichlorophenol in the freshwater fish *Carassius auratus*. Environ Toxicol Chem 24(9):2145–2153

Maity S, Roy S, Chaudhury S, Bhattacharya S (2008) Antioxidant responses of the earthworm Lampito mauritii exposed to Pb and Zn contaminated soil. Environ Pollut 151(1):1–7

Maity S, Roy S, Bhattacharya S, Chaudhury S (2011) Metallothionein responses in the earthworm *Lampito mauritii* (Kinberg) following lead and zinc exposure: a promising tool for monitoring metal contamination. Eur J Soil Biol 47:69–71

Mark EM (1961) Report of the secretary's commission on pesticides and their relationships to environmental health. U.S. Dept. of Health Education and Welfare, Washington, DC, p 269

Markad VL, Kodam KM, Ghole VS (2012) Effect of fly ash on biochemical responses and DNA damage in earthworm *Dichogaster curgensis*. J Hazard Mater 215–216:191–198

Mooney HA, Mack RN, McNeely JA, Neville LE, Schei PJ, Waage JK (2005) Invasive Alien species: a new synthesis, SCOPE, vol 63. Island Press, Washington, DC, p 368

Morgan JE, Morgan AJ (1998) The distribution and intracellular compartmentation of metals in the endogeic earthworm *Aporrectodea caliginosa* sampled from an unpolluted and a metal contaminated site. Environ Pollut 99:167–175

Morgan AJ, Mossekmans JFW, Charnock JM, Bennet A, Winters C, O'Reilly M, Fisher P, Andre J, Turner M, Gunning P, Kille P (2012) In situ metal imaging and Zn ligand-speciation in a soil-dwelling sentinel: complementary electron microprobe and synchrotron microbeam X-ray analyses. Environ Sci Technol 47:1073–1081

Muthukumaravel K, Amsath A, Sukumaran M (2008) Vermicomposting vegetable waste using cow dung. e-J Chem 5(4):810–813

Nahmani J, Hodson ME, Black S (2007) A review of studies performed to assess metal uptake by earthworms. Environ Pollut 145:402–424

Nath G, Singh K, Singh DK (2009) Effect of different combinations of animal dung and agro/kitchen wastes on growth and development of earthworm *Eisenia fetida*. Aust J Basic Appl Sci 4:3553–3556

Plaza C, Nogales R, Senesi N, Benitez E, Polo A (2007) Organic matter humification by vermicomposting of cattle manures alone and mixed with two-phase olive pomace. Bioresour Technol 9:5085–5089

Popović N, Strunjak-Perović I, Sauerborn Klobučar R, Barišić J, Babić S, Jadan M, Kepec S, Kazazić SP, Matijatko V, Beer Ljubić B, Car I, Klobučar GV, Čož-Rakovac R (2015) Impact of treated wastewater effluent on fish and toxicity biosensors. Sci Total Environ 538:23–37

Prakash M, Karmegam N (2010) Vermistabilization of press mud using *Perionyx ceylanensis* Mich. Bioresour Technol 101:8464–8468

Pramanik P, Chung YR (2011) Changes in fungal population of fly ash and vinasse mixture during vermicomposting by *Eudrilus eugeniae* and *Eisenia fetida*, documentation of cellulose isozymes in vermicompost. Waste Manag 31:1169–1175

Pramanik P, Ghosh GK, Ghosal PK, Banik P (2007) Changes in organic—C, N, P and K and enzyme activities in vermicompost of biodegradable organic wastes under limiting and microbial inoculants. Bioresour Technol 98:2485–2494

Reinecke AJ, Reinecke SA (1997) The influence of lead and manganese on spermatozoa of *Eisenia foetida* (Oligochaeta). Soil Biol Biochem 29:737–742

Richardson JB, Gorres JH, Jackson BP, Friedland AJ (2015) Trace metals and metalloids in forest soils and exotic earthworms in northern New England, USA. Soil Biol Biochem 85:190–198

Romero E, Plaza C, Senesi N, Nogales R, Polo A (2007) Humic acid-like fractions in raw and vermicomposted winery and distillery wastes. Geoderma 139:397–406

Saint-Denis M, Labrot F, Narbonne JF, Ribera D (1998) Glutathione, glutathione related enzymes, and catalase in the earthworm *Eisenia fetida andrei*. Arch Environ Contam Toxicol 35:602–614

Samali A, Nordgren H, Zhivotovsky B, Peterson E, Orrenius S (1999) A comparative study of apoptosis and necrosis in HepG2 cells: oxidant-induced caspase inactivation leads to necrosis. Biochem Biophys Res Commun 255:6–11

Sandrini JZ, Lima JV, Regoli F, Fattorini D, Notti A, Marins LF, Monserrat JM (2008) Antioxidant responses in the nereidid *Laeonereis acuta* (Annelida, Polychaeta) after cadmium exposure. Ecotoxicol Environ Saf 70:115–120

Sangwan P, Kaushik CP, Garg VK (2008) Feasibility of utilization of horse dung spiked filter cake in vermicomposters using exotic earthworm *Eisenia foetida*. Bioresour Technol 99 (7):2442–2448

Sarojini S, Ananthakrishnasamy S, Manimegala G, Prakash M, Gunasekaran G (2009) Effect of lignite fly ash on the growth and reproduction of earthworm *Eisenia fetida*. e-J Chem 6 (2):511–517

Savigny JC, Cuvier J (1826) Analysis of work of the royal academy of sciences during the year 1821, Physical part. Zoology. Mém Acad Sci Inst France (Hist) 5:176–184

Schreck E, Geret F, Gontier L, Treilhou M (2008) Neurotoxic effect and metabolic responses induced by a mixture of six pesticides on the earthworm *Aporrectodea caliginosa nocturna*. Chemosphere 71:1832–1839

Sharma K, Garg VK (2017) Vermimodificaion of ruminant excreta using *Eisenia fetida*. Environ Sci Pollut Res 24(24):19938–19945

Sharma K, Garg VK (2018) Comparative analysis of vermicompost quality produced from rice straw and paper waste employing earthworm *Eisenia fetida* (Sav.). Bioresour Technol 250:708–715

Singh S, Eapen S, D'Souza SF (2006) Cadmium accumulation and its influence on lipid peroxidation and antioxidative system in an aquatic plant, *Bacopa monnieri* L. Chemosphere 62:233–246

Singh S, Bhat SA, Singh J, Kaur R, Vig AP (2016) Vermistabilization of thermal plant fly ash using *Eisenia fetida*. J Ind Pollut Control 32(2):554–561

Sivakumar S, Kasthuri H, Prabha D, Senthilkumar P, Subbhuraam CV, Song YC (2009) Efficiency of compositing *Parthenium* plant and neem leaves in the presence and absence of an oligochaete, *Eisenia fetida*. Iran J Environ Health Sci Eng 6:201–208

Spurgeon DJ, Hopkin SP (1996) Effects of metal-contaminated soils on the growth, sexual development, and early cocoon production of the earthworm *Eisenia fetida*, with particular reference to zinc. Ecotoxicol Environ Saf 35:86–95

Stohs SJ, Bagchi D (1995) Oxidative mechanisms in the toxicity of metal ions. Free Radic Biol Med 18:321–336

Stürzenbaum SR, Kille P, Morgan AJ (1998) The identification, cloning and characterization of earthworm metallothionein. FEBS Lett 431:437–442

Stürzenbaum SR, Hockner M, Panneerselvam A, Levitt J (2012) Biosynthesis of luminescent quantum dots in an earthworm. Nat Nanotechnol 8:57–60

Sun YY, Yin Y, Zhang JF, Yu HX, Wang XR (2007) Bioaccumulation and ROS generation in liver of freshwater fish, goldfish *Carassius auratus* under HC Orange No. 1 exposure. Environ Toxicol 22:256–263

Suthar S (2006) Potential utilization of guar gum industrial waste in vermicompost production. Bioresour Technol 97:2474–2477

Suthar S (2007) Nutrient changes and biodynamics of epigeic earthworm *Perionyx excavatus* (Perrier) during recycling of some agriculture wastes. Bioresour Technol 98:1608–1614

Suthar S (2008a) Bioconversion of post harvest crop residues and cattle shed manure into value-added products using earthworm *Eudrilus eugeniae* Kinberg. Ecol Eng 32:206–214

Suthar S (2008b) Development of a novel epigeic-anecic-based polyculture vermireactor for efficient treatment of municipal sewage water sludge. Int J Environ Waste Manag 2 (1–2):84–101

Suthar S (2010) Pilot-scale vermireactors for sewage sludge stabilization and metal remediation process: comparison with small-scale vermireactors. Ecol Eng 36:703–712

Suthar S (2011) Production of earthworm *Allolobophora parva* (Eisen) in cattle dung. Ecol Eng 37:644–647

Suthar S (2012) Vermistabilization of wastewater sludge from milk processing industry. Ecol Eng 47:115–119

Suthar S, Pandey B, Gusain R, Gaur RZ, Kumar K (2017) Nutrient changes and biodynamics of *Eisenia fetida* during vermicomposting of water lettuce (Pistia sp.) biomass: a noxious weed of aquatic system. Environ Sci Pollut Res 24:199–207

Swarnam TP, Velmurugan A, Pandey SK, Roy SD (2016) Enhancing nutrient recovery and compost maturity of coconut husk by vermicomposting technology. Bioresour Technol 207:76–84

Tang R, Ding C, Ma Y, Wang J, Zhang T, Wang X (2017) Metabolic responses of *Eisenia fetida* to individual Pb and Cd contamination in two types of soils. Sci Rep 7:1–13

Ukpabi CF, Akubugwo EI, Ibiam UA, Agbafor KN, Ezeigbo RO, Wogu C (2013) Fertilizer application effect on heavy metal composition and biochemical responses of *Earthworm Eisenia fetida* (Savigny) in a sandy farm soil. Am J Biochem 3(3):74–79

Valavanidis A, Vlachogianni T, Fiotakis K (2009) 8-Hydroxy-2′-deoxyguanosine (8-OHdG): a biomarker of oxidative stress and carcinogenesis. J Environ Sci Health C, Environ Carcinog Ecotoxicol 27:120–139

Varma VS, Yadav J, Das S, Kalamdhad AS (2015) Potential of waste carbide sludge addition on earthworm growth and organic matter degradation during vermicomposting of agricultural wastes. Ecol Eng 83:90–95

Velki M, Hackenberger BK (2013) Biomarker responses in earthworm *Eisenia andrei* exposed to pirimiphos-methyl and deltamethrin using different toxicity tests. Chemosphere 90:1216–1226

Vodounnou DSJV, Kpogue DNS, Tossavi CE, Mennsah GA, Fiogbe ED (2016) Effect of animal waste and vegetable compost on production and growth of earthworm (*Eisenia fetida*) during vermiculture. Int J Recycling Organic Waste Agric 5:87–92

Wallwork J (1983) Annelids: the first coelomates. Earthworms biology. Edward Arnold, London

Wang L, Zhang Y, Lian J, Chao J, Gao Y, Yang F, Zhang L (2013) Impact of fly ash and phosphatic rock on metal stabilization and bioavailability during sewage sludge vermicomposting. Bioresour Technol 136:281–287

Wen H, Li Y, Xiang L, Zhao H, Chen L, Cai Q, Li H, Mo C, Zhou D, Wong M (2017) Toxicological effects of microcystin-LR on earthworm (*Eisenia fetida*) in soil. Biol Fertil Soils 53:849–860

Xu DM, Wen YZ, Wang KX (2010) Effect of chiral differences of metolachlor and its (S)-isomer on their toxicity to earthworms. Ecotoxicol Environ Saf 73:1925–1931

Xue YG, Gu XY, Wang XR, Sun C, Xu XH, Sun J, Zhang BG (2009) The hydroxyl radical generation and oxidative stress for the earthworm *Eisenia fetida* exposed to tetrabromobisphenol A. Ecotoxicology 18:693–699

Yadav A, Garg VK (2011a) Recycling of organic wastes by employing *Eisenia fetida*. Bioresour Technol 102:2874–2880

Yadav A, Garg VK (2011b) Vermicomposting—an effective tool for the management of invasive weed *Parthenium hysterophorus*. Bioresour Technol 102:5891–5895

Yang X, Song Y, Ackland ML, Liu Y, Cao X (2012) Biochemical responses of earthworm *Eisenia fetida* exposed to cadmium-contaminated soil with long duration. Bull Environ Contam Toxicol 89:1148–1153

Zhang QM, Zhang GL, Yin PJ, Lv YZ, Yuan S, Chen JQ, Wei BB, Wang CX (2015) Toxicological effects of soil contaminated with spirotetramat to the earthworm *Eisenia fetida*. Chemosphere 139:138–145

Chapter 16
Toxicity and Histopathological Effect of Distillery Industrial Sludge on the Earthworm *Eudrilus eugeniae*

Susila Sugumar, Selvakumar Muniraj, Tamilselvi Duraisamy, Ramarajan Selvam, and Vasanthy Muthunarayanan

Abstract Distillery sludge contains good nutrition because it is agro-based industrial waste and it has been recommended as a potential soil conditioner. Simultaneously, they contain heavy metals also. Hence, they need proper treatment before their usage as a soil conditioner and plant nutrient. However, the metal reduction in sludge was directly related to the bioaccumulation of metals in the tissues of composting earthworms. The inoculated earthworms showed a considerable level of Cd, Zn, Fe, Mn and Cu in their tissues and metallothionein protein was increased in the final when compared to initial. Histology and bioaccumulation factors also confirmed that the DSV1 (distillery sludge vermicompost set 1) inoculated earthworm's tissues were severely damaged than the DSV2 (distillery sludge vermicompost set 2) because of their concentration. Finally, we conclude that earthworms could be used effectively for removal of heavy metals from distillery industrial sludge.

Keywords Distillery industrial sludge · Vermicompost · Heavy metal · Histology

16.1 Introduction

In India, the Central Pollution Control Board (CPCB) has stated that distilleries are one of the industries that contribute to environmental pollution. The higher quantity of yeast sludge and spent malt grain add up to pollution. However, the sludge is reported to contain appreciable amount of nutrients, which can be used for promoting the growth of crops and yield. However, proper treatment of the sludge is essential because it contains not only nutrients but also heavy metals (Suthar

S. Sugumar · S. Muniraj · T. Duraisamy · R. Selvam · V. Muthunarayanan (✉)
Water and Solid Waste Processing Lab, Department of Environmental Biotechnology, Bharathidasan University, Tiruchirappalli, Tamil Nadu, India
e-mail: vasanthy@bdu.ac.in

S. A. Bhat et al. (eds.), *Earthworm Assisted Remediation of Effluents and Wastes*, https://doi.org/10.1007/978-981-15-4522-1_16

2008). Vermicomposting is a reliable method for converting biodegradable sludge into value-added products. Reports exist to prove that the earthworm could facilitate this conversion and also get the heavy metals accumulated in their tissues (Shahmansouri et al. 2005). Suthar and Singh (2008) have suggested the efficiency of earthworm to convert distillery sludge into fertilizer. In this context, Ponmani et al. (2014) have studied the possibility of vermicomposting paper mill sludge using *Eudrilus eugeniae*. Similarly, Biradar and Biradar (2015) have checked the possibility of vermicomposting different organic wastes, including jowar husk, rice straw, parthenium waste and red gram pod husk with cattle dung. From the earlier reports, it is clear that the pesticides and heavy metals are the major toxicants added to the soil ecosystem. But the treatment of such polluted sites for the reduction of heavy metals by conventional procedures is expensive and time-consuming.

However, one of the cost-effective biotechnological method for the degradation of solid wastes is the composting–vermicomposting strategy (Yadav et al. 2012; Rorat et al. 2016). During such process, the accumulation of toxic contaminants by the earthworm species and their transfer to higher trophic levels is also reported (Morgan et al. 1989). However, Rorat et al. (2017) have reported the bioaccumulation potential of *Eisena andrei* accumulating the polyaromatic hydrocarbons (PAH) and metal trace elements during the vermicomposting of sewage sludge. Suthar et al. (2014) attempted removing the heavy metals such as from the paper mill wastewater sludge along with cow dung using *Eisenia fetida*. Studies also report the efficiency of *Eudrillus eugeniae* to detoxify the soil collected from three abattoir sites along with a control site in Abeokuta located in Ogun state of Nigeria (Dedeke et al. 2016) by bio accumulating the heavy metals such as Cu, Zn, Pb, Cd and Mn. Although several studies have been done on composting of distillery and winery wastes (Inbar et al. 1992; Bertran et al. 2004), less work has been done on vermicomposting of distillery sludge (Nogales et al. 2005). Hence, this study was framed to check the vermicomposting potential of *Eudrillus eugeniae* and to bio accumulate the heavy metals from the distillery sludge after the addition of amendments.

16.2 Materials and Methods

16.2.1 Vermicompost Preparation

Experiments were carried out in the Department of Environmental Biotechnology, Bharathidasan University, Tiruchirappalli, Tamil Nadu. Distillery sludge was collected from Sri Ambiga Sugar Mills (Pvt. lit), Distillery Unit, Kanadhukandon, Virudachalam (Tk), Cuddalore (Dt), Tamil Nadu, India. Distillery sludge, was mixed with cow dung, vegetable waste and saw dust in different proportions as shown in Table 16.1. Distillery sludge with different proportion of waste and carbon, nitrogen source is termed as control (DSC1 and DSC2) and the same components with the added earthworms (*Eudrilus eugeniae*) were termed as DSV1 and DSV2.

Table 16.1 Quantity of wastes subjected to composting and vermicomposting

		Weight (g)			
Vermireactors	Components	DIS	VW	CD	SD
DSC1	DIS + VW + SD + CD	900	300	300	900
DSV1	DIS + VW + SD + CD + *E. eugeniae*	900	300	300	900
DSC2	DIS + VW + SD + CD	300	300	300	1500
DSV2	DIS + VW + SD + CD + *E. eugeniae*	300	300	300	1500

DSC distillery sludge compost, *DSV* distillery sludge vermicompost, *VW* vegetable waste, *SD* saw dust, *CD* cow dung

During the experiments, elimination of volatile substances was ensured by turning over the substrates manually once in 15 days. Initial decomposition was allowed for 15 days after which 50 non-ciliated earthworms were introduced into the vermi reactors. The reactors were maintained at 25 ± 3 °C in dark and the moisture content was about 60–80%. The quantity of waste was maintained constant throughout the study.

16.2.2 Metal Detection in Compost and Vermicompost

About 10 g of compost and vermicompost sample was taken from each treatment container for analysing the heavy metal as per Pedersen and Van Gestel (2001) and Kamitani and Kaneko (2007) using atomic absorption spectrophotometer (Perkin-Elmer model 3110).

16.2.3 Heavy Metals Analysis of Tissue Samples

The tissues of earthworm samples were digested in 2.5 conc. HNO_3 and heated to dryness in a hotplate. The dried earthworms were crushed, redissolved in 1.5 conc. HNO_3 and the solution was filtered and was made up to 50 mL with distilled water in a volumetric flask. This sample is further subjected for metal analysis using AAS.

16.2.4 Histology

The histopathological studies of the earthworm may signal a damaging effect of the toxicants to which the organism is exposed, hence this study includes the same. Following the experimental period, the gut contents ejected by the earthworms kept in the petridish were removed. The organisms were sacrificed following the freezing. Washing and dehydration was done further. Bovin's fluid was used for the tissue storage. Following the dehydration, the experimental tissue was dipped in xylene and embedded in paraffin wax. The material was further trimmed and the sections were made using Spencer's microtome. Deparaffinization and washing in absolute alcohol was done before staining with Delafield's haematoxylin. Later, the stained slide was washed till they appeared blue for the naked eye. Finally, the tissues stained with Eosin were dehydrated, cleared in xylene and mounted in DPX mountant (Sharma and Satyanarayan 2011).

16.2.5 Determination of Metallothionein

Due to various contaminants present in the soil ecosystem, organisms which dwell in the soil are dependent on the detoxification system. This happens in earthworms by eliciting certain biomolecules, which are called the biomarkers of soil pollution. One such biomarker is the metal-binding protein metallothionein as reported by Dedeke et al. (2016), Suriya et al. (2012), Homa et al. (2005), Dallinger-Marianne et al. (2000) and Gastaldi et al. (2007). Metallothionein is a protein of low molecular weight whose concentration is reported to increase in the tissues due to the exposure to heavy metals. The biomarker is estimated as per the procedure of Linde-Arias et al. (2008).

16.3 Results and Discussion

Earthworms were found to bioremediate the soil contaminated with pollutants such as heavy metals (Rorat et al. 2017). During the bioremediation of high density polyethylene (HDPE)-coated paper cups, the combined action of the earthworms with the microbes were reported (Arumugam et al. 2015). The concentration of heavy metals such as Fe, Cu, Cd, Zn and Mn ions was less in *Eudrilus eugeniae* worked vermicompost (Fig. 16.1a–e).

The concentration of Cd has reduced by about 82%, 98% and 96% in DSC1, DSV1 and DSV2, respectively, and about 70% in DSC2. Such higher accumulation of Cd is reported to be due to the higher uptake of the metal and very low excretion of the same (Spurgeon and Hopkin 1999). A similar trend is noticed for Mn, Fe, Zn and Cu. This may be attributed to the higher concentration of the sludge which could have been toxic to the organisms.

Figure 16.2a, b depicts the accumulation of heavy metals in earthworm tissues subjected to vermicomposting. Increase in the Mn concentration is noticed in DSV1 and DSV2 (12.63–156 mg/kg). A similar trend is noticed for all heavy metals. From the reduction in the heavy metal content of the worm-worked vermicompost sample, an increase in the same is noticed in the tissue samples.

Similarly, Shahmansouri et al. (2005) have studied the bioaccumulation of Iranian and Australian earthworms in sewage sludge vermicompost using *Eisenia*

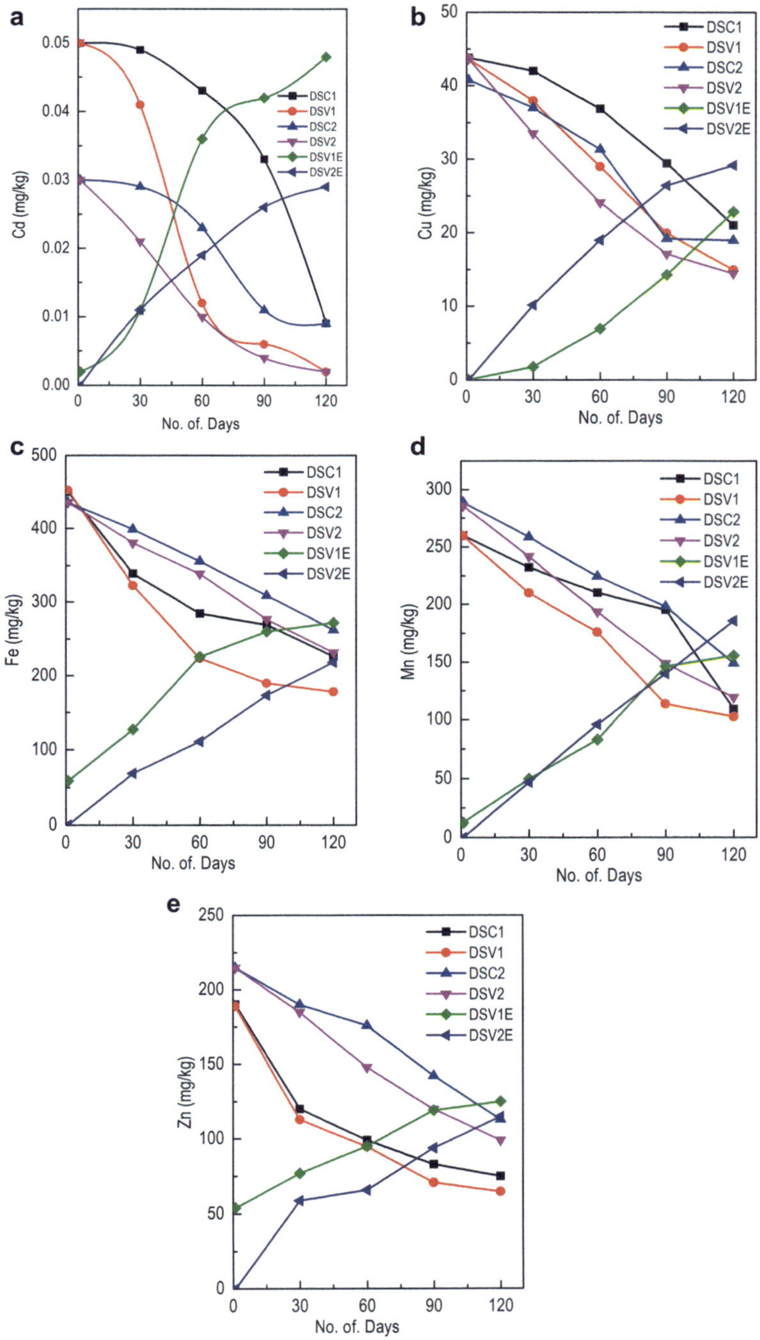

Fig. 16.1 (**a–e**) Changes in heavy metal concentration of composting and vermicomposting

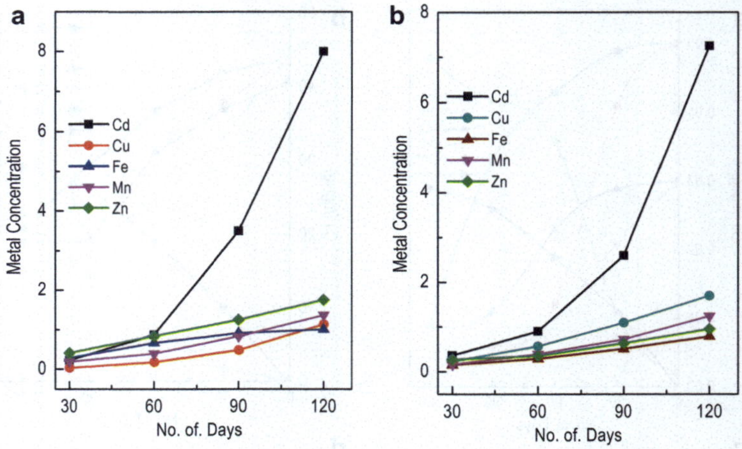

Fig. 16.2 (**a**, **b**) Concentration of heavy metal in earthworm tissues

fetida. And the results revealed a higher accumulation of micronutrients such as Zn and Cu. Rorat et al. (2017) ranked the bioaccumulation potential of *Eisenia andrei* in the following order, Cd > Cu > Zn > Ni > Cr > Pb when the organism worked on the sewage sludge. A similar result was observed by Yuvaraj et al. (2018), during the vermistabilization of paper mill sludge-spiked cow dung with Perionyx excavates. The reduction noticed in the heavy metal content was as follows: Cd (2.9–27.8%), Cu (0.22–42.3%), Pb (1.3–56.3%) and Cr (0.8–46.2%). Whereas, a 10–50% reduction of heavy metals in the fly ash + Cow dung mixture is noticed after vermicomposting with *E. fetida* (Gupta et al. 2005). The authors have suggested an inverse relationship that exists between the organic fraction and the reduction in the metal content. It is reported by Edwards and Bohlen (1996) that the gut microbes of earthworms facilitate the degradation of the wastes. During ingestion and decomposition of wastes, the metals along with the soluble fractions could be accumulated in the tissues, resulting in metal reduction. This concept could be validated on studying the amount of organic content and the rate of metal loss during the vermicomposting.

Suthar et al. (2014) have reported the reduction in the heavy metals such as Cd, Cr, Cu and Pb present in the paper mill sludge by about 32–37%, 47.3–80.9%, 68.8–88.4% and 95.3–97.5%, respectively, during the vermistabilization by *Eisenia fetida*. Singh and Kalamdhad (2013), Suthar et al. (2014), Manakova et al. (2014) and Lv et al. (2016) have reported heavy metal bioaccumulation in earthworm tissues resulting in the reduction of the heavy metal content in the experimental materials. Lemtiri et al. (2016) reported accumulation of Zn and Cu in the earthworm tissues following their introduction into the heavy metal contaminated soil for about 42 days.

Many researchers have reported the high quantity of heavy metals bioaccumulation in earthworm tissues resulting in an acceptable level of heavy

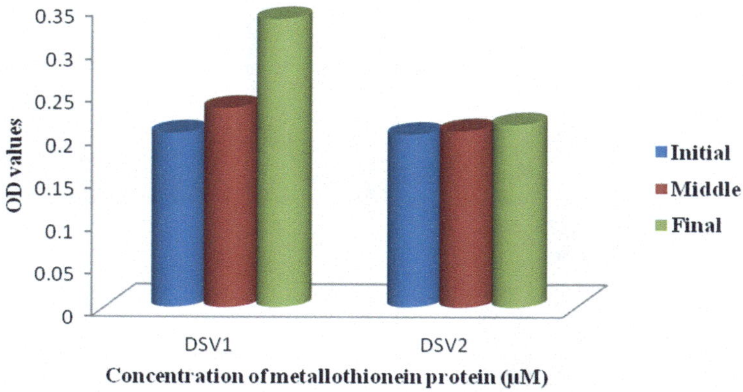

Fig. 16.3 Total metallothionein protein content in the tissues of earthworms exposed to distillery industrial sludge

metals in final materials (Lv et al. 2016; Manakova et al. 2014; Singh and Kalamdhad 2013; Suthar et al. 2014). According to Lemtiri et al. (2016), the earthworm *E. fetida* introduced to heavy metal contaminated soil for 42 days accumulated large quantities of heavy metals (Zn and Cu) in their body at the end of the experiment.

The amount of metallothionein protein in earthworm tissues exposed to DSV1 and DSV2 is represented in Fig. 16.3. The content has increased by 0.1325 mg in DSV1 and by 0.0115 mg in DSV2. Similarly, *L. rubellus* on exposure to Cd and Zn was proved to absorb negligible amount of the ions (Vijver et al. 2005). The metallothionein concentration was reported to be significantly higher in the earthworms identified from the abattoir sites of Nigeria namely, Gbonogun and Lafenwa (Dedeke et al. 2016).

The detoxification was reported as a result of their binding to organic or inorganic ligands. Such storage detoxication was reported in earthworms by Cancio et al. (1995). Sturzenbaum et al. (2001) reported the presence of granules in the chloragogenous tissues of earthworms near the intestine which contains ligands with a tendency to donate sulphur. Such ligands could be of metallothionein protein origin. Following this piece of study, it might be possible to infer that the heavy metal Cd, is responsible for the formation of metallothionein protein. Similarly, Demuynck et al. (2006) reported the presence of metallothionein in *E. fetida* following the exposure to Cd ion.

The histological reports of the earthworm show the damage caused to the cuticle and epidermis. Disintegration of the animal's circular and longitudinal muscle is also clear. Damage caused to chloragogen cells is also clear, along with the injury to intestinal epithelium. Foreign bodies accumulation of coelom vacuolization is seen. The DSV1 worked earthworm tissues and coelomic fluid were considerably damaged when compared to DSV2 worked earthworms. Further, the epithelial and intestinal walls were destructed after 120 days of exposure to waste (Fig. 16.4).

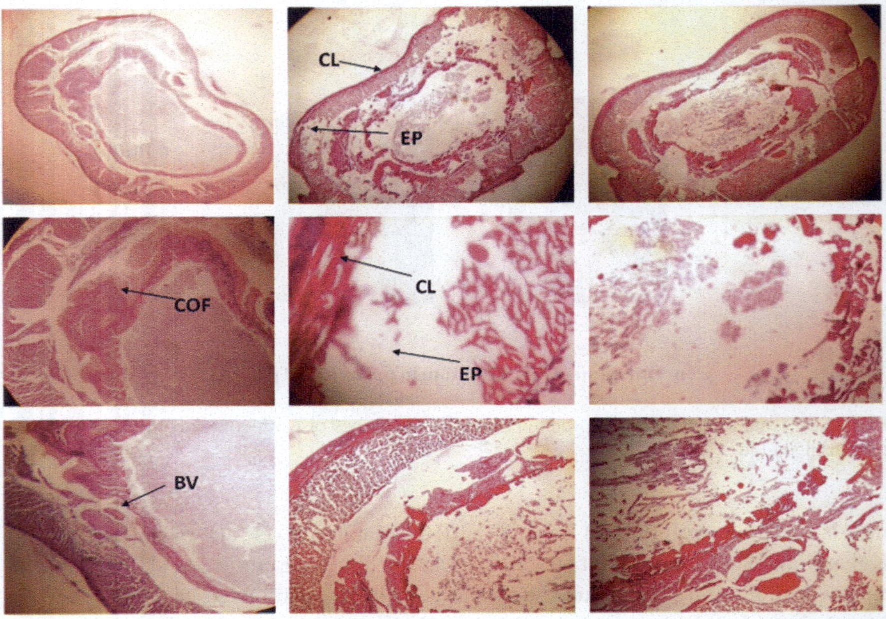

Fig. 16.4 Histological staining of earthworm tissues. *CL* chloragogen cells, *COF* coelomic fluid, *EP* epithelium, *BV* blood vessel

Similarly, Yuvaraj et al. (2018) have reported cell disintegration, irregular compartmentation of the cells and nucleus along with the irregularity of epidermal surface when *Perionyx excavatus* was employed for the vermistabilization of paper mill sludge-spiked cow dung in different ratios. However, the authors have reported that 1:1 ratio could be used for the management of paper mill sludge due to its compatibility. Kiliç (2011) and Lourenço et al. (2011) have confirmed the effect of heavy metal contaminated soils on the histopathology of earthworms. Further, Sharma and Satyanarayan (2011) reported the alteration in the histopathology and cell damage of Eudrillus eugeniae due to higher heavy metal concentration.

16.4 Conclusion

The study makes us understand that the earthworm tissues are affected by the sludge due to a wide range of concentrations. But the concentration of heavy metal ions such as Cd, Cu, Mn, Fe and Zn has decreased in the vermicompost, but has increased in the tissues substantially. The content of metallothionein protein and bioaccumulation factor content confirms the capacity of the animal to accumulate the ions in their tissues and thereby reducing the metal content in the compost. But it is clear that vermicomposting could be an efficient technology for the management of distillery sludge.

Acknowledgement The authors are grateful to the Ministry of Environmental Forest and Climate Change (MOEF&CC), Government of India, New Delhi for financial support and encouragement. They also acknowledge the DST-FIST, DST-PURSE, UGC-SAP and MHRD-RUSA for providing the instrumentation facilities in the department.

References

Arumugam K, Ganesan S, Muthunarayanan V, Vivek S, Sugumar S, Munusamy V (2015) Potentiality of Eisenia fetida to degrade disposable paper cups-an ecofriendly solution to solid waste pollution. Environ Sci Pollut Res 22(4):2868–2876

Bertran E, Sort X, Soliva M, Trillas I (2004) Composting winery waste: sludges and grape stalks. Bioresour Technol 95(2):203–208

Biradar PM, Biradar VA (2015) Influence of organic wastes on the biology of epigeic earthworm, Perionyx excavatus during different seasons. J Environ Biol 36(5):1095

Cancio I, Ap Gwynn I, Ireland MP, Cajaraville MP (1995) The effect of sublethal lead exposure on the ultrastructure and on the distribution of acid phosphatase activity in chloragocytes of earthworms (Annelida, Oligochaeta). Histochem J 27(12):965–973

Dallinger-Marianne J, Kohler-Vibeke HDHR, Weeks SRTJM (2000) Biomarkers in terrestrial invertebrates for ecotoxicological soil risk assessment. Rev Environ Contam Toxicol 164 (93):93–103

Dedeke GA, Owagboriaye FO, Adebambo AO, Ademolu KO (2016) Earthworm metallothionein production as biomarker of heavy metal pollution in abattoir soil. Appl Soil Ecol 104:42–47

Demuynck S, Grumiaux F, Mottier V, Schikorski D, Lemière S, Leprêtre A (2006) Metallothionein response following cadmium exposure in the oligochaete *Eisenia fetida*. Comp Biochem Physiol C Toxicol Pharmacol 144(1):34–46

Edwards CA, Bohlen PJ (1996) Biology and ecology of earthworms, vol 3. Springer Science & Business Media

Gastaldi L, Ranzato E, Capri F, Hankard P, Pérès G, Canesi L et al (2007) Application of a biomarker battery for the evaluation of the sublethal effects of pollutants in the earthworm *Eisenia andrei*. Comp Biochem Physiol C Toxicol Pharmacol 146(3):398–405

Gupta SK, Tewari A, Srivastava R, Murthy RC, Chandra S (2005) Potential of *Eisenia foetida* for sustainable and efficient vermicomposting of fly ash. Water Air Soil Pollut 163(1–4):293–302

Homa J, Olchawa E, Stürzenbaum SR, Morgan AJ, Plytycz B (2005) Early-phase immunodetection of metallothionein and heat shock proteins in extruded earthworm coelomocytes after dermal exposure to metal ions. Environ Pollut 135(2):275–280

Inbar Y, Hadar Y, Chen Y (1992) Characterization of humic substances formed during the composting of solid wastes from wineries. Sci Total Environ 113(1–2):35–48

Kamitani T, Kaneko N (2007) Species-specific heavy metal accumulation patterns of earthworms on a floodplain in Japan. Ecotoxicol Environ Saf 66(1):82–91

Kiliç GA (2011) Histopathological and biochemical alterations of the earthworm (*Lumbricus terrestris*) as biomarker of soil pollution along Porsuk River Basin (Turkey). Chemosphere 83 (8):1175–1180

Lemtiri A, Liénard A, Alabi T, Brostaux Y, Cluzeau D, Francis F, Colinet G (2016) Earthworms *Eisenia fetida* affect the uptake of heavy metals by plants Vicia faba and Zea mays in metal-contaminated soils. Appl Soil Ecol 104:67–78

Linde-Arias AR, Inácio AF, de Alburquerque C, Freire MM, Moreira JC (2008) Biomarkers in an invasive fish species, *Oreochromis niloticus*, to assess the effects of pollution in a highly degraded Brazilian River. Sci Total Environ 399(1–3):186–192

Lourenço J, Silva A, Carvalho F, Oliveira J, Malta M, Mendo S et al (2011) Histopathological changes in the earthworm *Eisenia andrei* associated with the exposure to metals and radionuclides. Chemosphere 85(10):1630–1634

Lv B, Xing M, Yang J (2016) Speciation and transformation of heavy metals during vermicomposting of animal manure. Bioresour Technol 209:397–401

Manakova B, Kuta J, Svobodová M, Hofman J (2014) Effects of combined composting and vermicomposting of waste sludge on arsenic fate and bioavailability. J Hazard Mater 280:544–551

Morgan JE, Norey CG, Morgan AJ, Kay J (1989) A comparison of the cadmium-binding proteins isolated from the posterior alimentary canal of the earthworms *Dendrodrilus rubidus* and *Lumbricus rubellus*. Comp Biochem Physiol C Comp Pharmacol 92(1):15–21

Nogales R, Cifuentes C, Benitez E (2005) Vermicomposting of winery wastes: a laboratory study. J Environ Sci Health B 40(4):659–673

Pedersen MB, van Gestel CA (2001) Toxicity of copper to the collembolan Folsomia fimetaria in relation to the age of soil contamination. Ecotoxicol Environ Saf 49(1):54–59

Ponmani S, Udayasoorian C, Jayabalakrishnan RM, Kumar KV (2014) Vermicomposting of paper mill solid waste using epigeic earthworm *Eudrilus eugeniae*. J Environ Biol 35(4):617

Rorat A, Suleiman H, Grobelak A, Grosser A, Kacprzak M, Płytycz B, Vandenbulcke F (2016) Interactions between sewage sludge-amended soil and earthworms—comparison between *Eisenia fetida* and *Eisenia andrei* composting species. Environ Sci Pollut Res 23(4):3026–3035

Rorat A, Wloka D, Grobelak A, Grosser A, Sosnecka A, Milczarek M et al (2017) Vermiremediation of polycyclic aromatic hydrocarbons and heavy metals in sewage sludge composting process. J Environ Manag 187:347–353

Shahmansouri MR, Pourmoghadas H, Parvaresh AR, Alidadi H (2005) Heavy metals bioaccumulation by Iranian and Australian earthworms (Eisenia fetida) in the sewage sludge vermicomposting. J Environ Health Sci Eng 2(1):28–32

Sharma VJ, Satyanarayan S (2011) Effect of selected heavy metals on the histopathology of different tissues of earthworm *Eudrillus eugeniae*. Environ Monit Assess 180(1–4):257–267

Singh J, Kalamdhad AS (2013) Assessment of bioavailability and leachability of heavy metals during rotary drum composting of green waste (Water hyacinth). Ecol Eng 52:59–69

Spurgeon DJ, Hopkin SP (1999) Comparisons of metal accumulation and excretion kinetics in earthworms (*Eisenia fetida*) exposed to contaminated field and laboratory soils. Appl Soil Ecol 11(2–3):227–243

Sturzenbaum SR, Winters C, Galay M, Morgan AJ, Kille P (2001) Metal ion trafficking in earthworms Identification of a cadmium-specific metallothionein. J Biol Chem 276 (36):34013–34018

Suriya J, Bharathiraja S, Sekar V, Rajasekaran R (2012) Metallothionein induction and antioxidative responses in the estuarine polychaeta Capitella capitata (Capitellidae). Asian Pac J Trop Biomed 2(2):S1052–S1059

Suthar S (2008) Bioconversion of post-harvest crop residues and cattle shed manure into value-added products using earthworm Eudrilus eugeniae Kinberg. Ecol Eng 32(3):206–214

Suthar S, Singh S (2008) Feasibility of vermicomposting in biostabilization of sludge from a distillery industry. Sci Total Environ 394(2–3):237–243

Suthar S, Sajwan P, Kumar K (2014) Vermiremediation of heavy metals in wastewater sludge from paper and pulp industry using earthworm Eisenia fetida. Ecotoxicol Environ Saf 109:177–184

Vijver MG, Wolterbeek HT, Vink JP, van Gestel CA (2005) Surface adsorption of metals onto the earthworm *Lumbricus rubellus* and the isopod Porcellio scaber is negligible compared to absorption in the body. Sci Total Environ 340(1–3):271–280

Yadav KD, Tare V, Ahammed MM (2012) Integrated composting–vermicomposting process for stabilization of human faecal slurry. Ecol Eng 47:24–29

Yuvaraj A, Karmegam N, Thangaraj R (2018) Vermistabilization of paper mill sludge by an epigeic earthworm Perionyx excavatus: mitigation strategies for sustainable environmental management. Ecol Eng 120:187–197

Chapter 17
Earthworm-Assisted Amelioration of Thermal Ash

Bhawana Sohal and Adarsh Pal Vig

Abstract Fly ash is the end product of burnt powdered coal. It has created various environmental problems and many health issues. So to control its negative impacts, earthworm-assisted technology (vermicomposting) was used for converting the fly ash into valuable manure. It was studied that it is not only changing the substrate from one form to another but also capable of improving its physicochemical properties such as pH, electrical conductivity (EC), and total organic carbon (TOC). It had been confirmed to increase all the nutrient contents such as nitrogen, phosphorus, and potassium (NPK), which are required for good agricultural purposes. It was also found to have the capacity to remove heavy metals from the fly ash samples. Thus, this chapter has mainly focused on the utilization and management of fly ash through earthworms.

Keywords Vermicomposting · Earthworm · Fly ash · Physicochemical properties

17.1 Introduction

An earthworm is a tube-like, segmented, unique hermaphrodite annelid of the soil fauna. It is capable of improving the whole ecosystem by reconstructing each structural and functional unit of the soil system. The process that involves the earthworms to mediate the subjected organic matter to natural manure is called vermicomposting. Mainly, it reduces the contaminants and enriches the soil with essential nutrients. It contributes natural supplements and microflora to the soil through its nutritious and neutral fecal pellets (vermicast). It is overloaded with all the essential nutrients for the healthy plant growth (Bombatkar 1996; Arancon et al. 2004). It is also responsible to enhance the microbial colonies in the soil, which in

B. Sohal · A. P. Vig (✉)
Department of Botanical and Environmental Sciences, Guru Nanak Dev University, Amritsar, Punjab, India

S. A. Bhat et al. (eds.), *Earthworm Assisted Remediation of Effluents and Wastes*,
https://doi.org/10.1007/978-981-15-4522-1_17

turn are helpful in conduit and aeration maintenance for the soil (Ouédraogo et al. 2001).

An earthworm belongs to the kingdom of Animalia, and they are multicellular eukaryotic organisms. The phylum is Annelida because its body is divided into different segments. It is the terrestrial worm of class Haplotaxida or Oligochaeta, as it consists of clitellum, which serves as the reproductive center. The family of the earthworm is Acanthodrilidae.

An earthworm is an only member of Oligochaeta that is very beneficial to the ecosystem and has the capacity to mold the ecology with respect to good agricultural practices. It has exhibited large diversity and shows up to more than 1800 species. The most commonly found genus Lumbricus shows a total of 13 native species. The most widely used earthworm species for vermicomposting are *Eisenia fetida* and *Eisenia andrei.* These species show large distribution with a high reproductive rate and short life span. They are good tolerant of a wide range of temperature and humidity (Domínguez 2004; Domínguez and Edwards 2011).

17.1.1 Ecological Classification of Earthworms

Earthworms are classified into four categories based on their different ecological pattern and behavior.

17.1.1.1 Compost Earthworms

These species are found mostly in the naturally enriched area of rotten vegetation. They are also called tiger worm as their name indicates, and they are very efficient in making compost because of their quick consumption of waste and rapid reproduction rate. They are also helpful to dispose of the waste as they can remove all the contaminants from the soil and also used to improve the properties and structure of the soil. The examples of compost earthworms are *Eisenia fetida* and *Dendrobaena veneta*. On the other hand, these species of earthworms are mostly preferred and used in vermicomposting due to their short life cycle, versatile temperature tolerance, high reproductive rate, easy availability, and ease of handling.

17.1.1.2 Epigeic Earthworms

It means surface feeders. Epigeic earthworms live on the leaf litter area of the soil surface. They feed on leaf litter but do not make burrows. They are mostly reddish-brown and non-stripy, and the species includes *Dendrobaena attemsi*, *Dendrodrilus rubidus*, and *Lumbricus festivus*.

17.1.1.3 Endogeic Earthworms

It is a category of deep feeders. They live deep into the soil and make deep, horizontal, and well-extended burrows to feed easily. Endogeic earthworms are mostly pinkish grey, and fever is found in green color. The most common species of this group are *Allolobophora chlorotica*, *Octolasion rosea*, and *Apporectodea icterica*.

17.1.1.4 Anecic Earthworms

They make a permanent vertical and long type of burrows for living and are usually brown in color with the paler tail. They drag the food into their burrows and are mostly found in grassland areas. They are the largest species of earthworm found in England. They cast on the surface of their particular habitat and required a long-time period for making vermicompost due to its long life cycle. They prefer cold temperatures and do not show any wide range of tolerance to the temperature. The examples of this category are *Aporrectodea longa*, *Lumbricus friendi*, and *Lumbricus terrestris*.

17.1.2 Fly Ash

Fly ash is the powdered minute particles (0.5–300 μm) and the end product of burnt pulverized coal material (Fig. 17.1). In 2017, it was found to produce 225 million tones of coal fly ash in each year and pollutes the environment very badly (Dixit and Lokesharappa 2011; Wikipedia n.d.). It is reported to contain various heavy metals that show a toxic potential, which may create many ailments in living beings like cancer (Singh et al. 2014a; Ćujić et al. 2016). If the fly ash is not properly managed (open dumping), it may combine with water and form slurry. The slurry fly ash, later on, becomes the supportive growth medium for many bacteria and mosquitoes and contaminates the water bodies (river, lakes and ponds), and underground water (Nawaz 2013).

17.1.2.1 Classification of Fly Ash

ASTM (American Society for Testing and Materials) C618 classified the fly ash into two categories on the basis of calcium oxide (CaO) concentrations. Afterward, IS3812-1981, fly ash has two grades. Grade I fly ash is composed of $SiO_2 + Al_2O_3 + Fe_2O_3$, which is greater than 70% and obtained from bituminous coal. It is also called Class C fly ash and contains more than 10% CaO. However, the source of grade II is lignite coal and consists of more than 50% of

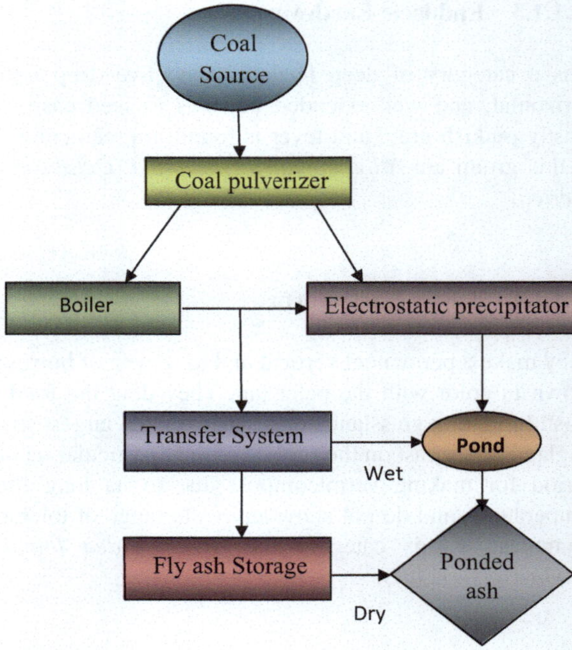

Fig. 17.1 Generation of fly ash in power plant

Table 17.1 Fly ash classification on the basis of different sources of coal

Types of coal	Mineral contents ($SiO_2 + Al_2O_3 + Fe_2O_3$)	Content of CaO	Classes of fly ash
Bituminous coal	70%	<10%	Class C (Grade I)
Lignite coal	50%	>10%	Class F (Grade II)

$SiO_2 + Al_2O_3 + Fe_2O_3$. It is also named as Class F Fly ash and has less than 10% CaO (Table 17.1). This chapter deals with the various types of earthworm-assisted technologies that help to mitigate the toxic effects of fly ash (the toxic waste product of thermal power plant).

The waste stabilization and management of various industries are highlighted in this chapter, and it is fully focused on the management of the end product of thermal power plant, i.e., coal fly ash. The heavy metal removal from the fly ash is done with the help of the vermicomposting process that involves earthworms, which focuses on proper utilization and conversion of the nutrient contents of fly ash into high-quality manure.

17.2 Different Potentials of Earthworms

17.2.1 Manure Formation

The vermicomposting is the earthworm-assisted process that is able to stabilize and detoxify all the toxic constituents of the waste into high-quality and nutrient-rich manure (Bhat et al. 2015a; Padmavathiamma et al. 2008). This manure is used to enhance the fertility of the soil as it enriches the soil with all the essential elements that are particularly needed by the healthy plant growth.

17.2.2 Removal of Heavy Metals

The earthworms are found to remove all the toxic heavy metals from all the types of industrial wastes into the nontoxic form of vermicompost (Brunetto et al. 2014). Many researchers have also studied that the earthworms are efficiently capable of removing almost all heavy metals like Cr, Co, Ni, Cd, and Pb.

17.2.3 Antimicrobial Activities

The earthworms were used as different traditional medicines in china to cure many infectious diseases. They are also used in making many medicines to cure diseases like pyorrhea postpartum weakness and postpartum hemorrhage. The earthworms have found to exhibit various antimicrobial activities (Cooper et al. 2004). It is possible due to the presence of newly found bacterial strain in the midgut of *Dendrobaena veneta* (Annelida), which is responsible for its antimicrobial activities (Marti et al. 2010).

17.2.4 Anticancer Activities

The various bioactive molecules are found in the coelomic fluid of the earthworms like anticancer agents (Lourdumary and Ramesh 2014). It is also responsible to cure many tumor-related ailments. It has been found that the coelomic fluid of earthworms contains many active components like cytolytic protein that is an important component of the human immune system, which is responsible to impart and enhance the anticancer activities in the earthworms (Bilej et al. 1995).

17.2.5 Earthworms as Bio-indicators

An earthworm can act as the bio-indicators to evaluate the heavy metal and pesticides in the soil (Paoletti et al. 1991). It is capable of remediating heavy metal waste during vermicomposting via accumulating them in their body (Gupta et al. 1997).

17.2.6 Earthworms as Eco-engineers

They act on toxic waste material and change it into nontoxic valuable manure with the help of microorganisms. In this way, it is helpful to maintain the nutrient cycle of the ecosystem (Edwards et al. 1996).

17.3 Earthworm-Assisted Waste Management Technologies

Various technologies such as open and closed composting have been invented by several researchers in the past decades to manage solid waste, which could only increase the soil fertility but could not completely manage the waste material. These traditional techniques, on the other hand, cause various types of pollution in an environment. Thus, earthworm-assisted technology to manage the organic wastes was introduced, which is called vermitechnology. It is found to manage all the types of organic wastes efficiently with the help of earthworms (Table 17.2). It converts

Table 17.2 Amelioration of different organic wastes using earthworms

S. no.	Types of wastes	Earthworm's species used	References
1.	Textile mill sludge and cow dung	*Eisenia fetida*	Kaushik and Garg (2003)
2.	Textile mill sludge with Poultry	*Eisenia fetida*	Garg and Kaushik (2005)
3.	Agro-industrial wastes	*Eisenia fetida*	Garg et al. (2006)
4.	Industrial wastes	*Eudrilus eugeniae*	Kitturmath et al. (2007)
5.	Fly ash—a lignite waste	*Lampito mauritii*	Ananthakrishnasamy et al. (2009)
6.	Thermal power plant fly ash	*Eisenia fetida*	Singh et al. (2016)
7.	Municipal solid waste	*Eisenia fetida* and *Lempito mauritii*	Sharma (2003)
8.	Bagasse waste	*Eisenia fetida*	Bhat et al. (2015a)
9.	Rice husk	*Eudrilus eugeniae*	Lim et al. (2012)
10.	Fly ash	*Eisenia fetida*	Bhattacharya and Chattopadhyay (2004)

the waste into a usable form (manure) without causing pollution and is totally eco-friendly. The earthworms were confirmed to play a major and important role in all the types of management by converting organic waste in black gold (Borowski 1995; Khwairakpam and Bhargava 2009). The different nature of wastes (solid and liquid) can be managed by the various earthworm-assisted techniques.

17.3.1 Vermicomposting Technology

This technology converts organic waste into high-quality manure with the help of earthworms. A vermicast is produced as the mutual action of the microorganisms and earthworms. It is responsible to increase the nutrient contents of the waste. Other secretions like coelomic fluid and mucilage of earthworms are responsible to stabilize the basic physicochemical properties of waste such as pH, EC, and TOC, which are good for soil health. It is very efficient for municipal and industrial solid-waste management (Bhat et al. 2015a; Sharma 2003).

17.3.2 Vermifiltration Technology

In this technology, liquid is extracted from the column of waste and earthworms. It is also called earthworm-assisted wastewater engineering. In this, water is added from the upper inlet of the column and collected from a lower outlet of the same. The collected liquid contains all the types of nutrients (NPK), earthworm secretion (enzymes and hormones) such as vermicompost. It is mainly used in floriculture. It is also capable of managing all the types of wastes and wastewater (Samal et al. 2017; Singh et al. 2019). It also provides all the required nutrients for the best plant growth.

17.3.3 Vermiremediation Technology

This technology uses earthworms to remove all the pollutants like heavy metals and other harmful chemicals from the soil. It is very useful to clean all the contaminated agricultural fields. It is also helpful to improve the soil health (Bhat et al. 2016; Shrestha et al. 2019).

17.3.4 Vermi-Agro-Production Technology

As the name indicates, this technology is useful in improving the total yield of the plants. In this technique, the earthworms were used as soil engineers to restore the soil fertility for high productivity (Islam et al. 2018; Ganeshnauth et al. 2018). The main purpose of this technology is to produce a high yield of the crops without using harmful chemicals such as pesticides and insecticides.

17.4 Impacts of Earthworms on Physicochemical Properties of Fly Ash

17.4.1 Electric Conductivity (EC)

The electric conductivity (EC) of the samples shows the ionic strength of the samples. An increase in electric conductivity of the sample indicates high ionic strength. The earthworms are capable of making the proper balance in EC of the sample by decreasing or increasing according to the nature of the sample. Many studies have shown that the earthworms are helpful to increase the electrical conductivity of the fly ash samples during vermicomposting. So, in this way, earthworms are helpful in maintaining the mineral contents of the fly ash, which is very important for good agricultural practices. The earthworms are also capable of decreasing the EC of the sample via increasing the precipitation and volatilization of mineral salts and ammonia during vermicomposting (Wong et al. 1995; Singh et al. 2010; Lim et al. 2012).

17.4.2 The Potential of Hydrogen (pH)

The neutral pH is showing the biological stabilization of the samples (Kiehl and Yamanouchi 1985). The good and healthy plant growth is the expression of the pH level of the soil. It was reported that a pH range from 5.0 to 9.0 supports the better growth of earthworms (Edwards 1995). The earthworms have the potential to balance the pH by adding their casts to the sample, which is very neutral in nature (Lee and Ramakrishnan 1985). They help to balance the pH of the acidic samples (low pH) by increasing the mineralization of organic materials and increasing the organic acid formation (Ndegwa et al. 2000). The earthworms decrease the pH of the fly ash via increasing aerobic digestion and various metabolic compound formation such as NH_3 (ammonia), CO_2 (carbon dioxide), and NO_3 (nitrate) during vermicomposting (Bhat et al. 2015b; Niyazi and Chaurasia 2014).

17.4.3 Total Organic Carbon (TOC)

TOC indicates the total carbon contained by the sample mixtures. Its high concentration can make the sample highly toxic for the soil biota (plants and microorganisms). So to manage the TOC content, most of the scientists have used the earthworms with the sample and incubate them for few (two) months (vermicomposting). It has found that the earthworms are capable of decreasing the TOC form the initial to final fly ash mixtures (Usmani et al. 2017; Singh et al. 2016). The earthworms help to encourage the microbial growth and their actions along with microbes bring the mineralization and decomposition of organic carbon that leads to carbon loss in the form of CO_2 in the feed mixtures during vermicomposting (Domínguez 2004; Suthar and Sharma 2013).

17.5 Impacts of Earthworms on NPK (Nitrogen, Phosphorus, and Potassium) Contents

17.5.1 Nitrogen Content

Nitrogen is an essential element for plant growth, and it is a basic structural component of the amino acids, which are further important components of proteins. It is available in the environment (air) in free form (N_2) and is not able to consume by plants. To make it consumable to plants, some microorganisms like bacteria are involved, which convert it into soluble forms like ammonia, nitrates, and nitrites. It is found that the earthworms are very useful in increasing the level of nitrogen in the post compost samples by mineralization and decomposition of the organic nitrogenous compound. It was found that the earthworms are helpful to make the nitrogen level high by the production of vermicast. Bhattacharya and Chattopadhyay (2004) also found the increasing trends of nitrogen during vermicomposting of fly ash. It is a rich source of nitrogen and also capable of producing the other stable products like ammonia, urea, and uric acid (Suthar 2009; Waseem et al. 2013). It was found that the nature and type of organic matter (cattle dung) used in vermicomposting is fully responsible for the number of nutrient contents of the vermicompost (Atiyeh et al. 2002; Bhat et al. 2015b; Kawaguchi and Nishi 2007).

17.5.2 Phosphorus Content

Phosphorus content was found to increase in the samples treated with earthworms during vermicomposting. It may be due to mineralization and decomposition of the organo-phosphorus compound by the earthworms (Hartenstein 1981; Hamilton and Mitchell 1997). The increasing content of phosphorus was reported in the water hyacinth feed mixture with the help of *E. fetida* (Gupta et al. 2007). Bhattacharya and Chattopadhyay (2004) and Ananthakrishnasamy et al. (2009) found that P content was increased at the end product of vermicomposting. It is found that the earthworms are helpful in increasing the phosphorus content during the vermicomposting of fly ash (Bhattacharya and Chattopadhyay 2002). It is the main and basic component of all types of genetic material (DNA and RNA), and it is common in all living beings. So, it is also helpful for plant and microbial growth, which is also good for soil health.

17.5.3 Potassium Content

Many researchers have found that the earthworms are helpful in increasing the potassium content in the fly ash during vermicomposting (Singh et al. 2016; Usmani et al. 2017). The microbial action inside the gut of the earthworm brings the decomposition and mineralization processes, which in turn increases the potassium content in the vermicompost (Kaviraj and Sharma 2003; Delgado and Mora 1995; Gunjal and Nikam 1992; Suthar 2006; Pattnaik and Reddy 2010). It is very useful in good plant growth as the potassium is helpful to regulate the osmotic variation according to changing seasons through stomatal movement.

17.6 Impacts of Earthworms on C/N Ratio

Earthworms have the potential to decrease the C/N ratio during vermicomposting, which is very important for the soil and plant health. They have capabilities to decrease the carbon and increase the nitrogen content. So, at the end product of earthworm-assisted vermicomposting, the C/N ratio decreases by the decomposition process of microorganisms (Wu et al. 2016; Wu et al. 2017).

17.7 Potential to Remove Heavy Metals (Detoxification)

Earthworms consist of chloragocytes cells and intestinal microflora, which have the capacity to remove the heavy metal concentrations. It is accomplished via the accumulation of these toxic heavy metals into the earthworm's tissues (Bhattacharya and Chattopadhyay 2006). It is also reported that as time increases, the rate of heavy metal reduction also increases (Shahmansouri et al. 2005). Heavy metal contents are found to decrease in water-hyacinth-based vermicompost (Gupta et al. 2007).

17.7.1 Copper (Cu)

A copper is one of the most important traces of essential element, and it is required in minor quantity. The excess range of this metal can be harmful to both plants and soil (Brunetto et al. 2014). It is found that the earthworms help to decrease the copper content from the fly ash (Tripathi et al. 1997; Gupta et al. 2005).

17.7.2 Cadmium (Cd)

Cadmium is considered a very toxic element for the soil system. It interferes in plant growth and damages the soil structure (Goswami et al. 2016). Suzuki et al. (1980) and Singh et al. (2016) found a sharp decline in the Cd content during the vermicomposting of fly ash.

17.7.3 Chromium (Cr)

Chromium exhibits in two states III and VI. The chromium III is non-toxic, whereas the latter is toxic. This chromium interconversion is highly required for healthy plant growth, which is accomplished with the help of earthworms (Kumar et al. 2017). The apparent reduction in the Cr content was seen in the fly ash vermicomposting (Singh et al. 2016).

17.7.4 Lead (Pb)

Pb is one of the most toxic elements for the life system. It is not only harming the ecosystem but also affects badly to each component of the living system deeply. It was found in many studies of vermicomposting that the earthworms are useful in the conversion of non-toxic stable products. Many researchers have studied that the earthworms can be capable of decreasing the Pb content from the fly ash mixture during vermicomposting (Gupta et al. 2005; Bhattacharya and Chattopadhyay 2006). Bozym (2017) found that *Eisenia fetida* is capable of converting the toxic Pb to nontoxic form.

17.7.5 Nickel (Ni)

Many studies have found that the earthworms have the potential to remove the Ni content from the waste material. It was seen to be decreased in post vermicompost samples of fly ash (Usmani et al. 2017). It is accomplished by the accumulation of Ni inside the body of earthworm (Jain et al. 2004; Vig et al. 2011).

17.7.6 Cobalt (Co)

Organo-cobalt-complex formations are enhanced during the vermicomposting due to the mutual action of earthworms and microorganisms. Patinvoh et al. (2017) also found the earthworm is able to reduce the cobalt content from the waste material. Singh et al. (2014b) and Mittra et al. (2005) found a high decline in Co content at the end product of vermicomposting of fly ash.

17.8 Conclusion

This chapter clearly indicates the best use and management of fly ash with the help of earthworms through vermitechnology. It also reveals the potential of the earthworms like improvement of chemical nature (pH, EC, TOC and N, P, K) of the fly ash and the removal of heavy metals from the studied substrate (fly ash) that were observed during vermicomposting.

References

Ananthakrishnasamy S, Sarojini S, Gunasekaran G, Manimegala G (2009) Fly ash—a lignite waste management through vermicomposting by indigenous earthworms *Lampito mauritii*. Am Eur J Agric Environ Sci 5:720–724

Arancon NQ, Edwards CA, Bierman P, Welch C, Metzger JD (2004) Influences of vermicomposts on field strawberries-1: effects on growth and yields. Bioresour Technol 93:145–153

Atiyeh RM, Arancon NQ, Edwards CA, Metzger JD (2002) The influence of earthworm-processed pig manure on the growth and productivity of marigolds. Bioresour Technol 81(2):103–108

Bhat SA, Singh J, Vig AP (2015a) Vermistabilization of sugar beet (*Beta vulgaris* L.) waste produced from sugar factory using earthworm *Eisenia fetida*: genotoxic assessment by *Allium cepa* test. Environ Sci Pollut Res 22:11236–11254

Bhat SA, Singh J, Vig AP (2015b) Potential utilization of bagasse as feed material for earthworm *Eisenia fetida* and production of vermicompost. SpringerPlus 4:11

Bhat SA, Bhatti SS, Singh J, Sambyal V, Nagpal A, Vig AP (2016) Vermiremediation and phytoremediation: eco approaches for soil stabilization. Austin Environ Sci 1(2):1006

Bhattacharya SS, Chattopadhyay GN (2002) Increasing bioavailability of phosphorous from fly ash through vermicomposting. J Environ Qual 31:2116–2119

Bhattacharya SS, Chattopadhyay GN (2004) Transformation of nitrogen during vermicomposting of fly ash. Waste Manag Res 22(6):488–491

Bhattacharya SS, Chattopadhyay GN (2006) Effect of vermicomposting on the transformation of some trace elements in fly ash. Nutr Cycl Agroecosyst 75(1–3):223–231

Bilej M, Brys L, Beschin A, Lucas R, Vercauteren E, Hanušová R, De Baetselier P (1995) Identification of a cytolytic protein in the coelomic fluid of Eisenia foetida earthworms. Immunol Lett 45(1–2):123–128

Bombatkar V (1996) The miracle called compost. The Other India Press, Pune

Borowski E (1995) Response of tomatoes to NO3-N or NH4-N applied to sand loam, and soil substrate. Annales Universitatis Mariae Curie Sklodowska 3:111–118

Bozym M (2017) Heavy metal content in compost and earthworms from home composters. Ochrona Srodowiska i Zasobow Naturalnych 28(4):1–4

Brunetto G, Miotto A, Ceretta CA, Schmitt DE, Heinzen J, Moraes MP, Canton L, Tiecher TL, Comin JJ, Girotto E (2014) Mobility of copper and zinc fractions in fungicide-amended vineyard sandy soils. Arch Agron Soil Sci 60:609–624

Cooper EL, Ru B, Weng N (2004) Earthworms: sources of antimicrobial and anticancer molecules. In: Complementary and alternative approaches to biomedicine. Springer, Boston, MA, pp 359–389

Ćujić M, Dragović S, Đorđević M, Dragović R, Gajić B (2016) Environmental assessment of heavy metals around the largest coal-fired power plant in Serbia. Catena 139:44–52

Delgado MA, Mora J (1995) Nonparametric and semiparametric estimation with discrete regressors. Econometrica 63:1477–1484

Dixit A, Lokesharappa (2011) Disposal and management of fly ash. In: International conference on 355 life science and technology, IPCBEE, vol 3. IACSIT Press, Singapore

Domínguez J (2004) State of the art and new perspectives on vermicomposting research. In: Edwards CA (ed) Earthworm ecology, 2nd edn. CRC Press, Boca Raton, FL, pp 401–424

Domínguez J, Edwards CA (2011) Biology and ecology of earthworm species used for vermicomposting. In: Edwards CA, Arancon NQ, Sherman RL (eds) Vermiculture technology: earthworms, organic waste and environmental management. CRC Press, Boca Raton, FL, pp 25–37

Edwards CA (1995) The commercial and environmental potential of vermicomposting: a historical overview. Biocycle, June: 62–63

Edwards CA, Bohlen PJ (eds) (1996) Biology and ecology of earthworm. Chapman & Hall, London, p 426

Ganeshnauth V, Jaikishun S, Ansari AA, Homenauth O (2018) The effect of vermicompost and other fertilizers on the growth and productivity of pepper plants in Guyana. In: Hussmann S (ed) Automation in agriculture-securing food supplies for future generations. Rijeka, Intech. https://doi.org/10.5772/intechOpen.73262

Garg VK, Kaushik P (2005) Vermistabilization of textile mill sludge spiked with poultry droppings by an epigeic earthworm Eisenia foetida. Bioresour Technol 96:1063–1071

Garg P, Gupta A, Satya S (2006) Vermicomposting of different types of waste using *Eisenia foetida*: a comparative study. Bioresourc Technol 97:391–395

Goswami LS, Pratihar S, Dasgupta P, Bhattacharyya P, Mudoi J, Bora SS, Bhattacharya KH (2016) Exploring metal detoxification and accumulation potential during vermicomposting of tea factory coal ash: sequential extraction and fluorescence probe analysis. Sci Rep 6:30402

Gunjal SS, Nikam TM (1992) Grape cultivation through earthworm farming. In: National seminar on organic farming, organized by Mahatma Phule Krishi Vidyapeeth, Rahuri, College of Agriculture, Pune, pp 18–19

Gupta SK, Singh SB, Sundararaman V (1997) Cadmium toxicity in earthworm, Mataphire posthums: ultra structural changes in secretary cells of clitellar epithelium. Indian J Exp Biol 35(7):780–786

Gupta SK, Tewari A, Srivastava R, Murthy RC, Chandra S (2005) Potential of Eisenia foetida for sustainable and efficient vermicomposting of fly ash. Water Air Soil Pollut 163(1–4):293–302

Gupta R, Mutiyar PK, Rawat NK, Saini MS, Garg VK (2007) Development of a water hyacinth based vermireactor using an epigeic earthworm *Eisenia foetida*. Bioresour Technol 98:2605–2610

Hamilton DP, Mitchell SF (1997) Wave induced shear stresses, plant nutrients and chlorophyll in the seven shadow lakes. Fresh Water Biol 38:159–168

Hartenstein R (1981) Production of earthworms as a potentially economical source of protein. Biotechnol Bioeng 23(8):1797–1811

Islam MA, Talukder MSU, Islam MS, Hossian MS, Mostofa MG (2018) Recycling of organic wastes through the vermicomposting process of cow dung and crop residues. J Bangladesh Acad Sci 42(1):1–9

Jain K, Singh J, Chauhan LKS, Murthy RC, Gupta SK (2004) Modulation of flyash-induced genotoxicity in *Vicia faba* by vermicomposting. Ecotoxicol Environ Saf 59:89–94

Kaushik P, Garg VK (2003) Vermicomposting of mixed solid textile mill sludge and cow dung with the epigeic earthworm Eisenia foetida. Bioresour Technol 90(3):311–316

Kaviraj, Sharma S (2003) Municipal solid waste management through vermicomposting employing exotic and local species of earthworms. Bioresour Technol 90:169–173

Kawaguchi S, Nishi S (2007) Nutritional and microbial parameters of earthworm cast, termite mound and surrounding bulk soil. J Fac Agric Kyushu Univ (Japan) 52:367–369

Khwairakpam M, Bhargava R (2009) Vermitechnology for sewage sludge recycling. J Hazard Mater 161:948–954

Kiehl JT, Yamanouchi T (1985) A parameterization for absorption due to the A, B, and γ oxygen bands. Tellus B Chem Phys Meteorol 37(1):1–6

Kitturmath MS, Giraddi RS, Basavaraj B (2007) Nutrient changes during earthworm-Eudrilus eugeniae mediated vermicomposting of agro industrial wastes, Karnataka. J Agric Sci 20 (3):653–654

Kumar R, Sharma RK, Singh AP (2017) Cellulose-based grafted biosorbents-journey from ligno-cellulose biomass to toxic metal ions sorption applications—a review. J Mol Liq 232:62–93

Lee PA, Ramakrishnan TV (1985) Disordered electronic systems. Rev Mod Phys 57(2):287

Lim SL, Wu TY, Sim EYS, Lim PN, Clarke C (2012) Biotransformation of rice husk into organic fertilizer through vermicomposting. Ecol Eng 41:60–64

Lourdumary AJ, Ramesh N (2014) Evaluation of anticancer activity of Indian earthworm Lampito mauritii. Int J Res Pharm Sci 4:2730

Marti R, Dabert P, Ziebal C, Pourcher AM (2010) Evaluation of Lactobacillus sobrius/amylovorus as a new microbial marker of pig manure. Appl Environ Microbiol 76(5):1456–1461

Mittra BN, Karmakar S, Swain DK, Ghosh BC (2005) Fly ash—a potential source of soil amendment and a component of integrated plant nutrient supply system. Fuel 84 (11):1447–1451

Nawaz I (2013) Disposal and utilization of fly ash to protect the environment. Int J Innov Res Sci Eng Technol 2(10):5259–5266

Ndegwa PM, Thompson SA, Das KC (2000) Effects of stocking density and feeding rate on vermicomposting of biosolids. Bioresour Technol 71:5–12

Niyazi R, Chaurasia S (2014) Vermiremoval of heavy metals from fly ash by employing earthworm Eisenia foetida. Int J Biol Sci Technol 6:1

Ouédraogo E, Mando A, Zombré NP (2001) Use of compost to improve soil properties and crop productivity under low input agricultural system in West Africa. J Agric Ecosyst Environ 84:259–266

Padmavathiamma PK, Li LY, Kumari UR (2008) An experimental study of vermi-biowaste composting for agricultural soil improvement. Bioresour Technol 99:1672–1681

Paoletti MG, Faveretto MR, Stinner BR, Purrington FF, Bater JE (1991) Invertebrates as bioindicators of soil use. Agric Ecosyst Environ 34:341–362

Patinvoh RJ, Osadolor OA, Horváth IS, Taherzadeh MJ (2017) Cost-effective dry anaerobic digestion in textile bioreactors: experimental and economic evaluation. Bioresour Technol 245:549–559

Pattnaik S, Reddy MV (2010) Assessment of municipal solid waste management in Puducherry (Pondicherry), India. Resourc Conserv Recycl 54:512–520

Samal K, Dash RR, Bhunia P (2017) Treatment of wastewater by vermifiltration integrated with macrophyte filter: a review. J Environ Chem Eng 5(3):2274–2289

Shahmansouri MR, Pourmoghadas H, Parvaresh AR, Alidadi H (2005) Heavy metals bioaccumulation by Iranian and Australian earthworms (*Eisenia fetida*) in the sewage sludge vermicomposting. Iran J Environ Health Sci Eng 2:28–32

Sharma KS (2003) Municipal solid waste management through vermicomposting employing exotic and local species of earthworms. Bioresour Technol 90:169–173

Shrestha P, Bellitürk K, Görres JH (2019) Phytoremediation of heavy metal-contaminated soil by switchgrass: a comparative study utilizing different composts and coir fiber on pollution remediation, plant productivity, and nutrient leaching. Int J Environ Res Public Health 16 (7):1261

Singh J, Kaur A, Vig AP, Rup PJ (2010) Role of *Eisenia fetida* in rapid recycling of nutrients from bio sludge of beverage industry. Ecotoxicol Environ Saf 73:430–435

Singh K, Bhartiya DK, Chauhan HK, Rai R, Singh RN (2014a) Bioaccumulation of cobalt and lead by earthworm Eisenia fetida from sewage sludge with different cattle dung during vermicomposting. Res J Sci Technol 6(4):175–179

Singh RK, Gupta NC, Guha BK (2014b) pH dependence leaching characteristics of selected metals from coal fly ash and its impact on groundwater quality. Int J Chem Environ Eng 5:218–222

Singh S, Bhat SA, Singh J, Kaur R, Vig AP (2016) Vermistabilization of thermal power plant fly ash using *Eisenia Fetida*. J Ind Pollut Control 32(2):554–561

Singh R, Samal K, Dash RR, Bhunia P (2019) Vermifiltration as a sustainable natural treatment technology for the treatment and reuse of wastewater: a review. J Environ Manag 247:140–151

Suthar S (2006) Potential utilization of guargum industrial waste in vermicompost production. Bioresour Technol 97:2474–2477

Suthar S (2009) Bioremediation of agricultural wastes through vermicomposting. Bioremediation 13(1):21–28. https://doi.org/10.1080/1088986080269051

Suthar S, Sharma P (2013) Vermicomposting of toxic weed-Lantana camara biomass: chemical and microbial properties changes and assessment of toxicity of end product using seed bioassay. Ecotoxicol Environ Saf 95:179–187

Suzuki KT, Yamamura M, Mori T (1980) Cadmium binding proteins induced in the earthworm. Arch Environ Contam Toxicol 9(4):415–424

Tripathi PSM, Singh G, Singh RC (1997) Prospects of bulk utilization of fly ash for soil amendment and management of degraded lands vis-à-vis environmental problems. In: First All India Peoples' Technology congress, Kolkata

Usmani Z, Kumar V, Mritunjay SK (2017) Vermicomposting of coal fly ash using epigeic and epi-endogeic earthworm species: nutrient dynamics and metal remediation. R Soc Chem Adv 7 (9):4876–4890

Vig AP, Singh J, Wani SH, Dhaliwal SS (2011) Vermicomposting of tannery sludge mixed with cattle dung into valuable manure using earthworm *Eisenia fetida* (Savigny). Bioresour Technol 102(17):7941–7945

Waseem MA, Giraddi RS, Math KK (2013) Assessment of nutrients and microflora in vermicompost enriched with various organics. J Exp Zool India 16:697–703

Wikipedia (n.d.) Fly Ash. The free encyclopedia. en.wikipedia.org/wiki/Fly_ash-12k

Wong JWC, Li SWY, Wong MH (1995) Coal fly ash as a composting material for sewage sludge: effects on microbial activities. Environ Technol 16(6):527–537

Wu S, Shen Z, Yang C, Zhou Y, Li X, Zeng G, Ai S, He H (2016) Effects of C/N ratio and bulking agent on speciation of Zn and Cu and enzymatic activity during pig manure composting. Int Biodeter Biodegr 119:429–436

Wu J, Zhao Y, Qi H, Zhao X, Yang T, Du Y, Zhang H, Wei Z (2017) Identifying the key factors that affect the formation of humic substance during different materials composting. Bioresour Technol 244:1193–1196

Part V
Soil

Chapter 18
Some Perspectives on Vermicompost Utilization in Organic Agriculture

Hupenyu A. Mupambwa, Balasuramani Ravindran, Ernest Dube, Noxolo S. Lukashe, Asteria A. N. Katakula, and Pearson N. S. Mnkeni

Abstract Over the past decades, agriculture has heavily relied on industrially produced inorganic sources of nutrients to fertilize crops for attaining the higher yields required to feed the growing world population. However, there is a growing realization that this type of agriculture tends to feed the crop rather than feeding the soil, which is a precious resource that is being degraded by the current forms of crop fertilization. Organic agriculture is now gaining momentum as a sustainable form of agriculture as it replicates nature by feeding the soil to feed the plant, thus increasing the resilience of soils to degradation. This chapter presents insights into how vermicomposting technology can be critical in driving the adoption of organic agriculture, whilst highlighting some of the areas where research is required to see this being a success.

H. A. Mupambwa (✉)
Sam Nujoma Marine and Coastal Resources Research Centre (SANUMARC), Sam Nujoma Campus, University of Namibia, Henties Bay, Namibia
e-mail: hmupambwa@unam.na

B. Ravindran
Department of Environmental Energy and Engineering, Kyonggi University Youngtong-Gu, Suwon, Gyeonggi-Do, South Korea

E. Dube
School of Natural Resources Management, Nelson Mandela Metropolitan University, George Campus, George, South Africa

N. S. Lukashe
Department of Soil Science, Stellenbosch University, Stellenbosch, South Africa

A. A. N. Katakula
Sam Nujoma Marine and Coastal Resources Research Centre (SANUMARC), Sam Nujoma Campus, University of Namibia, Henties Bay, Namibia

Crop Science Department, Ogongo Campus, University of Namibia, Ogongo, Namibia

P. N. S. Mnkeni
Faculty of Science and Technology, University of Arusha, Arusha, Tanzania

S. A. Bhat et al. (eds.), *Earthworm Assisted Remediation of Effluents and Wastes*,
https://doi.org/10.1007/978-981-15-4522-1_18

Keywords Earthworms · Mesophilic composting · Nutrient loss · Ionic nutrients · Soil quality

18.1 Introduction

The rapidly growing world population has increased food demand leading to intensification of agricultural and industrial activities to meet this demand (Mupambwa and Mnkeni 2018). Intensification of agricultural systems in the crop-based systems has, however, seen the overuse of and overdependence on synthetic inorganic fertilizers to provide nutrition to crops. The use of synthetic fertilizers, was the major driver of success for the green revolution; but of late, this has been realized to have its own challenges such as environmental pollution and causing soil degradation. Against this background, Rahmann et al. (2017) suggested that future food production and consumption challenges will require to focus on the following: (1) producing enough, healthy and safe food for the growing world population; (2) reducing environmental pollution associated with feeding this population; (3) employing renewable energy and recycled nutrients in food production; (4) adapting and improving agriculture resilience to climate change; and (5) protecting soil health and its function. The growing realization of the negative impacts of inorganic fertilizers to soil health and their economic impact on agriculture has stimulated interest in the use of soil health compatible organic nutrient sources like vermicomposts (Arancon et al. 2004, 2008). However, the low macronutrient concentrations of organic nutrient sources relative to inorganic fertilizers has been among the major limitations to the sustained use of organic fertilizers like vermicomposts in agriculture. In this chapter, we give in-depth insights to how improved organic fertilizers like vermicomposts can be critical in driving the future sustainability of crop-based agriculture.

18.2 What Is Organic Agriculture: The Basics

The idea of organic agriculture is widely believed to have been pioneered by Rudolf Steiner (1861–1925) through a series of lectures that gave birth to the movement of biodynamic agriculture (Rahmann et al. 2017; Paull and Hennig 2011). Rudolf Steiner ran this first course on organic agriculture in 1924 to a group of farmers in the Polish village of Koberwitz now known as Kobierzyce (Paull and Hennig 2011). According to literature, Rudolf Steiner's lectures were in partial response to the growing chemicalization of agriculture, which was being principally driven by industrialization of the Haber–Bosch process developed by Fritz Haber and Carl Bosch in 1909, allowing for the production of nitrogen fertilizer (Paull and Hennig 2011). Generally, the system of organic agriculture, which arose at the beginning of the twentieth century, has gone through three stages outlined by Rahmann et al.

(2017) as organic 1.0, organic 2.0 and organic 3.0. According to Rahmann et al. (2017), organic 1.0 is that period led by organic pioneers who developed the vision of organic agriculture (OA); organic 2.0 described a recent period which saw the growth and marketing of OA, founding of organic research institutes culminating in the formation of the International Federation of Organic Agriculture Movements (IFOAM), whilst organic 3.0 addresses the future challenges and aims at pushing for OA at a global scale.

The FAO/WHO (1999) Codex Alimentarius Commission has by far given the most complete definition of organic agriculture, describing it as 'a holistic production management system which promotes and enhances agro-ecosystem health, including biodiversity, biological cycles and soil biological activity. It emphasises the use of management practices in preference to the use of off-farm inputs, taking into account that regional conditions require locally adapted systems. This is accomplished by using, where possible, agronomic, biological and mechanical methods, as opposed to using synthetic materials, to fulfil any specific function within the system'. Due to the environmental challenges associated with modern agricultural practices, the period between 1980 and 1990 saw a great revival in organic agriculture (Kirchmann et al. 2008). In simpler terms, organic refers to anything derived from living matter and in crop-based agriculture, the term organic agriculture loosely refers to a farming system where plant nutrition is provided for by decomposing materials of usually of animal or plant origin such as manure and leaf litter. To understand more the difference between organic and inorganic plant nutrition, it is important to note that plants can only absorb nutrients in their ionic form. For example, for the primary fertilizer elements; nitrogen is absorbed as ammonium (NH_4^+) or nitrate (NO_3^-); phosphorus is absorbed as either $H_2PO_4^-$ or HPO_4^{2-}; whilst potassium is absorbed as K^+. In inorganic fertilization, mineral nutrients that are either mined or chemically produced and these mineral salts dissociate in water when applied to soil thus availing their ionic nutrients for plant uptake instantly. However, organic fertilizers due to their minerals being organically bound require soil organisms to facilitate their mineralization over time to convert these nutrients into their ionic, plant available forms (Masunga et al. 2016).

As outlined by Lotter (2003), 'in the United States of America, the organic rule in crop production, which is the focus of this chapter, states that an organic crop producer must manage soil fertility through use of crop rotations, cover crops and use plant and animal materials or low solubility natural minerals. These practices must improve or maintain soil organic matter content, manage deficient or excess plant nutrients and control erosion to the extent that these functions are applicable to the operation. The producer must use preventive practices to manage crop pests, weeds and diseases through use of crop rations, precise nutrient management, sanitary measures and cultural practices that enhance crop health. Furthermore, animal manures used must be composted or be applied in soil 120 days before crop harvest'. It is also important to note that organic agriculture can also refer to animal production, which limits use of synthetic products during production, though this is not the focus of this chapter. In crop-based organic and inorganic agriculture,

Table 18.1 Comparison between organic and inorganic nutrient management in soils for crop production

Organic nutrients in soil fertility management		Inorganic nutrients in soil fertility management		
Advantages	Disadvantages	Advantages	Disadvantages	References
Improves soil structure and overall soil health.	Nutrient content not standardized.	Nutrients readily available.	High cost of production.	Naguib (2011), Hazra (2016), Rahmann et al. (2017), Lotter (2003)
Enhances soil microbial activity.	Lower macronutrient content.	Fast crop response.	Can result in possible crop damage.	
Sustainable and environmentally friendly.	Not easy to provide recommendations for use.	Ease of application.	Susceptible to leaching and thus causing environmental pollution.	
Low cost.	Usually bulky and expensive to transport.	Nutrient content standardized and predictable.	Feeds the plant rather than the soil, thus does not enhance soil health.	
Slow release of nutrients.	Can transmit weed seeds; crop and soil diseases if not properly cured.	High macronutrient content.		
Broad range of micronutrients		Can be prepared in different forms for application.		
Grows more resilient crops with pest and disease resistance.				
Offers an opportunity to re-cycling nutrients.				

several researchers have presented the advantages and disadvantages of both systems and these are presented in Table 18.1.

Though the pros and cons of organic and inorganic agriculture seem to favour organic agriculture, its uptake is still slow especially in African countries. This is clearly reflected in the reconstructed world map by Paull and Hennig (2011), which depicts proportionally the area under organic agriculture in each country (Fig. 18.1).

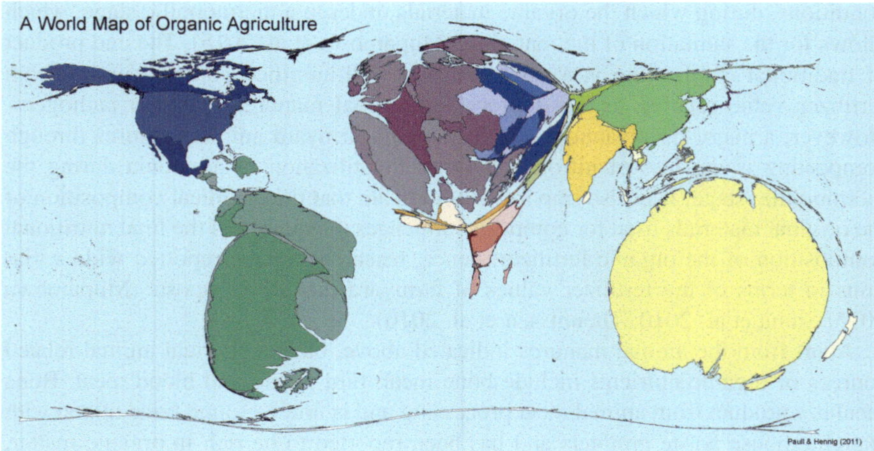

Fig. 18.1 A reconstruction of the world map by Paull and Hennig (2011) based on the application under organic agriculture in countries

18.3 Sources of Organic Fertilizers in Agriculture

As indicated previously, organic fertilizers are those that are derived from living materials, which means these are mainly of plant or animal origins. However, though organic fertilizers are mostly described as being derived from living materials, there are mineral sources that have been accepted as nutrient sources in organic agriculture and these include dolomitic limestone, sulphur, greensand, Epsom salts and rock phosphate (Green 2019; Lavison 2013). For this chapter, the sources of organic fertilizers discussed include farmyard manure composts; urban waste and sewage sludge; cover crops and green manure; compost leachates and animal urine; mineral fertilizers and vermicomposts.

18.3.1 Farmyard Manure Composts

The use of animal manures such as cow dung, sheep, pig, poultry, goat and horse manures as sources of nutrients in agriculture dates back to several centuries ago (Loss et al. 2019). However, though rich in nutrients, these animal manures are not suitable for direct application for plant nutrition as most of the nutrients present are organically bound and require mineralization to convert the organic nutrients into inorganic forms required for plant nutrition. Composting is one of the most popular methods that have been used to convert farmyard manures into effective nutrient sources (Mupambwa et al. 2015). The composting process involves the accelerated aerobic degradation of organic matter by microorganisms under controlled conditions, during which the organic materials undergo a thermophilic stage, which allows for the sanitation of the materials (Mupambwa et al. 2015). The end product of traditional composting is a stabilized and well-humified compost with a high fertilizer value relative to the unprocessed animal manures, with no pathogens. However, a major disadvantage of mineralizing farmyard animal manures through composting is the loss of nitrogen through volatilization of ammonia during the thermophilic stage. It is also important to indicate that the chemical composition of the original materials used for composts influences the quality of the final nutritional composition of the organic fertilizer; hence, researchers have reported wide variations in terms of the fertilizer values of farmyard manure composts (Mupambwa 2015; Sinha et al. 2010; Theunissen et al. 2010).

Apart from the animal manures indicated above, other important animal-related sources of organic nutrients include bone meal, bird guano and blood meal. Bone meal is a product from animal meat processing and is finely ground bone mixed with slaughterhouse waste products and has been reported to be rich in organic matter, nitrogen, phosphorus and calcium (Jeng et al. 2006). Meanwhile, guano which is of bird or bat origin is made up of deposits consisting of excrements, eggshells and carcasses of dead bird and bats. Guano has been reported to be a very high source of the essential macro-elements nitrogen, phosphorus and potassium (Schnug et al. 2018). Lastly, blood meal, which is dried and powdered blood collected from animal slaughterhouses is an excellent source of organic nitrogen due to its the high protein composition.

18.3.2 Urban or Municipal Waste and Sewage Sludge

Urban wastes and sewage sludge are all waste products resulting from the increase in world population, urbanization and industrial activities. Due to the increase in these municipal wastes and sewage sludge, options such as incineration, pyrolysis and direct soil applications have been evaluated (Ayuso et al. 1996). Application of the composted municipal wastes presents an interesting option for the sustainable management and use of these wastes though the major challenge is that it could lead to the accumulation of phytotoxic levels of heavy metals in soils (Hemmata et al. 2010). The interest of using municipal wastes and sewage sludge has been driven by their high content of organic matter, essential macronutrients and other micronutrients required for plant growth (Mdolo 2016). However, the presence of potentially toxic heavy metals and pathogens in these wastes has seen their strict regulation when used in organic soil fertility management as the heavy metals taken up by the plants can easily bioaccumulate in other organisms higher up in the trophic levels. This has seen the use of municipal wastes and sewage sludge being strongly regulated by different bodies in many countries (Mdolo 2016; Alvarenga et al. 2015).

18.3.3 Cover Crops or Green Manure

As previously outlined, in organic soil fertility management, there is a need to supply nutrients from mineralized organic sources. Cover crops and green manures have thus been widely promoted in organic agriculture and these are cereals, annual or perennial grasses or legumes used to cover the soil for a certain period and then ploughed under to contribute organic matter and nutrients to the soil (Gaskell and Smith 2007). Cover crops play an important function in soil fertility management as they capture residual nutrients during crop production and sequester it in plant biomass for subsequent supply to the next crops during decomposition. Furthermore, other green manure crops are grown mainly due to their ability to fix atmospheric nitrogen into organic nitrogen which is an important nutrient source in organic agriculture.

18.3.4 Mineral Organic Fertilizers

There is a shift towards classifying nutrient sources that are of *natural* origin as suitable sources of nutrients in organic agriculture. This has seen the acceptance of natural unprocessed mineral nutrient sources such as rock phosphate (source of phosphorus), dolomitic limestone (raises the pH of soils that are low in magnesium), sulphur (lowers soil pH), greensand (supplies potassium and several micronutrients) and Epsom salts (provide rapid release of magnesium) in organic agriculture (Mupambwa and Mnkeni 2018; Lavison 2013; Edwards et al. 2010; Chianu et al. 2012). Mineral fertilizers are used mainly for soil pH control, as well as for the supply of nutrients like phosphorus, potassium and magnesium but none of these provides any nitrogen (Green 2019). Unlike synthetic inorganic fertilizers, the nutrients on mineral fertilizers require natural soil chemical processes to make them bio-available for plant uptake; hence, the rate of nutrient release is also slow when directly applied to soil. However, processes such as vermicomposting can be used to further enhance the bio-availability of nutrients in the mineral fertilizers (Mupambwa and Mnkeni 2018).

18.3.5 Compost Leachates and Human Urine

Unlike solid nutrient sources identified before, several researchers have looked at the possibility of extracting nutrients from solid composts by leaching them with water thus producing liquid nutrient sources called leachates or using liquid excreta (urine) from animals and humans (Gutierrez-Miceli et al. 2008; Mnkeni et al. 2008). The leachates from composts have been reported to be high in macro- and micronutrients and also contain essential humic acids which are important in regulating several plant development processes such as nutrient absorption (Gutierrez-Miceli et al. 2011). Furthermore, these leachates have also been shown to be effective in controlling several plant diseases. Apart from using leachates from composts, researchers have also evaluated the potential of using human urine as a nutrient source in crop production (Mnkeni et al. 2005, 2008; Karak and Bhattacharyya 2011). Human urine can be a valuable source of nutrients and has been used since ancient times to enhance crop growth. It has the advantage that most of the nutrients in it are in ionic form with plant availability that compares very well with that of synthetic fertilizers (Mnkeni et al. 2008). Karak and Bhattacharyya (2011) in a review reported that human urine can be an important source of the macro-elements N, P and K. However, despite its potential as a nutrient source in organic agriculture, the adoption of human urine as a nutrient source has been very limited by its social stigma as a human waste and its high salinization potential when continually applied to arable land.

18.3.6 Vermicomposts

One important source of nutrients in organic agriculture is vermicomposts, which has started to gain more attention compared to traditional composts due to its superiority in terms of chemical, biological and physical quality. Vermicomposts, which are the main focus of this chapter, are a product of the action of mainly epigeic earthworm species and microorganisms in driving the mineralization of organic nutrients in organic materials (Mupambwa and Mnkeni 2018). Unlike traditional thermophilic composting, vermicomposting only involves mesophilic composting which is critical in avoiding the loss of nitrogen as ammonia as nitrifying bacteria are able to survive under these mesophilic temperatures. Vermicomposts can be made from different organic materials similar to those used for composting; hence, their quality is also largely dependent on the quality of the organic materials used. Vermicomposts have also been reported to be provide not only nutrients for plant growth but are also rich in humic acids and plant growth hormones (Singh et al. 2008). Though vermicomposts are being widely promoted in organic agriculture, the greatest limitation to their commercial adoption is the wide variation in the quality of vermicomposts, which is dependent on the method of preparation, earthworm species used and stocking density, and quality of organic materials, among other factors. Research on optimization and standardization of the production procedures of vermicomposts is critical, as this can allow for the production of vermicomposts with predictable nutrient composition.

18.4 Vermicomposting

18.4.1 The Process

The term *vermi* in vermicomposting is derived from the Latin word 'vermis', which means a worm. However, vermicomposting refers to a composting process that is done by epigeic, anecic, or endogeic earthworm species, which have a natural ability to colonize and degrade organic wastes (Bhat et al. 2018; Das et al. 2016). According to Gomez-Brandon and Dominguez (2014), vermicomposting has thus been defined as 'bio-oxidative process in which detrivorous earthworms interact with microorganisms and other fauna within the decomposer community, thus accelerating the stabilization of organic matter (OM) and greatly modifying its physical and biochemical properties'. Though there are more than 4000 known species of earthworms classified according to the three groups, only a few belonging to the epigeic class have been shown to be effective for the vermicomposting process mainly due to their high feed consumption rates, high reproduction rates and high tolerance to a wide range of environmental conditions (Bhat et al. 2017; Dominguez and Edwards 2011). Among the widely used vermicomposting earthworm species which belong to the epigeic earthworm class include *Eisenia fetida, Eisenia andrei, Eudrilus eugeniae* and *Perionyx excavates* (Mupambwa et al. 2016; Atiyeh et al. 2000; Ravindran et al. 2015). Also, though having a low vermicomposting potential, endogeic species *Metaphire posthuma* and *Drawida barwelli* (Das et al. 2016; Bhat et al. 2017) together with anecic species *Lampito mauritii; Apporrectodea trapezoids* and *Lumbricus terrestris* (Bhat et al. 2017; Anbalagan and Manivannan 2012) have also been used for vermicomposting.

Earthworms have been shown to enhance the composting process directly by mechanically breaking down larger organic wastes using their gizzards, thus increasing the substrate surface area and consequently altering microbial activity, in processes that have been collectively called gut-associated processes (Swati and Hait 2017; Gomez-Brandon and Dominguez 2014). Therefore, apart from the mechanical action of the earthworms, the microorganisms within the earthworm gut and the compost are then responsible for producing various bio-compounds like enzymes and organic acids, which are responsible for accelerating the biodegradation and nutrient mineralization within vermicompost (Dominguez et al. 2010; Aira et al. 2007), as schematically shown in Fig. 18.2.

According to Dominguez et al. (2010), microorganisms are the most abundant and diverse members of the vermidegradation food web, with the earthworms being secondary higher level consumers existing together with microorganisms, which feed and disperse the microbes. The crucial link between earthworms and microbes during the vermicomposting process has driven several researchers into trying to deliberately modify the microbial composition of vermicomposts by inoculating vermicomposts with specialized microbial cocktails such as effective microorganism (EM), phosphate solubilizing microbes (PSM), N_2-fixing bacteria (Mupambwa et al. 2016; Busato et al. 2012; Kumar and Singh 2001). Such microbial cocktails are

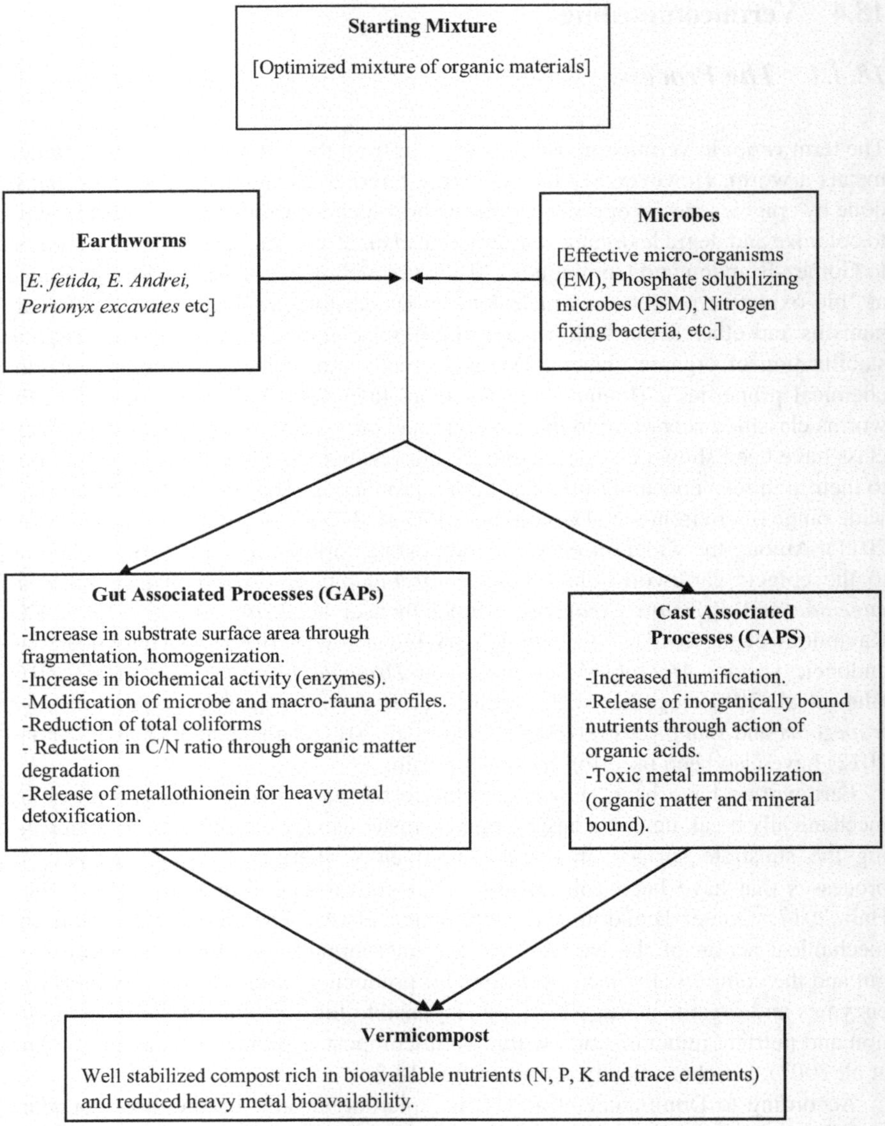

Fig. 18.2 Schematic representation of earthworm and microbe roles during vermicomposting of organic materials. (Source: Mupambwa and Mnkeni 2018)

believed to play a crucial role in cast-associated processes (CAPs), which take place during the ageing of the vermicompost, following transit of the organic materials through the earthworms' gut.

During the vermicomposting process and mainly in the CAPs, low molecular weight organic acids which include oxalic, formic, citric, acetic acids among others are produced (Busato et al. 2012). These organic acids have been observed to play a crucial role in nutrient mineralization during vermicomposting, but have, however, received limited research attention (Bolan et al. 1994; Kumari et al. 2008). Though these organic acids are not produced directly by the earthworms, the earthworm-microbe driven process of organic decomposition during vermicomposting produces these organic acids which are crucial in nutrient mineralization, particularly in releasing inorganically bound nutrients (Busato et al. 2012). These organic acids have been reported to cause this mineralization via three processes which include competition for adsorption sites, dissolution of adsorbents and changes in the surface charges of adsorbents (Bolan et al. 1994). So, nutrient release during vermicomposting can be viewed from both a biological and chemical perspective. Much of the research has, however, focused on the biological aspects of vermicomposting related to nutrient mineralization, with limited attention being given to how changes in other chemical properties such as organic acids influence nutrient mineralization.

18.4.2 Nutrient Enrichment of Vermicomposts

Though vermicomposts are being promoted as organic fertilizers, their adoption as commercial fertilizers in agriculture is limited by their low macronutrient concentrations relative to inorganic fertilizers. For example, commercial inorganic fertilizers can have as much as 46% N, 48% P, whilst vermicomposts have been reported to have around 0.96% N and 0.21% P when made from cow dung and banana wastes (Padmavathiamma et al. 2008). In another study, Singh et al. (2008) reported macronutrient levels of N (0.92%), P (1.21%) and K (1.45%) for a vermicompost prepared from vegetable waste and cow dung. This means that, to achieve the same nutrient level as inorganic fertilizers, organic sources of nutrients need to be applied at higher volumes per land area or amended with inorganic fertilizers sources as suggested by Singh et al. (2008) and Arancon et al. (2006). However, organic fertilizers have other extra benefits such as improving soil physical, chemical and biological properties, slow release of nutrients, reducing physiological disorders in vegetables and being environmentally friendly if properly managed, compared to inorganic fertilizers (Singh et al. 2008; Das et al. 2016). As a solution to the low nutrient content of organic fertilizers, several researchers have looked at the potential of amending vermicomposts with other nutrient-rich inorganic materials (Kumar and Singh 2001). One important macronutrient whose concentration can be enhanced through amending vermicomposts with inorganic materials is phosphorus (P), and this has been the subject of several studies (Bhattacharya and Chattopadhyay 2002; Edwards et al. 2010; Adhami et al. 2014).

Apart from nitrogen, phosphorus is the most important plant nutrient that plays critical roles in almost all major bio-chemical processes within plants such as photosynthesis, energy transfer and molecular biosynthesis (Roy 2017; Sharma et al. 2013). Although P is abundant in soils, it is mostly insoluble and precipitated in different forms which are not easily available for plant uptake (Sharma et al. 2013). Due to this, available P levels in soils have to be supplemented in most agricultural soils by adding chemical P fertilizers to improve crop productivity, which represents a major cost in agriculture and poses great environmental risks (Alamgir et al. 2012). The economic and environmental challenges associated with the use of chemical P fertilizers such as eutrophication and diminishing P reserves have resulted in a renewed interest in alternative sources of P like phospho-composts (Alamgir et al. 2012). Recent research has focused on two inorganic materials that have been shown to have potential in improving the phosphorus nutrition of organic vermicompost, namely rock phosphate (Kumar and Singh 2001; Aria et al. 2010; Busato et al. 2012) and fly ash (Malik and Thapliyal 2009; Bhattacharya et al. 2012; Mupambwa et al. 2015). What has created this interest is the high total P content within fly ash and rock phosphate, which can be converted into bioavailable forms through processes like vermicomposting. Phosphate rock or rock phosphate is a term that describes any naturally occurring geological material that contains high concentrations of phosphate bearing minerals with a phosphate content of 15–20% (Edwards et al. 2010). There are various types of these rock phosphates with varying chemical, mineralogical and physical properties depending on origin (Van Kauwenbergh 2010). Fly ash, on the other hand, is a product of coal combustion, which is captured through electrostatic precipitators at thermal power stations, and there are different types depending on the origin of the coal materials (Mupambwa et al. 2015; Ukwattage et al. 2013).

Apart from inorganic materials that are being promoted as amendments, biochar, which is a highly aromatic product of pyrolysis, has recently received great interest as highly stabilized organic additive for vermicomposting. Lehmann et al. (2006) reported that biochar was shown to yield high aromatic materials with high stability when added to soil and that it enhances the carbon sequestration potential of vermicomposts, thus mitigating climate change. Different biochar qualities can be produced from different feedstock and pyrolysis temperature (Zhao et al. 2013). Barthod et al. (2018) indicated that higher pyrolysis temperatures during biochar preparation leads to greater aromatization of the material, decreasing its surface area, cation exchange capacity and content of volatile compounds. Biochar has a potential to enhance vermicompost microbial activity due to its porous structure through moisture and aeration control. However, biochar application rate higher than 20% is not recommended since it may hinder organic matter biodegradation. Research is required to identify the optimum biochar amendment rate for different vermicomposts (Xiao et al. 2017).

18.5 Factors to Consider in the Production and Use of Organic Fertilizers

In line with principles of sustainable agriculture, farmers are called upon to improve the soil on cropped lands through organic fertilizer inputs. Inherent poor fertility of most soils in the tropics is a major constraint to sustainable smallholder crop production. Synthetic fertilizers often lack essential micronutrients crucial for crop growth on the nutrient poor soils cultivated by smallholder farmers. Application of organic fertilizers like vermicomposts on agricultural land can complement inorganic fertilizers because they not only add mineral nutrients but also improve soil organic matter. This in turn positively influences soil quality by reducing compaction, erosion processes and the toxicity of pollutants while also improving soil structure, porosity, aeration, water infiltration, water holding capacity and nutrient availability. Cocktails of beneficial bacteria and fungi found in organic fertilizers convert nutrients into easily accessible forms and aid biological activity. Slow nutrient release by vermicomposts reduces the risk of toxicity to plants and the environment.

The tropics are experiencing extreme weather events due to climate change, and these include high annual temperatures and rainfall variability. Farmers here are facing many challenges such as unpredictable rains, soil degradation and new or different pests and diseases. Organic farming can help to mitigate the harms of climate variability, mostly by storing carbon in the soil and improving plant health. In organic matter rich soils, roots develop well to exploit the supply of mineral nutrients and to take up water to meet crop requirements for growth. Organic matter also provides a buffer between the soil and air interface, thereby protecting the roots from desiccation and steep gradient changes in temperature. Organic fertilizers increase the return of carbon to the soil, and this raises productivity whilst favouring carbon storage. Additionally, organic fertilizers help farmers to adapt to climate change because the high cation exchange capacity (CEC) from soil organic matter can help to reduce nutrient losses in the event of floods. Soils and plants under organic farming are more resilient to pests, floods, droughts and land degradation processes.

Most organic fertilizers are products of a composting or vermicomposting process. Composting organic waste to make fertilizers does not only reduce pests and toxic compounds composition, but it also improves the handling characteristics of nutrient rich organic waste by reducing its volume and weight while increasing product uniformity and ease of application. Producing good organic fertilizer is a blend of art and science. Composting techniques can be as simple as stockpiling and turning, or more complex as in vermicomposting or active aeration or in-vessel systems. In all essence, good aeration is the key to achieving good compost. Bad composting can produce as many potential problems with odour and greenhouse gas emissions. The mixing technique (method and frequency) significantly influences chemical and biological parameters, speed of degradation, as well as the quality of the product. Decomposition is microbial dependent and thus influenced by several factors including C/N ratio, temperature, soil moisture and aeration (Mupambwa and Mnkeni 2018). Organic fertilizer quality can be widely variable depending on several factors such as parent organic materials, vermicomposting technique and maturity, as well as the interactions between these factors. Such variations in quality, hence yield response are a major reason why farmers have at times regarded organic fertilizers as an undependable source of nutrients (Mupambwa et al. 2019).

Materials low in certain essential plant nutrients will not create an organic fertilizer rich in those nutrients. For example, organic fertilizers derived from livestock feeding on a depleted veld tend to be poor in nutritional quality, especially P. On the contrary, organic fertilizers derived from composted feedlot wastes typically contain more P because the base ration ingredients are high in P. In fact, the quality of organic fertilizers that are based on animal manures may also vary considerably depending on the type of animal, its age and physical condition. Handling methods, application technique and timing, including field conditions post application, are also critical factors that influence crop response to organic fertilizers. In the sub-sections below, we review variables that influence quality of organic fertilizers and crop response to provide standards for production and application of organic fertilizers.

18.5.1 Application Rates for Organic Fertilizers

Plants require a proper balance of all the essential nutrients for normal growth and optimum yield. Organic fertilizers usually contain all the nutrients required for plant growth, albeit in varying quantities. Prior to application onto the soil, it is important to collect a sample that accurately represents the quality of the organic fertilizer and have it properly analysed. The nutrient availability of an organic fertilizer needs to be determined so that these organic resources can be effectively utilized. Application of excessive amounts could result in environmental pollution and plant toxicity, whereas under-application results in poor crop performance. It should be noted that not all the nutrients in organic fertilizers are plant-available. Some need to be mineralized in the soil before being absorbed by plants. The plant-availability spectrum of nutrients may increase due to mineralization or decrease due to immobilization, depending on soil properties. Organic fertilizers with finer texture, or those more matured, tend to be characterized by lower C/N ratios, as well as faster decomposition and nutrient release. Farmers with limited access to soil laboratories could consider a greyish black colour, fine texture, longer composting time and earthly smell as indicators of better organic fertilizer quality.

Nutrient levels in organic fertilizers do not precisely match crop nutrient needs. For example, crops generally require N and P in a ratio of 10:1. The N:P ratio in most organic fertilizers is often much less, and this often translates into an excess application of P when the N requirements are met solely through composts. Hence, applying an organic fertilizer to meet the crop's needs for one nutrient may lead to over- or under-application of other nutrients. Getting the maximum value out of composts thus requires applying the composts at proper rates and frequency, as well as balancing the rest of nutrients with inorganic fertilizers. Proper application of rates is crucial in reducing adverse effects such as build-up of toxic elements. The optimal amount of organic fertilizer to be applied in a field is dependent on the composition, available soil nutrients, the crop to be produced and environmental conditions. Efficient, economic, and environmentally sustainable organic fertilizer application should aim to achieve a balance between nutrient application and crop nutrient demand. The first step in determining application rates for the organic fertilizers is establishing target yields.

Organic fertilizer application software packages have been developed by commercial testing laboratories for calculating application rates. The computer-based software simplifies application rate calculations, allows record keeping and provides applicator calibration information. They also allow users to keep records for each field on nutrient application rates, soil test values, crop yield and nutrient removal. Such records are useful for longer term nutrient budgeting. The use of global positioning systems, site-specific soil and yield capability information within a field can further enhance the ability to match nutrient applications to crop. Lastly, proper calibration of a spreader will help insure correct rate and uniform application.

All organic fertilizers continue to decompose during storage. This results in nutrient transformations and losses, especially N. Aeration, higher temperatures and higher N concentration increase the risk for N loss before collection and during storage. Turning has a similar stimulating effect on decomposition. Nutrient and carbon losses during storage tend to be lower for more compact and anaerobic storage systems and for organic fertilizers with higher C/N ratios. There will be higher nutrient losses if the organic fertilizer is not handled properly in transition from production to field.

18.5.2 Application Methods for Organic Fertilizers

Organic fertilizers should be placed in soil, rather than surface-applied, to increase crop root access to nutrients and to reduce potential losses by runoff and volatilization. Under dry conditions, surface-applied organic fertilizer nutrients may be isolated from crop roots by remaining in the dry surface layer. With heavy rain and high runoff, surface-applied organic fertilizers are subject to loss by runoff and erosion. Surface applications of organic fertilizers are accompanied by inevitable losses of N in the form of ammonia gas. Applying organic fertilizers close to the time the crop will need the nutrients will help to reduce losses. Timing and method of organic fertilizer application determines the efficiency of nutrient recycling. It must be spread uniformly to achieve consistent results. Applying and incorporating compost too early for the crop, as in the autumn on saturated soils in high rainfall areas, could result in significant N leaching and groundwater contamination.

When using organic fertilizers, knowledge of the mineralization rate under field conditions is needed. Since mineralization is microbe driven, it is influenced by several factors, including temperature, soil moisture, soil properties and organic fertilizer characteristics. Laboratory incubation studies can be important for evaluating N mineralization mechanisms but relating incubation results to mineralization under actual field conditions can be difficult. Mineralization increases with increasing temperature under conditions found in agricultural soils. Mineralization is greatest when soil moisture is near field capacity and declines with soil drying. Growing degree day, which is a function of temperature, has been used to predict N availability from applied composts. In this case, it is important to understand post field temperature and moisture conditions post compost application. Fields should preferably be managed to minimize ammonia loss following application, hence tillage practices post compost application should be minimal.

18.5.3 Economic Benefits of Organic Fertilizers

To encourage the use of vermicomposts on both conventional and organic agricultural production, it is important that there be a thorough understanding of the benefits. Koelsch (2017) states that, organic fertilizers such as vermicomposts derive their economic value, from improvements in soil quality, increases in yield and replacement of purchased nutrients required for crop production. Compared to commercial inorganic fertilizers, organic fertilizers contain organic carbon, which is the key to maintaining soil health, including the characteristics of cation exchange capacity, soil tilth and water holding capacity. Targeting fields requiring supplemental P produces the largest economic value from organic sources. As mentioned previously, most nutrients in organic fertilizers are slowly released over a longer period than is obtained from most inorganic fertilizers. Soil organic matter has an added benefit of preventing nutrients from leaching or loss through increasing the soil CEC. The organic matter in organic fertilizers plays an important role in nutrient cycling, which provides food for soil organisms and facilitates the formation of aggregates that maintain the soil as a habitat favourable for beneficial soil organisms. The value of nutrients in organic fertilizers like vermicomposts is therefore more than the accumulated value of the individual nutrients. The residual effects of composts last for at least three seasons; therefore, farmers could benefit from its residual fertility in subsequent seasons providing nutrients to crops for several years. Also, addition of organic fertilizers can help to buffer the soil against pH changes, hence saving on liming costs on acidic soils.

18.5.4 Organic Fertilizer (Vermicompost) Quality

Though vermicomposts can be optimized in terms of degradation efficiency and nutrient content, there has been very limited information on product standards and indicators or compost quality parameters to govern their marketing and field application (Fan et al. 2016; Brinton 2000). This is unlike inorganic fertilizers that have to

Table 18.2 Typical physiochemical characteristics of selected composts and vermicompost in USA and India (Source: Mupambwa and Mnkeni 2018)

	USA[a]	India[b]	
Compost parameter	Municipal compost	City compost	Vermicompost
pH	6.0–7.5	6.5–7.5	–
Soluble salts (dS/m)	5.0 or below	4.0 or below	–
C:N ratio	–	<20	–
Nutrients			
N (%)	1.0 or above	0.8 or above	1.0 or above
P (%)	1.0 or above	0.17 or above	0.35 or more
K (%)	–	0.33 or above	0.66
Organic matter content (%)	50–60	21	31
Water holding capacity (dry basis) (%)	100 or more	–	–
Moisture content (%)	40–50	15-25	15-25
Particle size			
Bulk density	0.47–0.59 g/cm^3	<1.0 g/cm^3	0.7–0.9 g/cm^3
Particle size	Pass through 1 mm screen or smaller	90% must pass through a 4 mm IS screen	
Growth screening	Must pass seed germination test and plant growth assay	–	–
Stability	Stable to highly stable	–	–
Pathogens	–	Nil	–

[a]Brinton (2000)
[b]http://ncof.dacnet.nic.in/Training_manuals/Training_manuals_in_English/BF_and_OF_in_FCO.pdf. Downloaded 19 Dec 2017

carry a mandatory label declaring their nutritional composition. Specification of compost quality is important as it enables compost users to more effectively manage their 'plant growing systems' (Cerda et al. 2018; Fan et al. 2016; Brinton 2000). Brinton (2000) provided an in-depth analysis of the status of compost standards in North America and Europe, but other countries including India have also developed such guidelines. Table 18.2 shows compost parameters used in the USA and India for guiding the agronomic use and management of composts. Generally, the development of compost standards has placed greater emphasis on environmental than on agronomic considerations. Accordingly, compost quality is commonly categorized based on heavy metals, non-organic content, maturity and stability (Cerda et al. 2018). Table 18.3 shows typical maximum permissible values of selected heavy metals in composts adopted in the indicated countries. The wide country-to-country variation on acceptable permissible levels of heavy metals indicates the need for ongoing international engagement to establish permissible metal levels for critical compost quality parameters, which have greater universal acceptance. It is noteworthy that only India has attempted to develop standards for vermicomposts. Therefore,

Table 18.3 Typical permissible heavy metal concentrations (mg/kg) in organic fertilizers in different countries (Source: Mupambwa and Mnkeni 2018)

	USA[a]	EU range[b]	EU standard[a]	Canada[a]	India[c]		Malaysia[d]
Element (mg/kg)	Municipal compost and bio-solids	Municipal compost	Municipal compost	Municipal compost	City compost	Vermicompost	Municipal compost
Pb	300	70–1000	150	150	100	100	300
Cu	1500	70–600	100	100	300	–	
Zn	2800	210–4000	400	500	1000	–	
Cr	1200	70–200	100	210	50	50	200
Ni	420	20–200	50	62	50	50	150
Cd	39	0.70–10.0	2.0	3	5	5	2
Hg	17	0.70–10.0	0.5	0.8	0.15	–	–
As	41	–	–	13	–	–	–
Se	36	–	25	2	–	–	–
B	–	–	300	–	–	–	–
Mo	–	–	10	–	–	–	–

[a]Wood's End Research Laboratory (2005)
[b]Brinton (2000)
[c]http://ncof.dacnet.nic.in/Training_manuals/Training_manuals_in_English/BF_and_OF_in_FCO.pdf
[d]Fan et al. (2016)

the optimization of vermicomposting needs to be coupled with the development of standards for this amendment whose importance is steadily growing.

18.6 Soil Quality Under Organic Soil Fertility Management

Sustainable management of agricultural land simultaneously aims at maintaining or enhancing food production, reducing the level of production risk, protecting the potential of natural resources and preventing degradation of soils and water quality, while being economically viable and socially acceptable (Schjonning et al. 2004). Organic farming meets the above listed points through maintaining and improving soil fertility by adopting crop rotations including leguminous fertility building crops and plants with different rooting depths (Watson et al. 2002). Furthermore, organic wastes such as animal manures, by-products of several kinds of wastes and composted residues can be used as amendments to increase soil fertility, since they are important sources of nutrients for growing crops and means for enhancing the overall soil quality (Diacono and Montemurro 2010).

Organic waste materials used as fertilizers are known to have low bioavailable nutrient contents relative to the total content, which can make producers to prefer synthetic inorganic fertilizers. With recent advances in biotechnology, scientifically produced bio-fertilizers (composts) have been reported to be more nutritive and productive than those produced traditionally by farmers in earlier days. As a result, organic farming systems with the aid of nutritive bio-fertilizer like vermicompost are more recommended as they can give very 'high food productivity' with significantly 'higher nutritional quality' (Sinha et al. 2011). Similar to organic based materials, vermicomposted materials have important functions in enhancing soil quality which the synthetic inorganic fertilizers do not have. Organic-based fertilizers provide low cost sorbents that can bind contaminants, subsequently minimizing natural water contamination, minimize eutrophication whilst adding organic matter to the soil, which can enhance chemical, biological and physical properties of the soil (Watson et al. 2002; Ferreras et al. 2006; Antonious 2016). Several studies have documented the effects of vermicompost application in improving soil quality (Song et al. 2015). The following subsections discuss the influence of organic waste materials processed through vermicomposting in enhancing soil quality indicators.

18.6.1 Soil Chemical Properties

Vermicomposts are a nutritive organic fertilizer or soil conditioner with ability to improve soil quality even in degraded soils. They contain most nutrients in plant-available (bioavailable) forms, both the macro- and micronutrients (Rajiv et al. 2009). The addition of vermicomposts in soils improves cation exchange capacity

(CEC), which is important as it translates to more nutrients held in the soils and reduced acidity (Sinha et al. 2011). Reduced soil acidity can also be influenced by the neutralizing capacity of most organic fertilizers which can contain about 5% calcium carbonate in dry matter. The neutralizing capacity of vermicompost has been reported in a study by Lukashe et al. (2019) in acidic mine-waste-contaminated soils amended with vermicompost enriched with fly ash. In this study, the original pH values in the amended soils ranged between acidic 3.7 and 5.3 and were increased to values close to neutrality ranging between 6.8 and 7.6, following application of the vermicomposts.

Neutral soil pH values promote nutrient availability, and vermicomposts are known to provide balanced nutritional release patterns in terms of available N, soluble K, exchangeable Ca, Mg and P, which are subsequently used by plants (Singh et al. 2013). In another study, Zhao et al. (2017) applied vermicomposts at 3000 kg/ha to soil reporting an increase in total phosphorus, available phosphorus, soil organic matter, extractable nitrogen, microbial N and reduced NH_4^+–N. The improved fertility of the soil as measured by the mentioned increased parameters was positively correlated with improved cucumber yields (Zhao et al. 2017). A decrease in pH with addition of vermicompost in soils has also been reported by Manivannan et al. (2009), and this was attributed to acidifying effects of organic acids and/or the increased permeability and leaching of salts. Even so, it is noteworthy that even though there was a decrease in pH, the change was very slight from 7.15 to 7.08 and 7.14 to 6.97 in clay loam and sandy loam soil, respectively, still maintaining the soil pH values at neutrality.

18.6.2 Soil Physical Properties

Vermicompost application in soils mainly modifies physical properties such as the water holding capacity, structure, porosity, aeration, infiltration and resistance to erosion. Aeration, water retention and improved soil structure are naturally influenced by soil aggregate stability. Aggregate stability is influenced by polysaccharides from organic matter, which can act as cementing agents between soil particles in the soil (Mupambwa and Wakindiki 2012). Vermicompost products have been reported to be also enriched with polysaccharides, which could have an influence in aggregate stability when applied in the soil (Lim et al. 2014). Parthasarathi et al. (2008) observed increased porosity in clay-loam and sandy-loam soils amended with vermicompost, and the increased porosity was also attributed to the presence of polysaccharides in vermicompost. The increased porosity in the study subsequently resulted in improved water holding capacity in soils with vermicompost at 100% relative to soils without vermicompost.

18.6.3 Soil Biological Properties

Biological indicators, especially assessment of microbial activity and biomass, are sensitive to soil management practices. As a result, any changes in these properties may precede detectable changes in other soil properties (chemical and physical properties), thus providing early signal of improvement or warning of degradation (Gomez et al. 2006; Diacono and Montemurro 2010). Vermicasts generally contain enormous quantitiesof bacteria, actinomycetes and fungi. The increased population can possibly be attributed to increased surface area for microbial activity as a result of organic matter fragmentation within the earthworm's gizzard. Vermicompost is also a source of organic matter, which is a food source for beneficial soil microbes, and it aids in improving microbial population and diversity. The enhanced beneficial microbial diversity influences transformation, release and cycling of essential nutrients (Sinha and Herat 2012; Soni and Sharma 2016). Indicators of soil microbiological activity include biomass C, basal respiration, ratio of biomass C to total organic C, enzyme activity and metabolic quotients (Arancon et al. 2006). A comparative study was conducted by Arancon et al. (2006) between vermicompost and inorganically fertilized soils with the application of food waste vermicompost, resulting in significantly greater dehydrogenase activity and microbial N relative to the soils with inorganic fertilizer.

18.7 Crop Responses to Vermicomposts and Other Organic Fertilizers

Vermicomposts, whether used as soil amendments or as component of plant growing media, have the ability to improve seed germination, enhance seedling growth, development and overall plant productivity (Peyvast et al. 2007). Positive results have been reported on both media quality and crop development, where vermicomposts were used as a combination of artificial media (Atiyeh et al. 2001; Arancon et al. 2006). The increased growth and yield are attributed to the improved soil or media properties such as increased nutrient availability for plant uptake, microbial activity and plant growth regulators, which are produced by activity of microorganisms such as fungi, bacteria and actinomycetes (Singh et al. 2008; Azarmi et al. 2008). It is noteworthy that addition of vermicompost in soils does not only improve activity of beneficial microorganisms but it also suppresses growth disease causing microorganisms and population of plant parasitic nematodes (Singh et al. 2008). Plant growth is also enhanced by the ability of the vermicompost to supply essential plant nutrients when required. Vermicompost provides a large particulate surface area that provides microsites for strong retention of nutrients in the soil (Singh et al. 2013).

Kumari and Ushakumari (2002) used a rock phosphate enriched vermicompost to amend a soil planted with cow peas (*Vigna unguiculata*) and evaluated yield and nutrient uptake. Though the rock phosphate was applied after the vermicomposting process, it was interesting to observe that the combination of the vermicompost and the rock phosphate showed the best results on grain yield, plant N and P. Compared to the following treatments, no manure, vermicompost alone, and inorganic P fertilizer, the rock phosphate-enriched vermicompost resulted in a grain yield 49.5% greater and plant P 93.5% greater compared to the average of all the other treatments (Kumari and Ushakumari 2002). It would have been interesting to observe the P mineralization had the RP been incorporated during the vermicomposting process rather than after, as it was done in this study. Also, the influence of the vermicompost on the soil properties like heavy metal accumulation was not evaluated in this study.

Mihreteab et al. (2016) used a compost prepared from grass clippings, palm, olive and ornamental tree prunings with or without rock phosphate enrichment. It was noteworthy in this study that the rock phosphate amended compost applied to supply 0.59 g P/kg-compost produced seedlings that were comparable to the substrates substituted with 1.18 P as rock phosphate per kg compost. These results clearly demonstrated the apatite mineralization potential of the composting, which could have even been more if vermicomposting had been done. What is interesting in all vermicomposting research is that though several researchers have indicated positive nutrient results following rock phosphate or fly-ash-enriched vermicomposting, research that could compliment these studies, in terms of actual soil studies, are still rare. Such a study was performed by Bhattacharya et al. (2012) using a fly ash-enriched vermicompost on red and lateritic soils over a period of two years. Unlike the normal belief that fly ash-enriched vermicompost will contribute higher levels of heavy metals, this study reported slightly changed or even reduce heavy metal levels in soils in the second year. Such information is encouraging, as it points to the possibility of using materials like fly ash in organic soil fertility management systems, though thorough studies that support this possibility are still essential. Using fly ash as a P source, Mupambwa et al. (2017) used a cow dung–waste paper vermicompost amended with 33% fly ash as a substitute for pine bark planting media and evaluated its ornamental horticulture potential. Compared to the pine bark alone, substitution of the fly ash-based vermicompost between 25% and 50% resulted in the best marigold germination and flower growth. However, this greenhouse study did not also determine the heavy metal uptake in the ornamental marigold.

18.8 Organic Hydroponics Powered by Vermicomposts: Opportunities and Challenges

In the face of climate change, most regions have seen erratic rainfall patterns which have greatly affected crop production, especially within the resource-poor subsistence farmers. One technology that is being promoted as a solution to improve crop production under arid conditions is hydroponics. Hydroponics is a technology that allows the production of crops in soilless systems using a balanced liquid nutrient solution. In hydroponic systems, the source of nutrients is mainly chemical inorganic fertilizers, which are usually very expensive and beyond reach of resource-poor farmers, especially in developing countries (Aboutalebi et al. 2013; Mowa 2015). Research focusing on generating optimized organic nutrient sources utilizing abundant animal (goat, cow and donkey) manures that can be effectively used in hydroponics can be crucial in enhancing adoption of hydroponics even amongst the resource-poor farmers. Research on organic nutrient sources in hydroponics is currently very limited. Table 18.4 summarizes some of the research that has been done so far.

As highlighted in Table 18.4, though several researchers have attempted to make use of different organic nutrient sources in hydroponics, there is still paucity of information on the development of an optimized organic nutrient formula. Despite being a great challenge, a holistic approach that considers the following (1) the nutritional composition of the organic materials used, (2) optimization and standardization of organic nutrient solution method of preparation and (3) development of effective models can allow researchers to predict final nutrient composition of organic nutrient solutions, as proposed by Son and Okuya (1991). Research that optimizes organic hydroponic solution preparation and utilization is required to allow for adoption of these cheaper nutrient solutions.

In generating organic nutrient sources, one technology highlighted already with ability to enhance nutrient mineralization, creating a nutrient-rich final product, is vermicomposting (Bhat et al. 2017; Garcia-Sanchez et al. 2017). However, during vermicomposting research, most researchers have focused on the nutrient composition of the solid vermicompost itself (Malinska et al. 2017; Mupambwa et al. 2016), with limited interest in the vermi-leachate. During vermicomposting, excess water called vermi-leachate or vermi-wash can be collected. Unlike nutrient solutions produced directly from organic substances such as biogas digestates, vermi-leachates have great potential as effective nutrient sources in soilless crop production, due to their superior quality. These vermi-leachates are superior as they contain significant concentrations of macro- and micronutrients, several enzymes, plant growth hormones and can also increase plant disease resistance (Verma et al. 2017; Quaik et al. 2012). However, very limited studies have been carried out to evaluate the potential organic nutrient value of vermi-leachates, particularly in hydroponics (Quaik et al. 2012). Furthermore, the nutrient quality of vermi-leachates is highly dependent on factors such as chemical composition of original organic materials, vermicomposting process and moisture content of the vermicompost. Integrated research that involves optimization of the vermicomposting process, identifying optimum organic materials and vermicompost leaching frequency can allow for the development of commercially effective vermi-leachates as liquid fertilizers (Quaik and Ibrahim 2013). Research in this direction will be crucial in driving the adoption of cheap nutrient sources for hydroponics among resource-poor in developing countries.

Table 18.4 Summary of research done evaluating the potential of organic nutrient sources in hydroponic vegetable production (Source: Mupambwa et al. 2019)

Organic nutrient sources	Method of organic nutrient solution preparation	Nutrient content of solution	Observations and research gaps	Reference
Chicken manure, goat manure, cow manure, goat manure compost, and conventional hydroponic fertilizer	A 58.8 g of manure was immersed in 1 L of water whilst enclosed in a filter bag and left to brew for 7 days. The goat manure compost was prepared by thermophilic composting for 6 months, after which the compost was treated as in above.	Not determined.	The control with conventional hydroponic fertilizer yielded the highest results across all parameters measured. However, it was observed that composted goat manure yielded the highest among all other organic nutrient sources. This study points to the need for optimized mineralization of organic nutrient sources before they can be used in preparation of hydroponic nutrient solutions.	Mowa (2015)
Biogas digestate from a biogas digester treating food and vegetable waste; whilst a control of commercial hydroponic solution was also included.	Concentrated biogas digestate was diluted with de-ionized water to achieve a dilution factor of 5, 10, 15, 20, 30, and 50.	The diluted biogas digestate contained the following nutrient concentrations (mg/L): NH_4—475; NO_3—68.2; total P—28.1; total K—558; total Ca—171; total Mg—256; total Na—319.	The results clearly showed that growth of silverbeet was significantly higher in the control compared to all other treatments. Within the organic nutrient sources, the treatment with a dilution of 20 resulted in the highest growth and yield parameters, higher than even the 50% treatment. At 20% dilution, the nitrate and P	Krishnasamy et al. (2012)

(continued)

Table 18.4 (continued)

Organic nutrient sources	Method of organic nutrient solution preparation	Nutrient content of solution	Observations and research gaps	Reference
			levels were 87.8% and 92.7% lower than the control treatment, respectively. The fast depletion of nutrients within the organic nutrient sources points to a need to investigate influence of frequency of nutrient change on growth and yield.	
Fish waste, a by-product of the production of dried bonito fish flakes, with a conventional hydroponic fertilizer being a control treatment. Oyster shell lime was also used to supplement minor nutrients in all treatments.	The fish waste was added at a rate of 300 g/ 200 L of water and inoculated with bark compost as a microbial inoculant for nutrient mineralization. The mixture was then aerated for 50 days before use.	The solution contained 123 mg/L of nitrates with no ammonium. The added oyster shells provided 11.9 mmol Mg; 9.67 mmol Fe; 2.17 mmol B; 1.53 mmol Mn; 0.077 mmol Zn; 0.0113 mmol Cu; 0.005 mmol Mo; and 62.7 mmol Ca.	The tomato fruit yield, Brix value and ascorbic acid content did not differ significantly between the organic and inorganic nutrient sources. Interestingly, for lettuce, the organic fertilizer gave a significantly higher yield and root weight compared to the conventional hydroponic fertilizer. This study shows that using a high N organic material like fish waste, can be critical in developing a workable organic hydroponic nutrient solution.	Shinohara et al. (2011)
Biogas slurry from an unnamed organic source.	The biogas slurry was diluted with water 5.22 times with water, then supplemented with different P	The biogas slurry contained the following nutrients: N—0.73 g/L; P—0.028 g/L; K—0.74 g/L; NH_4—	The biogas slurry alone showed a significantly lower lettuce shoot biomass compared to other	Liu et al. (2011)

(continued)

Table 18.4 (continued)

Organic nutrient sources	Method of organic nutrient solution preparation	Nutrient content of solution	Observations and research gaps	Reference
	and Fe combinations to achieve seven different treatments.	0.7 g/L; NO_3—8.62 mg/L; Mg—227 mg/L.	treatments that had been supplemented with P and Fe. Unlike supplementing the nutrient solution with Fe or P singly, the combination of Fe and P supplementation significantly increased all parameters measured in this study. However, this study did not indicate why the slurry was diluted 5.22 times originally, which could have not been the optimum dilution level.	
Seaweed extracts that were prepared from two species (*Ulva rigida* and *Fucus spiralis*) collected in Morocco. A control of Hoagland's solution was also used.	Fresh seaweed was harvested, cleaned with tap water, and shredded into smaller pieces. One kilogram of each sea weed was then separately grounded, then boiled with 1 L of water for an hour, then filtered to remove solids. The filtrates were then diluted using water to make 25% extract of *Fucus spiralis* and 25% of *Ulva rigida.* These solutions were further mixed at	Extracts from *Ulva rigida* had the following elemental composition: Na (ppm)—42.4; Ca (ppm)—91.9; K (ppm)—55.8; and total nitrogen (%)—1.28. Extracts from *Fucus spiralis* had the following elemental composition: Na (ppm)—72.5; Ca (ppm)—51.2; K (ppm)—142.9; and total nitrogen (%)—0.56.	When beans was grown in a commercial hydroponic solution and foliar applied with the 2 seaweed extracts, the chlorophyll content was significantly higher than the unsprayed control. On leaf protein content, the control behaved similar to the 100 seaweed extracts when it was used a hydroponic solution. It was interesting to observe that the 2 seaweeds	Latique et al. (2013)

(continued)

Table 18.4 (continued)

Organic nutrient sources	Method of organic nutrient solution preparation	Nutrient content of solution	Observations and research gaps	Reference
	different levels with or without Hoagland's solution to achieve the treatments in this study.		showed significantly different results on leaf protein content with 25% Hoagland's solution and 75% of *Fucus spiralis* being optimum, whilst 50% of Hoagland's solution and 50% *Ulva rigida* was optimum.	

18.9 Conclusions

The concept of organic agriculture is not new, though recent realizations of its benefits relative to inorganically fertilized agriculture are driving the renewed interest in this concept. Organic agriculture feeds the soil so that it can feed the crop rather than inorganically fertilized agriculture, which focuses solely on feeding the crop. This chapter presented information on how organic agriculture driven by vermicomposts can be a crucial driver leading to the increased adoption of organic agriculture at a commercial scale. However, much of the evidence presented has been based on small-scale experiments, and there is still need for large-scale field evaluation of the vermicomposting technologies presented. Furthermore, there is still also a need for the creation of proper standards of production and utilization of organic fertilizers similar to those existing in inorganically fertilized agriculture. Overly, vermicomposting presents a biotechnology-driven process that can be adopted in organic agriculture as an effective nutrient source.

References

Aboutalebi A, Jahromi MG, Farahi MH (2013) Evaluation of growth and yield of organically-grown basil (*Ocimum basilicum* L.) in soilless culture. J Food Agric Environ 11(1):299–301

Adhami E, Hosseini S, Owliaie H (2014) Forms of phosphorus of vermicompost produced from leaf compost and sheep dung enriched with rock phosphate. Int J Recycl Org Waste Agric 3:5

Aira M, Monroy F, Dominguez J (2007) Earthworms strongly modify microbial biomass and activity triggering enzymatic activities during vermicomposting independently of the application rates of pig slurry. Sci Total Environ 385:252–261

Alamgir M, McNeill A, Tang C, Marschner C (2012) Changes in soil P pools during legume residue decomposition. Soil Biol Biochem 49:70–77

Alvarenga P, Mourinha C, Farto M, Santos T, Palma P, Sengo J, Morais M, Cunha-Queda C (2015) Sewage sludge, compost and other representative organic wastes as agricultural soil amendments: benefits versus limiting factors. Waste Manag 40:44–52

Anbalagan M, Manivannan S (2012) Capacity of fly ash and organic additives to support adequate earthworm biomass for large scale vermicompost production. J Res Ecol 1:1–5

Antonious GF (2016) Soil amendments for agricultural production. In: Organic fertilizers—from basic concepts to applied outcomes. https://doi.org/10.5772/63047

Arancon NQ, Edwards CA, Atiyeh R, Metzger JD (2004) Effects of vermicomposts produced from food waste on the growth and yields of greenhouse peppers. Bioresour Technol 93:139–144

Arancon NQ, Edwards CA, Bierman P (2006) Influences of vermicomposts on field strawberries: part 2. Effects on soil microbiological and chemical properties. Bioresour Technol 97:831–840

Arancon NQ, Edwards CA, Babenko A, Cannon J, Galvis P, Metzger JD (2008) Influences of vermicomposts, produced by earthworms and microorganisms from cattle manure, food waste and paper waste, on the germination, growth and flowering of petunias in the greenhouse. Appl Soil Ecol 39:91–99

Aria MM, Lakzian A, Haghnia GH, Berenji AR, Besharati H, Fotovat A (2010) Effect of *Thiobacillus*, sulfur, and vermicompost on the water-soluble phosphorus of hard rock phosphate. Bioresour Technol 101:551–554

Atiyeh RM, Dominguez J, Subler S, Edwards CA (2000) Changes in biochemical properties of cow manure during processing by earthworms (*Eisenia andrei*, Bouche) and the effects on seedling growth. Pedobiologia 44:709–724

Atiyeh RM, Edwards CA, Subler S, Metzger JD (2001) Pig manure vermicompost as a component of a horticultural bedding plant medium: effects on physicochemical properties and plant growth. Bioresour Technol 78:11–20

Ayuso M, Pascual JA, García C, Hernández T (1996) Evaluation of urban wastes for agricultural use. Soil Sci Plant Nutr 42:105–111

Azarmi R, Giglou MT, Taleshmikail RD (2008) Influence of vermicompost on soil chemical and physical properties in tomato (Lycopersicum esculentum) field. Afr J Biotechnol 7:2397–2401

Barthod J, Rumpel C, Dignac M (2018) Composting with additives to improve organic amendments. A review. Agron Sustain Dev 38:17

Bhat SA, Singh J, Vig AP (2017) Instrumental characterization of organic wastes for evaluation of vermicompost maturity. J Anal Sci Technol 8:2

Bhat SA, Singh J, Vig AP (2018) Earthworms as organic waste managers and biofertilizer producers. Waste Biomass Valor 9:1073–1086. https://doi.org/10.1007/s12649-017-9899-8

Bhattacharya SS, Chattopadhyay GN (2002) Increasing bioavailability of phosphorus from fly ash through vermicomposting. J Environ Qual 31(6):2116–2119. https://doi.org/10.2134/jeq2002.2116

Bhattacharya SS, Iftikar W, Sahariaha B, Chattopadhyay GN (2012) Vermicomposting converts fly ash to enrich soil fertility and sustain crop growth in red and lateritic soils. Resour Conserv Recycl 65:100–106

Bolan NS, Naidu R, Mahimairaja S, Baskaran S (1994) Influence of low molecular weight organic acids on the solubilization of phosphates. Biol Fertil Soils 18:311–319

Brinton WF (2000) Compost quality standards and guidelines: an international view. New York State Association of Recyclers. Wood End Research Laboratory, New York

Busato JG, Lima LS, Aguiar NO, Canellas LP, Olivares FL (2012) Changes in labile phosphorus forms during maturation of vermicompost enriched with phosphorus-solubilizing and diazotrophic bacteria. Bioresour Technol 110:390–395

Cerda A, Artola A, Font X, Barrena R, Gea T, Sanchez A (2018) Composting of food wastes: status and challenges. Bioresour Technol 248:57–67
Chianu J, Chianu J, Mairura (2012) Mineral fertilizers in the farming systems of sub-Saharan Africa. A review. Agron Sustain Dev 32(2):545–566
Das D, Bhattacharyya P, Ghosh BC, Banik P (2016) Bioconversion and biodynamics of *Eisenia foetida* in different organic wastes through microbially enriched vermiconversion technologies. Ecol Eng 86:154–161
Diacono M, Montemurro F (2010) Long-term effects of organic amendments on soil fertility. A review. Agron Sustain Dev 30:401–422
Dominguez J, Edwards CA (2011) Relationships between composting and vermicomposting. In: Edwards CA, Arancon NQ, Sherman RL (eds) Vermiculture technology: earthworms, organic waste and environmental management. CRC Press, Boca Raton, FL, pp 1–14
Dominguez J, Aira M, Gomez-Brandon M (2010) Vermicomposting: earthworms enhance the work of microbes. In: Insam H, Franke-Whittle I, Goberna M (eds) Microbes at work: from wastes to resources. Springer, New York
Edwards CA, Walker RL, Maskell P, Watson CA, Rees RM, Stockdale EA, Knox OGG (2010) Improving bioavailability of phosphate rock for organic farming. In: Genetic engineering, bio-fertilisation, soil quality and organic farming. Springer, Dordrecht
Fan YV, Lee CT, Klemes JJ, Bong CPC, Ho WS (2016) Economic assessment system towards sustainable composting quality in the developing countries. Clean Techn Environ Policy 18:2479–2491. https://doi.org/10.1007/s10098-016-1209-9
FAO/WHO (1999) Codex Alimentarius Commission approves guidelines for organic food. http://www.fao.org/unfao/bodies/coag/coag15/x0075e.htm. Accessed 25 Nov 2019
Ferreras L, Gomez E, Toresani S, Firpo I, Rotondo R (2006) Effect of organic amendments on some physical, chemical and biological properties in a horticultural soil. Bioresour Technol 97:635–640
Garcia-Sanchez M, Tausnerova H, Hanc A, Tlustos P (2017) Stabilization of different starting materials through vermicomposting in a continuous-feeding system: changes in chemical and biological parameters. Waste Manag 62:33–42
Gaskell M, Smith R (2007) Nitrogen sources for organic vegetable crops. HortTechnology 17 (4):431–441
Gomez-Brandon M, Dominguez J (2014) Recycling of solid organic wastes through vermicomposting: microbial community changes throughout the process and use of vermicompost as a soil amendment. Crit Rev Environ Sci Technol 44(12):1289–1312
Gomez E, Ferreras L, Toresani S (2006) Soil bacterial functional diversity as influenced by organic amendment application. Bioresour Technol 97:1484–1489
Green J (2019) Examples of organic fertilizers. https://homeguides.sfgate.com/examples-organic-fertilizers. Accessed 5 Dec 2019
Gutierrez-Miceli FA, Garcia-Gomez RC, Rincon RR, Abud-Archila M, Oliva L, Maria A, Marcos Joaquin GC, Luc D (2008) Formulation of a liquid fertilizer for sorghum (*Sorghum bicolor* (L.) Moench) using vermicompost leachate. Bioresour Technol 99:6174–6180
Gutierrez-Miceli FA, Maria A, Oliva L, Paula MN, Benigno RS, Alvarez-Solis JD, Luc D (2011) Optimization of vermicompost and worm-bed leachate for the organic cultivation of radish. J Plant Nutr 34(11):1642–1653
Hazra G (2016) Different types of eco-friendly fertilizers: an overview. Sustain Environ 1(1):54–70. ISSN: 2470-637X (Print) and 2470-6388 (Online)
Hemmata A, Aghilinategh N, Rezainejad Y, Sadeghi M (2010) Long-term impacts of municipal solid waste compost, sewage sludge and farmyard manure application on organic carbon, bulk density and consistency limits of a calcareous soil in central Iran. Soil Tillage Res 108:43–50
Jeng AS, Haraldsen TK, Grønlund A, Pedersen PA (2006) Meat and bone meal as nitrogen and phosphorus fertilizer to cereals and rye grass. Nutr Cycl Agroecosyst 76:183–191
Karak T, Bhattacharyya P (2011) Human urine as a source of alternative natural fertilizer in agriculture: a flight of fancy or an achievable reality. Resour Conserv Recycl 55:400–408

Kirchmann H, Thorvaldsson G, Bergstrom L, Gerzabek M, Andrén O, Eriksson L, Winninge M (2008) Fundamentals of organic agriculture—past and present. In: Kirchmann H, Bergström L (eds) Organic crop production—ambitions and limitations. Springer, Dordrecht, pp 13–38

Koelsch R (2017) What is the economic value of manure. University of Nebraska. http://www.water.unl.edu. Accessed 15 Jul 2018

Krishnasamy K, Nair J, Bauml B (2012) Hydroponic system for the treatment of anaerobic liquid. Water Sci Technol 65(7):1164–1171

Kumar V, Singh KP (2001) Enriching vermicompost by nitrogen fixing and phosphate solubilizing bacteria. Bioresour Technol 76:173–175

Kumari MSS, Ushakumari K (2002) Effect of vermicompost enriched with rock phosphate on the yield and uptake of nutrients in cowpea *(Vigna unguiculata.* Walp). J Trop Agric 40:27–30

Kumari A, Kapoor KK, Kundu BS, Mehta RK (2008) Identification of organic acids produced during rice straw decomposition and their role in rock phosphate solubilization. Plant Soil Environ 54(2):72–77

Latique S, Halima C, Mounir M, El K (2013) Seaweed liquid fertilizer effect on physiological and biochemical parameters of bean plant (Phaesolus vulgaris var Paulista) under hydroponic system. Eur Sci J 9(30)

Lavison RK (2013) Factors influencing the adoption of organic fertilizers in vegetable production in Accra. MPhil thesis, University of Ghana

Lehmann J, Gaunt J, Rondon M (2006) Bio-char sequestration in terrestrial ecosystems–a review. Mitig Adapt Strat Glob Change 11:403–427. https://doi.org/10.1007/s11027-005-9006-5

Lim SL, Wu TY, Lim PN, Shak KPY (2014) The use of vermicompost in organic farming: overview, effects on soil and economics. J Sci Food Agric 95:1143–1156

Liu WK, Yang QC, Lian Feng D, Cheng RF, Zhou WL (2011) Nutrient supplementation increased growth and nitrate concentration of lettuce cultivated hydroponically with biogas slurry. Acta Agric Scand Sec B Soil Plant Sci 61(5):391–394. https://doi.org/10.1080/09064710.2010.482539

Loss A, da Rosa Couto R, Brunetto G, da Veiga M, Toselli M, Baldi E (2019) Animal manure as fertilizer: changes in soil attributes, productivity and food composition. Int J Res Granthaalayah 7(9):307–331

Lotter DW (2003) Organic agriculture. J Sustain Agric 21(4):59–128

Lukashe NS, Mupambwa HA, Green E, Mnkeni PNS (2019) Inoculation of fly ash amended vermicompost with phosphate solubilizing bacteria (Pseudomonas fluorescens) and its influence on vermi-degradation, nutrient release and biological activity. Waste Manag 83:14–22

Malik A, Thapliyal A (2009) Eco-friendly fly ash utilization: potential for land application. Crit Rev Environ Sci Technol 39(4):333–366

Malinska K, Golanska M, Caceres R, Rorat A, Weisser P, Slezak E (2017) Biochar amendment for integrated composting and vermicomposting of sewage sludge—the effect of biochar on the activity of *Eisenia fetida* and the obtained vermicompost. Bioresour Technol 225:206–214

Manivannan S, Balamurugan K, Parthasarathi K, Gunasekaran G, Ranganathan LS (2009) Effect of vermicompost on soil fertility and crop productivity – Beans (Phaseolus vulgaris). J Environ Biol 30:275–281

Masunga RH, Uzokwe VN, Mlaya PD, Odeh I, Singh A, Buchan D, De Neve S (2016) Nitrogen mineralization dynamics of different valuable organic amendments commonly used in agriculture. Appl Soil Ecol 101:185–193

Mdolo P (2016) Sewage sludge and municipal waste: potential sources of phosphorus for land restoration. United Nations University Land Restoration Training Programme [final project]. http://www.unulrt.is/static/fellows/document/Mdolo2016.pdf

Mihreteab HT, Ceglie FG, Aly A, Tittarelli F (2016) Rock phosphate enriched compost as a growth media component for organic tomato (*Solanum lycopersicum* L.) seedlings production. Biol Agric Hort 32(1):7–20

Mnkeni PNS, Austin A, Kutu FR (2005) Preliminary studies on the evaluation of human urine as a source of nutrients for vegetables in the Eastern Cape Province, South Africa. In: Ecological sanitation: a sustainable integrated solution. Proceedings of the third international ecological sanitation conference, Durban, South Africa, 23–26 May 2005, pp 418–426. http://conference2005.ecosan.org/papers/mnkeni_et_al.pdf

Mnkeni PNS, Kutu FR, Muchaonyerwa P, Austin LM (2008) Evaluation of human urine as a source of nutrients for selected vegetables and maize under tunnel house conditions in the Eastern Cape, South Africa. Waste Manag Res 26:132

Mowa E (2015) Organic manure for vegetable production under hydroponic conditions in arid Namibia. Int Sci Technol J Namibia 5:3–12

Mupambwa HA (2015) Optimization of bio-conversion and nutrient release from coal fly ash amended cow dung–waste paper composts using earthworms (*Eisenia fetida*) and effective micro-organisms. PhD thesis, University of Fort Hare, South Africa

Mupambwa HA, Mnkeni PNS (2018) Optimizing the vermicomposting of organic waste materials for production of nutrient rich organic fertilizers: a review. Environ Sci Pollut Res 25:10577–10595. https://doi.org/10.1007/s11356-018-1328-4

Mupambwa HA, Wakindiki IIC (2012) Influence of mono- and biculture cover cropping systems on soil polysaccharides and aggregate stability in a sandy loam Oakleaf soil in Eastern Cape, South Africa. S Afr J Plant Soil 29(3-4):165–171

Mupambwa HA, Dube E, Mnkeni PNS (2015) Fly ash composting to improve fertiliser value—a review. S Afr J Sci 111(7/8), 6 p

Mupambwa HA, Ravindran B, Mnkeni PNS (2016) Potential of effective micro-organisms and Eisenia fetida in enhancing vermi-degradation and nutrient release of fly ash incorporated into cow dung–paper waste mixture. Waste Manag 48:165–173

Mupambwa HA, Lukashe SN, Mnkeni PNS (2017) Suitability of fly ash vermicompost as a component of pine bark growing media: effects on media physicochemical properties and ornamental marigold (*Tagetes* spp.) growth and flowering. Comp Sci Util 25(1):48–61

Mupambwa HA, Hausiku MK, Nciizah AD, Dube E (2019) The unique Namib desert-coastal region and its opportunities for climate smart agriculture: a review. Cogent Food Agric 5:1645258

Naguib NYM (2011) Organic vs chemical fertilization of medicinal plants: a concise review of researches. Adv Environ Biol 5(2):394–400

Padmavathiamma PK, Li LY, Kumari UR (2008) An experimental study of vermi-biowaste composting for agricultural soil improvement. Bioresour Technol 99:1672–1681

Parthasarathi K, Balamurugan M, Ranganathan LS (2008) Influence of vermicompost on the physico-chemical and biological properties in different types of soil along with yield and quality of the pulse crop-blackgram. Iran J Environ Health, Sci Eng 5:51–58

Paull J, Hennig B (2011) A world map of organic agriculture. Eur J Soc Sci 24:3

Peyvast G, Olfati JA, Madeni S, Forghani A (2007) Effect of vermicompost on the growth and yield of spinach (Spinach oleracea L.). J Food, Agri Environ 6:133–135

Quaik S, Ibrahim MH (2013) A review on potential of vermicomposting derived liquids in agricultural use. Int J Sci Res Publ 3(3)

Quaik S, Embrandiri A, Rupani PF, Ibrahim MH (2012) Potential of vermicomposting leachate as organic foliar fertilizer and nutrient solution in hydroponic culture: a review. In: Second international conference on environment and bioscience, IPCBEE, vol 44. IACSIT Press, Singapore. https://doi.org/10.7763/IPCBEE.2012.V44.10

Rahmann G, Ardakani MR, Bàrberi P, Boehm H, Canali S, Chander M, David W et al (2017) Organic agriculture 3.0 is innovation with research. Org Agric 7:169–197

Rajiv S, Sunil H, Dalsukhbhai V, Krunalkumar C (2009) Earthworms vermicompost: a powerful crop nutrient over the conventional compost & Protective soil conditioner against the destructive chemical fertilizers for food safety and security. Am Eurasian J Agric Environ Sci 5(S):01–55

Ravindran B, Contreras-Ramos SM, Sekaran G (2015) Changes in earthworm gut associated enzymes and microbial diversity on the treatment of fermented tannery waste using epigeic earthworm *Eudrilus eugeniae*. Ecol Eng 74:394–401

Roy ED (2017) Phosphorus recovery and recycling with ecological engineering: a review. Ecol Eng 98:213–227

Schjonning P, Elmhol, S, Christensen BT (2004) Soil quality management–synthesis. In: Schjonning P, Elmholt S, Christensen BT (eds) Managing soil quality: challenges in modern agriculture. CABI Publishing, Wallingford, UK, pp 368

Schnug E, Jacobs F, Stöven K (2018) Guano: the white gold of the seabirds, seabirds, Heimo Mikkola. IntechOpen, London. https://doi.org/10.5772/intechopen.79501. https://www.intechopen.com/books/seabirds/guano-the-white-gold-of-the-seabirds

Sharma SB, Sayyed RZ, Trivedi MH, Gobi TA (2013) Phosphate solubilizing microbes: sustainable approach for managing phosphorus deficiency in agricultural soils. SpringerPlus 2:587

Shinohara M, Aoyama C, Fujiwara K, Watanabe A, Ohmori H, Uehara Y, Takano M (2011) Microbial mineralization of organic nitrogen into nitrate to allow the use of organic fertilizer in hydroponics. Soil Sci Plant Nutr 57(2):190–203. https://doi.org/10.1080/00380768.2011.554223

Singh R, Sharma RR, Kumar S, Gupta RK, Patil RT (2008) Vermicompost substitution influences growth, physiological disorders, fruit yield and quality of strawberry (*Fragaria ananassa* Duch.). Bioresour Technol 99:8507–8511

Singh R, Singh R, Soni S, Singh SP, Chauhan UK, Kalra A (2013) Vermicompost from biodegraded distillation waste improves soil properties and essential oil yield of Pogostemon cablin (patchouli) Benth. Appl Soil Ecol 70:48–56

Sinha RK, Herat S (2012) Organic farming: producing chemical-free, nutritive and protective food for the society while also protecting the farm soil by earthworms and vermicompost – reviving the dreams of sir Charles Darwin. Agricl Sci Resh J 25:217–239

Sinha RK, Agarwal S, Chauhan K, Valani D (2010) The wonders of earthworms and its vermicompost in farm production: Charles Darwin's 'friends of farmers', with potential to replace destructive chemical fertilizers from agriculture. Agric Sci 1:76–94

Sinha RK, Hahn G, Singh PK, Suhane RK, Anthonyreddy A (2011) Organic farming by vermiculture: producing safe, nutritive and protective foods by earthworms (Charles Darwin's Friends of Farmers). Am J Exp Agric 1:363–399

Son JE, Okuya T (1991) Prediction of electrical conductivity of nutrient solution in hydroponics. J Agric Meteorol 47(3):159–163

Song X, Liu M, Wu D, Griffiths BS, Jiao J, Li H, Hu F (2015) Interaction matters: Synergy between vermicompost and PGPR agents improves soil quality, crop quality and crop yield in the field. Appl Soil Ecol 89:25–34

Soni R, Sharma A (2016) Vermiculture technology: a novel approach in organic farming. Indian Hortic J 6:150–154

Swati A, Hait S (2017) Fate and bioavailability of heavy metals during vermicomposting of various organic waste—a review. Process Saf Environ Prot 109:30–45

Theunissen J, Ndakidemi PA, Laubscher CP (2010) Potential of vermicompost produced from plant waste on the growth and nutrient status in vegetable production. Int J Phys Sci 5(13):1964–1973

Ukwattage NL, Ranjith PG, Bouazza M (2013) The use of coal combustion fly ash as a soil amendment in agricultural lands (with comments on its potential to improve food security and sequester carbon). Fuel 109:400–408

Van Kauwenbergh SJ (2010) World phosphate rock reserves and resources. International Fertilizer Development Center (IFDC), Muscle Shoals, Alabama, USA

Verma S, Singh A, Pradhan SS, Singh RK, Singh JP (2017) Bio-efficacy of organic formulations on crop production—a review. Int J Curr Microbiol Appl Sci 6(5):648–665

Watson CA, Atkinson D, Gosling P, Jackson LR, Rayns FW (2002) Managing soil fertility in organic farming systems. Soil Use Manag 18:239–247

Wood's End Research Laboratory (2005) Interpreting waste and compost tests. J Woods End Res Lab 2(1):1–4. www.woodsend.org. Downloaded 19 Dec 2017

Xiao R, Awasthi MK, Li R, Park J, Pensky SM, Wang Q, Wang JJ, Zhang Z (2017) Recent developments in biochar utilization as an additive in organic solid waste composting: a review. Bioresour Technol 246:203–213. https://doi.org/10.1016/j.biortech.2017.07.090
Zhao L, Cao X, Mašek O, Zimmerman A (2013) Heterogeneity of biochar properties as a function of feedstock sources and production temperatures. J Hazardous Mat 256–257:1–9
Zhao H, Li T, Zhang Y et al (2017) Effects of vermicompost amendment as a basal fertilizer on soil properties and cucumber yield and quality under continuous cropping conditions in a greenhouse. J Soils Sediments 17:2718–2730. https://doi.org/10.1007/s11368-017-1744-y

Chapter 19
Earthworm Communities and Soil Structural Properties

Sharanpreet Singh, Jaswinder Singh, Adarsh Pal Vig, Falwinder Verma, and Surindra Suthar

Abstract Earthworms are the important soil organism belonging to phylum Annelida and class Oligochaeta. They are also known as ecosystem engineers for their important role in influencing the physical, chemical and biological properties of the soil. Earthworms stimulate microbial activity, mix and aggregate the soil, water infiltration rate, soil water content and water-holding capacity, etc. They also increase litter decomposition, soil organic matter dynamics, nutrient cycles, promote plant growth and reduce some soil-borne diseases. On the other hand, soil properties like organic matter, pH, EC, temperature, soil texture, etc., also plays a crucial role in earthworm abundance in the soil. In this chapter, an attempt has been made to highlight the earthworm communities and their dependence on soil physico-chemical properties.

Keywords Earthworm · Soil properties · Microbial activity · Physico-chemical characteristics

S. Singh · A. P. Vig
Department of Botanical and Environmental Sciences, Guru Nanak Dev University, Amritsar, India

J. Singh (✉)
Department of Zoology, Khalsa College Amritsar, Amritsar, India

F. Verma
I.K. Gujral Punjab Technical University, Jalandhar, India

S. Suthar
School of Environment and Natural Resources, Doon University, Dehradun, India

S. A. Bhat et al. (eds.), *Earthworm Assisted Remediation of Effluents and Wastes*,
https://doi.org/10.1007/978-981-15-4522-1_19

19.1 Introduction

The soil is the most important asset to the farmer because all agricultural activities invariably take place on the soil. Therefore, the protection of the soil against depletion is a good step towards sustainable management of its biological properties that determine long-term stability and productivity. It is generally believed that soil organisms benefit soil productivity but not much is known to small-scale farmers about the organisms that live in the soil and the functioning of the soil ecosystem. Of the soil organisms, earthworms are among the most important soil organisms because of their crucial roles in the soil (Ebewore and Ovharhe 2016). They are most ancient soil dwellers known since the Pre-Cambrian and Ordovician period, i.e. 650–570 million years ago (Lee 1985; Miller and Harley 1999). An earthworm is a tube or cylinder shaped with a segmented body found in the phylum Annelida and class Oligochaeta. The class oligochaetes are further divided into two groups on the basis of habitat and size: microdrili (which includes small aquatic worms of terrestrial family Enchytraeidae) and megadrili (which larger terrestrial worms and their aquatic representatives). The megadrilli group mainly consists of those earthworm individuals who have large numbers and biomass in the soils of temperate and tropical regions of the world. Earthworms are omnivorous in nature and consume dead organic matter present in the soil. Charles Darwin gave the importance of earthworms to soil fertility. They are also known as ecosystem engineers for their important role in influencing the physical, chemical and biological properties of the soil (Singh et al. 2016a). Many studies have also supported the beneficial role of the earthworm to the sustainable function of ecosystems (Barrios 2007).

In the present study, an attempt has been made to highlight the earthworm communities and their dependence on soil physico-chemical properties. The main objective of the present study was to know the benefits of earthworm in the soil and how the soil properties influence earthworm communities.

19.2 Earthworm Classification on the Basis of Ecology

The earthworm usually prefers to live within the soil. But, some species of earthworms also live above ground habitats, i.e. leaf litter layers, organic material accumulated at the base of epiphytes and even in the intertidal zone (Lee 1985). However, overall, the soil is the main habitat of earthworms to live, but depending on the ecological nature, each earthworm species has its own feeding habits on different food resources. Hence, earthworms are classified into three different ecological categories depending on the soil horizons in which they are most commonly found (Table 19.1) (Bouche 1977; Lee 1985).

1. Anecic: Anecic earthworms are large sized with uniform pigmentation at their anterior and posterior end. They are mostly phytophagus and play an important

Table 19.1 Ecological category, habitat and food preference of earthworms

Ecological category	Subcategory	Habitat	Food preference
Epigeic	Epigeic Epi-anecic Epi-endogeic	Litter Surface soil	Leaf litter, microbes
Anecic	Anecic	Burrows	Leaf litter and soil
Endogeic	Polyhumic Mesohumic Endo-anecic Oligohumic	Upper 0–20 cm of soil Upper 0–20 cm of soil 0–50 cm of soil 15–20 cm of soil	Soil with high organic matter Soil from 0–10 cm Soil from 20–40 cm Soil from 0–10 cm

Source: Iman Nejad and Rahmani (2005) and Kooch and Jalilvand (2008)

role in mixing the nutrients present in deep layers to the surface of the soil by casting. They feed on the surface residue along with the soil after pulling the same into their burrows (Anon 2001), but they egest their vermicast at the surface of the soil. Examples: *Lampito mauritii, Aporrectodea longa, Lumbricus terrestris.*

2. Epigeic: Earthworms having epigeic ecological category are usually small in size with uniform body colour. They live on the upper surface of litter and are also able to tolerate a variable range of temperature. These types of earthworms do not prefer the food having low organic matter content (Anon 2001; Duiker and Stehouwer 2008). They also play an important role in humus formation. They have a high reproductive rate, hatching rate and feeding rate. Thus, epigeic earthworm species are good for vermicomposting. Examples: *Eisenia fetida, Dendrobaena rubida, Perionyx excavatus* and *Eudrilus eugeniae.*
3. Endogeic: Endogeic earthworms are large- or small-sized worms with weak pigmentation on their bodies. These are geophagus and are mostly found in the soil layer with the organic and mineralized matter. They move within the soil after creating the three-dimensional horizontal burrows by consuming the soil (Duiker and Stehouwer 2008). They also egest their vermicast within the soil. Examples: *Metaphire posthuma, Octochaetona thurstoni, Allolobophora caliginosa, Aporrectodea rosea* and *Octolasion cyaneum.*

Lee (1985) classified the earthworm's species on the basis of their feeding habits into two groups:

1. Detritivorous earthworms: Earthworms mainly feed on plant litter or dead roots or mammalian dung at or near the soil surface. Example: *Eisenia fetida, Lampito mauritii, Perionyx excavatus* and *Eudrilus eugeniae.*
2. Geophagus earthworms: They mainly feed on organically rich soil beneath the soil surface. Example: *Metaphire posthuma, Octochaetona thurstoni.*

19.3 Roles of Earthworms Communities in Soil Ecosystem

Earthworms stimulate microbial activity, mix and aggregate the soil, water infiltration rate, soil water content and water-holding capacity, etc. They also increase litter decomposition, soil organic matter dynamics, nutrient cycles, promote plant growth and reduce some soil-borne diseases (Chandran et al. 2012; Elmer 2012). Earthworms are functionally very important and diverse, and are therefore potentially useful for the management of biodiversity and ecosystem services (Blouin et al. 2013).

19.3.1 Stimulate Microbial Activity in the Soil

Earthworms derive their nutrition with the help of microorganisms. Many more microorganisms are present in their casts than in organic matter that they consume. The earthworm's ability to decompose organic matter has been attributed to the microbial species which inhabit in their digestive tract or the structures they create, which in effect contribute to making up the drilosphere, a microbial activity hotspot (Medina-Sauza et al. 2019). According to Kale and Karmegam (2010), there is a mutualistic relationship between the microorganisms and earthworms (Fig. 19.1). As the organic matter passes through their intestines, it is fragmented and inoculated with microorganisms. Increased microbial activity facilitates the breakdown of organic matter and their conversion is readily taken up by plants (Postma-Blaauw et al. 2006; van Groenigen et al. 2014). Four mechanisms that are generally assumed to be responsible for earthworm-–microbial interactions (Brown 1995) are the following: (1) Soil ingestion stimulates the growth of some microorganisms through the addition of mucus and brings microorganisms in contact with organic residues. (2) Incorporation of organic matter into the soil creates hot spots of microbial activity. (3) Earthworms modify soil structure, creating habitats favourable to microbial activity. (4) Earthworms are responsible for horizontal and vertical transport of microorganisms, which are either transported on earthworm body or in their

Fig. 19.1 Mutualistic relationship of earthworm and microorganisms in soil

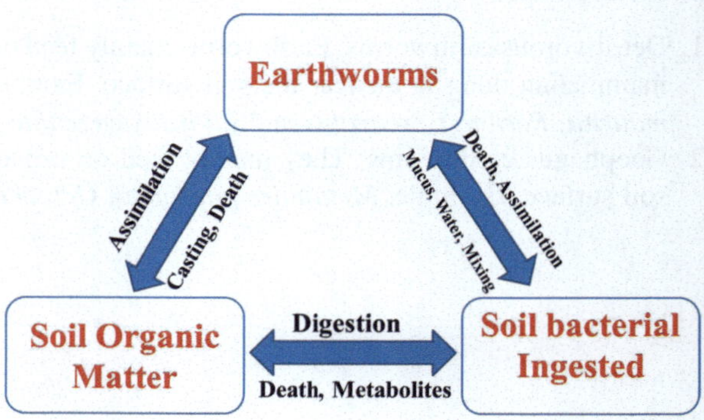

gut, ingested with soil or litter (Bertrand et al. 2015). Earthworms are considered to be important ecological mediators capable of affecting soil functions and microbial activity (Lipiec et al. 2016), generating an energy-rich mucus which triggers microorganisms by priming (Medina-Sauza et al. 2019) and signalling molecules with hormone-like effects and influencing plant gene expression (Puga-Freitas and Blouin 2015).

The impact of earthworms on the microbial communities can be neutral, negative or positive on the basis of species of earthworms. De Menezes et al. (2018) reported the neutral effects of earthworms on soil bacterial communities, showing that the introduction of the endogeic *Aporrectodea trapezoid* did not affect the microbial communities of the entire soil. Hoeffner et al. (2018) observed the positive effects on bacterial richness and diversity in the burrows as well as vermicompost of four *Lumbricus* epi-anecic species compared to bulk soil. They had reported that bacterial diversity increased in the final vermicompost as compared to the initial substrate. On the other hand, negative effects of earthworms on bacterial diversity were observed in earthworm casts by Koubová et al. (2015), which showed that the diversity of bacterial species (estimated from cultural bacteria) decreased during the passage through the intestine of the *Eisenia fetida*. This decline in microbial diversity after soil intake was due to the increased dominance of several bacterial groups in earthworm casts (Furlong et al. 2002; Gopal et al. 2017). More studies have shown that the type of food eaten by earthworms seems to have little effect on the diversity of bacterial communities in casts (Medina-Sauza et al. 2019).

19.3.2 Development of Soil Structure and Fertility

The structure of the soil is a critical factor in most soil functions, including soil fertility (Bertrand et al. 2015). The interaction of different physical forces such as soil water content, surrounding vegetation, presence or absence of organic matter, soil tillage and soil microorganisms form the structure of soil (Sharma et al. 2017). Among the soil organisms, earthworms have a prominent effect on the soil structure due to the reorganization of soil as it feeds and moves in the soil (Martin and Marinissen 1993). They also affect the mechanical and hydraulic properties of soil through their burrowing activities which produce macropores that have a significant impact on water absorption and are therefore important for water-supplying crops and controlling surface runoff and erosion (Bertrand et al. 2015). Earthworms also improve the soil structure by forming soil aggregates which help in soil erosion control (Julka 2008). A large proportion of soil passes through the guts of earthworm, as they turn over the top 15 cm of soil in 10–20 years. Earthworms contribute to soil organic matter dynamics in agroecosystems by consuming an estimated 2–15 mg/ha/year of organic residues and process as much as 10% of the topsoil each year (Whalen and Parmelee 2000; Topoliantz et al. 2000). Earthworms are estimated to produce casts of approximately 40–100 tons/ha/year and thus contribute

Table 19.2 Comparison of different properties of soil and vermicast

S. no.	Parameters	Soil	Vermicast
1.	pH	5.96	8.09
2.	EC (mS/cm)	0.33	0.18
3.	Moisture (%)	24.9	53.5
4.	Water-holding capacity (%)	36.1	110.3
5.	Total carbon (%)	3.10	18.10
6.	Total nitrogen (%)	0.30	0.87
7.	C/N ratio (%)	10.34	20.81
8.	Phosphorus (%)	<0.01	0.01
9.	Potassium (%)	0.09	0.13
10.	Calcium (%)	1.05	2.63
11.	Sodium (%)	0.01	0.02

Source: Adhikary (2012)

tremendously to the creation of stable soil aggregates (Pietz et al. 1984; Sharma et al. 2017).

The earthworm cast has five times more available nitrogen, seven times more available potassium, three times more phosphorus and two times more calcium as compared to surrounding soil (Julka 2008; Bhadauria and Saxena 2010). The vermicast is also rich in various plant nutrients, which are much less in the soil (Table 19.2).

19.3.3 Increases Water Infiltration

Earthworms enhance porosity as they move through the soil. Some species make permanent burrows deep into the soil. These burrows can persist long and can be the major conduct for soil drainage and also help to minimize soil water erosion. The horizontal burrowing in the top several inches of soil increases overall porosity or drainage. Earthworms remove the surface thatch material that can block water from entering the soil (Zaller et al. 2011; Spurgeon et al. 2013). Clements et al. (1991) showed that after 10 years of earthworm inoculation, the water infiltration rate increased from 15 to 27 mm/h. In Mediterranean soils, water percolation was found to be positively correlated with earthworm biomass, burrow length and burrow surface. The increase in the infiltration rate due to anecic earthworm burrows reduced soil erosion by 50–75% (Shuster et al. 2002; Jouquet et al. 2007). Endogeic species also increases soil macroporosity and water infiltration, which tends to reduce runoff. However, it has been shown that some endogeic species also produce small-sized casts, which favour surface sealing and contribute to soil erosion (Blanchart et al. 2004). The presence of earthworms, particularly of anecic ecological category forms vertical burrows which facilitates larger pores accompanying water and contributes to infiltration when water is delivered in large quantities (Fischer et al. 2014). These also influence the mechanical and hydraulic properties of soil through their burrowing activities, producing macropores that have a significant impact on water absorption and are therefore critical for water-supplying crops, as well as managing surface runoff and erosion (Bertrand et al. 2015). The improved infiltration rates due to burrows created by the earthworms also reduced the soil erosion by 50% (Shuster et al. 2002). On the other hand, many earthworms having an endogeic ecological category have also been shown to produce small casts that support surface sealing and contribute to soil erosion (Blanchart et al. 2004; Bertrand et al. 2015). Such discrepancies between the effects of earthworm on soil structure, water absorption and soil erosion are likely due to the fact that this impact depends on (1) the precipitation system, (2) earthworm abundance, (3) earthworm organisms and (4) the amount of organic matter available on soil surface.

19.3.4 Improve Water-Holding Capacity

Water is the key productivity limiting factor in all terrestrial ecosystems because all vegetation needs huge amounts of water for its growth and fruit production (Munnoli and Bhosle 2011). It is therefore of the utmost importance to 'save water'as much as possible and has been adopted as a motto of the world. Earthworms have the capacity to increase the water-holding capacity of soils by fragmenting organic matter and increasing soil porosity. Vermicast increases the softness of soil thus requiring less tillage and irrigation (Guild 1955). Ernst et al. (2008) showed that earthworm belonging to different ecological groups (Epigeic, Anecic, Endogeic) affects soil water characteristics. The anecic *Lumbricus terrestris* and the endogeic *Aporrectodea caliginosa* enhanced drying in the 0–15 cm soil layer by increasing soil aeration and subsequently evaporation through their burrows. In contrast, the epigeic *Lumbricus rubellus* tended to favour water storage in the topsoil. This is probably due to the fact that *Lumbricus rubellus* deposited plants litter at the soil surface rather than burying it, which prevents evaporation. *Aporrectodea caliginosa* induced higher water infiltration rates and faster water discharges to the subsoil than other species, probably because its burrows are temporary and are continually being rebuilt (Bertrand et al. 2015). Munnoli and Bhosle (2011) studied the capacity of vermicompost egested by earthworms in monoculture (vermicompost egested by one earthworm species) and polyculture (vermicompost egested by more than one earthworm species) in the vermireactors by using sugar industry wastes. They reported that vermicompost of monoculture experiment can hold 110–170% water, while polyculture can hold 140–210% of the water in the soil. There experiment clearly states that mixture or combination of different earthworm species in the soil has more water-holding capacity as compared to the field having single earthworm species or without any earthworms.

19.4 Earthworm Communities and Soil Physico-Chemical Properties

19.4.1 Moisture

Earthworms are the soil invertebrates that have very little protection against the fluctuations in moisture (Edwards and Bohlen 1996). The diversity, abundance and biomass of earthworm species at a particular site are principally dependent on soil moisture (Perreault and Whalen 2006; Mariotte et al. 2016). The body of earthworm consists of 75–90% water, which helps them to respire due to the cutaneous respiration mechanism. So, moisture is the critical factor for earthworm survival and its activity in soil is also influenced by soil moisture (Edwards and Bohlen 1996; Dewi and Senge 2015). If the soil is dry, then earthworms move to the deeper layers of the soil or revert to diapause stage, during which earthworms roll their body in a knot with slimy substance lined over the same which avoids moisture loss (Whalen 2004; Mariotte et al. 2016). The anecic earthworm species makes permanent vertical burrows within the soil and also has the ability to enter into a true diapause stage during the dry period (Mariotte et al. 2016). This downward movement of earthworm due to less soil moisture makes sampling more difficult including with extraction solution such as formalin and allyl isothiocyanate. During the dry summer season, earthworm with epigeic ecological category was more effected due to less moisture availability on the upper surface, while on the other hand, endogeic species have the ability to live deep in the soil and can get required moisture (Eggleton et al. 2009). The high level of moisture is also harmful to the earthworms and even also caused death due to excessive water logging (Duiker and Stehouwer 2008). The high moisture level also moistens the soil which affects the nutrient status of the soil and hence affects the earthworm species (Fragoso and Lavelle 1992; Briones and Schmidt 2017). The high moisture in the soil also has a compact structure with poor aeration which also has a negative effect on the earthworm population (Birkas et al. 2004; Capowiez et al. 2009). The requirement of moisture for earthworms varies from species to species and each species has its own moisture requirements for earthworm populations from different regions of the world (Dewi and Senge 2015). For example, *Aporrectodea caliginosa* goes into the diapauses stage as moisture in the soil goes below 25–30% in Europe (Edwards and Bohlen 1996). On the other hand, in Argentina, *Aporrectodea caliginosa* remains active in the soils with moisture content as low as 15% (Ljungstrom et al. 1973). Zaller and Arnone (1999) studied the activity of earthworm in the soil with different earthworm density and moisture conditions (ambient and additional rain). The activity of earthworms was measured on the basis of cast production by the earthworms. They observed that vermicast production was increased with increasing earthworm density with ambient moisture conditions. But as additional rain was given to the earthworm, there was no increase in cast production and above-ground biomass which was due to a decrease in earthworm density below the ground due to the increase in moisture level. According to Ivask et al. (2006), soil moisture has a strong influence on the earthworm distribution as compared to the type of soil. Wever et al. (2001) studied the growth response of earthworm, *Aporrectodea tuberculata*, at different moisture levels, i.e. 10%, 15%, 20% and 25%. They observed that with an increase in moisture content, the growth rate of earthworm was increased and the maximum growth rate was observed in 25% moisture level. They assessed the growth rate on the basis of biomass and maturity of earthworm in different moisture conditions. Andriuzzi et al. (2015) reported that earthworm tries to avoid high moisture conditions during intense rainfall by producing more burrows and vice versa. Thus, change in the pattern of rainfall influences the burrowing behaviour of earthworms in the soil. Singh et al. (2016b) also studied that the earthworm diversity and abundance changed with change in soil moisture level. Thus, soil moisture can directly influence the numbers and biomass of earthworms at any location.

19.4.2 pH

The earthworms' population is very sensitive to the hydrogen ion concentration which is an important factor that affects the species, numbers and distribution of earthworms within the soil. Several researchers have indicated that maximum earthworm species prefer to live in a soil having a neutral pH and also can resist pH from 5.0 to 8.0 (Edwards and Bohlen 1996). According to Bouche (1972), earthworms can also survive in pH less than 4 but their burrowing activity is decreased and the same has been increased beyond pH 5.4. Various studies reported that the pH of any particular site varies according to organic carbon (OC) and organic matter (OM) content. Gillman (1985) reported that with a 1% increase in OC content in the soil, pH declined by 1 unit and vice versa. Thus, long-term accumulation of OC and OM in the soil might result in the acidification and addition of inorganic and organic fertilizers further decreases the pH to an acidic level which ultimately disturbs the earthworm species within the soil (Solomou et al. 2013).

19.4.3 Temperature

The average worldwide surface temperature has increased by 0.8 °C since 1880 and is expected to rise by 1.5 °C during the following century (IPCC 2014). This increase in the average surface temperature of the earth day by day, impacts all living organisms (Brose et al. 2012). This increase in temperature has no effect on species which favours high temperature but has a negative effect on those species which prefer cold or normal temperature for their activities (Hering et al. 2009; Somero 2010). The earthworm is also depending on the optimal temperature which is required for its growth, metabolism, respiration and reproduction. The fecundity rate in earthworm is greatly affected by different temperature conditions. The optimum temperature requirement by *Lumbricus terrestris* for production of the cocoon is 15 °C and the time period for the hatching of cocoon also increases with an increase in temperature (Butt 1993). Lee (1985) also observed that the optimum temperature requirement for the growth of indigenous populations of Lumbricidae in Europe ranged from 10 to 15 °C. Satchell (1967) concluded that an earthworm performs their best activities at night due to less soil temperature on the surface. They can move down into the soil to escape these adverse temperature. Raja et al. (2017) also observed the positive correlation between soil temperature and earthworm abundance which showed that earthworm abundance increased with an increase in temperature and vice versa. While according to Duiker and Stehouwer (2008) and Dewi and Senge (2015), the optimum temperatures for earthworm's growth ranged from 0 to 35 °C, Tripathi and Bhardwaj (2004) observed that earthworm can survive best at 20 °C in laboratory conditions. Lalander et al. (2015) observed that earthworm moves to the edges and top levels of vermicomposting bed if the temperature of the vermibed is more than the threshold temperature. Edwards et al. (1998) studied the life cycle of *Perionyx excavatus* in a variety of organic wastes under various population density pressures and temperature between 15 and 30 °C. They observed that an increase in temperatures up to 30 °C accelerated the growth of earthworms but decreased the time of sexual maturity. Edwards (1998), in another study, evaluated the optimal conditions for breeding of *Eisenia fetida* with different ranges of animal and vegetable wastes under aerobic condition with a temperature range of 15–20 °C, moisture 80–90%, ammonia content <0.5 mg/g, salt content <0.5% and pH in the range of 5–9. He found that the population density of earthworms per unit volume or weight of waste was very important in affecting the rate of growth and reproduction.

With an increase in temperature, the metabolism rate of the earthworm is also increased which requires more food stock for the same. Tripathi et al. (2011) studied the effects of varying temperatures (12–44 °C) on the specific activity of cytoplasmic malate dehydrogenase (cMDH), mitochondrial malate dehydrogenase (mMDH) and lactate dehydrogenase (LDH) in earthworms. They concluded that the specific activity of the above-said enzymes decreases in the earthworm with an increase in temperature and vice versa. They also reported that this change in enzyme activity is dependent on the ecological category of the earthworm species. They observed more variations in the enzyme activity in earthworm with epigeic ecological nature while less variation was observed in the endogeic ecological nature of earthworm species. The less variation in endogeic earthworm species was due to less temperature fluctuation within the soil.

The abundance of earthworms is very less during the winter periods which acts as a harsh condition for earthworm survival. According to Robinson et al. (2018), there was a decrease in the mean weight and growth in the invertebrate soil fauna with a change in soil temperature, but the impacts of temperature on the invertebrate population varied from species to species. Kanianska et al. (2016) observed that no earthworm species have been reported from the sites having high soil temperature and low precipitation. Thus, appropriate soil temperature is a limiting factor for earthworm survival in the soil. Haokip and Singh (2012) observed that earthworm abundance and diversity should be at a maximum level when the threshold level of temperature was in the soil. They observed that the abundance of earthworm decreased from summer to winter when there is a decrease in soil surface temperature.

19.4.4 Soil Texture

Soil texture has a considerable effect on the feeding and burrowing activities of earthworms, i.e. soil type and soil texture because of its effects on other soil properties such as moisture, nutrient and cation exchange capacity (Edwards and Bohlen 1996; Briones and Schmidt 2017). Soil texture also plays an important role in the amount of water available in the soil. Soil having high loam content had greater numbers of earthworms as compared to soil having high clay or sand content (Edwards and Bohlen 1996; Holmstrup et al. 2011). Some shapeless soil with high sand content forms a compressed structure which inhibits air and water movement in the soil and hence affects earthworm species (Zhang et al. 2007; Morris et al. 2010; Briones and Schmidt 2017). El-Duweini and Ghabbour (1965) also observed less population of *Aporrectodea caliginosa* in Egypt as the proportion of gravel and sand increased in the soil. On the other hand, Nordstrom and Rundgren (1974) observed a significant positive correlation between soil clay content and earthworm abundance. Lapied et al. (2009) reported that soil rich in clay content has high organic matter content, which favours earthworm frequency and abundance in the soil. In contrast to this, they also observed low organic matter content in sandy soil. It was observed that large content of clay in soil normally supports high population numbers of an earthworm because of higher water maintenance by the clay, which is an essential factor for earthworm survival (Baker et al. 1998; Curry 2004; Briones and Schmidt 2017). Thus earthworm prefers soil having high clay content. On the other hand, Hendrix et al. (1992) observed a positive correlation between earthworm biomass and the amount of silt content.

The clay content also keeps the organic matter content unchanged by deposition of organic matter on the surface of clay, which reduces the decomposition of organic matter (Rice 2002). Hendrix et al. (1992) also observed a positive correlation between the silt content in the soil and earthworm abundance. Similarly, Baker et al. (1992) observed weak positive correlations between clay content in the soil and earthworm abundance. Coarse sand can be a negative factor either because the abrasive action of sand grains damages their skin or these soils dry out more easily (Kanianska et al. 2016). The sandy soils have lower ground water level horizons which makes an anoxic circumstance for earthworm while the upper horizons dry out very rapidly as a result of discharge and thus both conditions with sandy soil are lethal for earthworms (Sankar and Patnaik 2018). Yvan et al. (2012) also reported that the variables such as soil bulk density or soil texture may influence earthworm growth and activity. Nordstrom and Rundgren (1974) reported that the abundance of *Aporrectodea rosea* in alluvial soil was due to higher clay content.

Fischer et al. (2014) reported that the infiltration of water in the soil directly depends on the soil texture and earthworm abundance. They observed high water infiltration within the soil having high sand content with large earthworm population and vice versa. This increase in water infiltration was due to the formation of burrows in the soil by earthworm which acts as a channel for water infiltration. They also observed that a decrease in the earthworm abundance at the site also disturbed the water infiltration. They also reported that this water infiltration also changes with the ecological nature of the earthworm species.

19.4.5 Organic Carbon and Organic Matter

Organic carbon (OC) and Organic matter (OM) are the main food reserve for earthworms (Brown et al. 2004). Earthworm populations are usually associated with plant species in the forest or native grasslands which provide them organic matter produced by the above-or below-ground litter (Marhan and Scheu 2005; Campana et al. 2002; Nachtergale et al. 2002). Most species of earthworms have the capability to discriminate between diverse kinds of forest litter. Thus, the litter composition is the main factor that greatly affects the abundance and biomass of earthworms. The type of vegetation, quantity and quality of their food supply also affect the earthworm populations. Earthworms have the capability to feed on a wide variety of organic materials and also accumulate a sufficient amount of nourishment from organic matter from soil or microorganisms even in adverse conditions to survive. Soil organic carbon is a critical factor in the distribution and abundance of earthworms at a particular site (Chan and Barchia 2007). The availability of organic matter in the soil directly influences the earthworm's abundance and distribution. The soil having a low content of organic matter does not support a high abundance of earthworms (Edwards and Bohlen 1996). The types and quantity of food available affect not only the biomass and number of earthworm populations but also their rate of growth and fecundity (Edwards and Bohlen 1996). Various studies have been reported a positive relationship between earthworm abundance and OC and they showed that with an increase in OC and OM content within the soil, the abundance and density of earthworm also increased and vice versa (Briones and Schmidt 2017). According to Ayuke et al. (2011), the agricultural soil has a lower content of OC due to agricultural management practices, while high OC content was observed in undisturbed soil, which may be the reason for less earthworm diversity and abundance in agricultural fields. On the other hand, Kibet et al. (2016) and Piccoli et al. (2016) reported that OC and OM content was much higher in the field with crop residue left as such in the field as compared to the field without any crop residue. The type of vegetation and its leaf litter around the earthworm species sampling site is also an important factor. The vegetation which supports low understory biomass C:N and C:P ratio supports high microbial activity, which in turn provides more nutrient to earthworm species around that vegetation (De Wandeler et al. 2016). The type of collection site, i.e. disturbed (tillage) or non-disturbed (non-tillage) are also affecting the OM content. Bartz et al. (2014) observed the positive relationship between the age of non-disturbed land and OM content, i.e. non-disturbed land has high OM content and vice versa. The site with minimum tillage or zero tillage has high earthworm abundance due to the retention of high OM content (Bartz et al. 2014).

19.5 Conclusion

Earthworms play a crucial role in the soil by altering the water-holding capacity, water infiltration, soil microfauna, etc., but the physico-chemical properties of soil such as pH, electrical conductivity, moisture, temperature, soil texture, organic matter, etc., influence the earthworm communities in the soil. The vermicast egested by the earthworms acts as an organic manure, which helps in plant growth and development. The farmers should also be educated regarding the role of earthworms in the soil and also to use earthworm friendly agricultural practices.

References

Adhikary S (2012) Vermicompost, the story of organic gold: a review. Agric Sci 3(7):905

Andriuzzi WS, Pulleman MM, Schmidt O, Faber JH, Brussaard L (2015) Anecic earthworms (*Lumbricus terrestris*) alleviate negative effects of extreme rainfall events on soil and plants in field mesocosms. Plant Soil 397(1–2):103–113

Anon (2001) Standards for Organic Food Production, Feb 2001 (amended to Nov 2001). UKFROFS Reference OB4

Ayuke FO, Pulleman MM, Vanlauwe B, de Goede RG, Six J, Csuzdi C, Brussaard L (2011) Agricultural management affects earthworm and termite diversity across humid to semi-arid tropical zones. Agric Ecosyst Environ 140(1–2):148–154

Baker G, Buckerfield J, Grey-Gardner R, Merry R, Doube B (1992) The abundance and diversity of earthworms in pasture soils in the Fleurieu Peninsula, South Australia. Soil Biol Biochem 24 (12):1389–1395

Baker GH, Carter PJ, Curry JP, Cultreri O, Beck A (1998) Clay content of soil and its influence on the abundance of Aporrectodea trapezoides Duges (Lumbricidae). Appl Soil Ecol 9 (1–3):333–337

Barrios E (2007) Soil biota, ecosystem services and land productivity. Ecol Econ 64(2):269–285

Bartz MLC, Brown GG, da Rosa MG, Klauberg Filho O, James SW, Decaëns T, Baretta D (2014) Earthworm richness in land-use systems in Santa Catarina, Brazil. Appl Soil Ecol 83:59–70

Bertrand M, Barot S, Blouin M, Whalen J, de Oliveira T, Roger-Estrade J (2015) Earthworm services for cropping systems. A review. Agron Sustain Dev 35(2):553–567

Bhadauria T, Saxena KG (2010) Role of earthworms in soil fertility maintenance through the production of biogenic structures. Appl Environ Soil Sci 2010:816073

Birkas M, Jolánkai M, Gyuricza C, Percze A (2004) Tillage effects on compaction, earthworms and other soil quality indicators in Hungary. Soil Tillage Res 78(2):185–196

Blanchart E, Albrecht A, Brown G, Decaens T, Duboisset A, Lavelle P et al (2004) Effects of tropical endogeic earthworms on soil erosion. Agric Ecosyst Environ 104(2):303–315

Blouin M, Hodson ME, Delgado EA, Baker G, Brussaard L, Butt KR et al (2013) A review of earthworm impact on soil function and ecosystem services. Eur J Soil Sci 64(2):161–182

Bouche MB (1972) Lombriciens de France: écologie et systématique. Annales de Zoologie écologique et animale 72(2), special number

Bouche MB (1977) Strategies lombriciennes. In: Lohm U, Persson T (eds) Soil organisms as components of ecosystems, vol 25. Biol. Bull., Stockholm, pp 122–132

Briones MJI, Schmidt O (2017) Conventional tillage decreases the abundance and biomass of earthworms and alters their community structure in a global meta-analysis. Glob Chang Biol 23 (10):4396–4419

Brose U, Dunne JA, Montoya JM, Petchey OL, Schneider FD, Jacob U (2012) Climate change in size-structured ecosystems. Phil Trans R Soc B 367:2903–2912

Brown GG (1995) How do earthworms affect microfloral and faunal community diversity? Plant Soil 170(1):209–231

Brown GG, Edwards CA, Brussaard L (2004) How earthworms affect plant growth: burrowing into the mechanisms. Earthworm Ecol 2:13–49

Butt KR (1993) Reproduction and growth of three deep-burrowing earthworms (Lumbricidae) in laboratory culture in order to assess production for soil restoration. Biol Fertil Soils 16 (2):135–138

Campana C, Gauvin S, Ponge JF (2002) Influence of ground cover on earthworm communities in an unmanaged beech forest: linear gradient studies. Eur J Soil Biol 38(2):213–224

Capowiez Y, Cadoux S, Bouchant P, Ruy S, Roger-Estrade J, Richard G, Boizard H (2009) The effect of tillage type and cropping system on earthworm communities, macroporosity and water infiltration. Soil Tillage Res 105(2):209–216

Chan KY, Barchia I (2007) Soil compaction controls the abundance, biomass and distribution of earthworms in a single dairy farm in south-eastern Australia. Soil Tillage Res 94(1):75–82

Chandran MS, Sujatha S, Mohan M, Julka JM, Ramasamy EV (2012) Earthworm diversity at Nilgiri biosphere reserve, Western Ghats, India. Biodivers Conserv 21(13):3343–3353

Clements RO, Murray PJ, Sturdy RG (1991) The impact of 20 years' absence of earthworms and three levels of N fertilizer on a grassland soil environment. Agric Ecosyst Environ 36 (1–2):75–85

Curry JP (2004) Factors affecting the abundance of earthworms in soils. Earthworm Ecol 9:113

De Menezes AB, Prendergast-Miller MT, Macdonald LM, Toscas P, Baker G, Farrell M et al (2018) Earthworm-induced shifts in microbial diversity in soils with rare versus established invasive earthworm populations. FEMS Microbiol Ecol 94(5):fiy051

De Wandeler H, Sousa-Silva R, Ampoorter E, Bruelheide H, Carnol M, Dawud SM et al (2016) Drivers of earthworm incidence and abundance across European forests. Soil Biol Biochem 99:167–178

Dewi WS, Senge M (2015) Earthworm diversity and ecosystem services under threat. Rev Agric Sci 3:25–35

Duiker S, Stehouwer R (2008) Earthworms. Penn State College of Agricultural Sciences, University Park, PA

Ebewore SO, Ovharhe OJ (2016) Crop farmers' perception of the role of earthworm in soil improvement in delta state, Nigeria. Sch J Agric Vet Sci 3(1):72–78

Edwards CA (1998) The use of earthworms in the breakdown and management of organic wastes. In: Earthworm ecology. CRC Press LLC, Boca Raton, FL, pp 327–354

Edwards CA, Bohlen PJ (1996) Biology and ecology of earthworms, 3rd edn. Chapman & Hall, London

Edwards CA, Dominguez J, Neuhauser EF (1998) Growth and reproduction of Perionyx excavatus (Perr.) (Megascolecidae) as factors in organic waste management. Biol Fertil Soils 27 (2):155–161

Eggleton P, Inward K, Smith J, Jones DT, Sherlock E (2009) A six year study of earthworm (Lumbricidae) populations in pasture woodland in southern England shows their responses to soil temperature and soil moisture. Soil Biol Biochem 41(9):1857–1865

El-Duweini AK, Ghabbour SI (1965) Population density and biomass of earthworms in different types of Egyptian soils. J Appl Ecol 2:271–287

Elmer WH (2012) Using earthworms to improve soil health and suppress diseases. The Connecticut Agricultural Experiment Station, New Haven, CT

Ernst G, Zimmermann S, Christie P, Frey B (2008) Mercury, cadmium and lead concentrations in different ecophysiological groups of earthworms in forest soils. Environ Pollut 156 (3):1304–1313

Fischer C, Roscher C, Jensen B, Eisenhauer N, Baade J, Attinger S et al (2014) How do earthworms, soil texture and plant composition affect infiltration along an experimental plant diversity gradient in grassland? PLoS One 9(6):e98987

Fragoso C, Lavelle P (1992) Earthworm communities of tropical rain forests. Soil Biol Biochem 24 (12):1397–1408

Furlong MA, Singleton DR, Coleman DC, Whitman WB (2002) Molecular and culture-based analyses of prokaryotic communities from an agricultural soil and the burrows and casts of the earthworm Lumbricus rubellus. Appl Environ Microbiol 68(3):1265–1279

Gillman GP (1985) Influence of organic matter and phosphate content on the point of zero charge of variable charge components in oxidic soils. Soil Res 23(4):643–646

Gopal M, Bhute SS, Gupta A, Prabhu SR, Thomas GV, Whitman WB, Jangid K (2017) Changes in structure and function of bacterial communities during coconut leaf vermicomposting. Antonie Van Leeuwenhoek 110(10):1339–1355

Guild WJML (1955) Earthworms and soil structure. In: Kevan DKME (ed) Soil zoology (Proceedings of Nottingham School of Agricultural Science). Butterworth, London, pp 83–98

Haokip SL, Singh TB (2012) Diversity and distribution of earthworms in a natural reserved and disturbed sub-tropical forest ecosystem of Imphal-West, Manipur, India. Int Multidiscip Res J 2 (2):28–34

Hendrix PF, Mueller BR, Bruce RR, Langdale GW, Parmelee RW (1992) Abundance and distribution of earthworms in relation to landscape factors on the Georgia Piedmont, USA. Soil Biol Biochem 24(12):1357–1361

Hering D, Schmidt-Kloiber A, Murphy J, Lücke S, Zamora-Munoz C, López-Rodríguez MJ et al (2009) Potential impact of climate change on aquatic insects: a sensitivity analysis for European caddisflies (Trichoptera) based on distribution patterns and ecological preferences. Aquat Sci 71 (1):3–14

Hoeffner K, Monard C, Santonja M, Cluzeau D (2018) Feeding behaviour of epi-anecic earthworm species and their impacts on soil microbial communities. Soil Biol Biochem 125:1–9

Holmstrup M, Lamandé M, Torp SB, Greve MH, Labouriau R, Heckrath G (2011) Associations between soil texture, soil water characteristics and earthworm populations in grassland. Acta Agric Scand B Soil Plant Sci 61(7):583–592

Iman Nejad E, Rahmani R (2005) Relation between number and biomass of earthworms with tree species and soil properties. Doctoral dissertation, MSc thesis, Gorgan University (in Persian)

IPCC (2014) Climate change 2014: synthesis report. In: Pachauri RK, Meyer LA (eds) Contribution of working groups I, II and III to the fifth assessment report of the intergovernmental panel on climate change. IPCC, Geneva, 151 p

Ivask M, Kuu A, Truu M, Truu J (2006) The effect of soil type and soil moisture on earthworm communities. Agric Sci 17:7–11

Jouquet P, Bottinelli N, Mathieu J, Orange D, Podwojewski P, des Tureaux TH, Duc TT (2007) Impact of land-use change on earthworm diversity and activity: the consequences for soil fertility and soil erosion. In: Gebbie L, Glendinning A, Lefroy-Braun R, Victor M (eds) Proceedings of the 2nd international conference on sustainable sloping lands and watershed management: linking research to strengthen upland policies and practices, pp 130–141

Julka JM (2008) Know your earthworms. Foundation for Life Sciences and Business Management, Solan, 51 pp

Kale RD, Karmegam N (2010) The role of earthworms in tropics with emphasis on Indian ecosystems. Appl Environ Soil Sci 2010:414356, 1–16

Kanianska R, Jaďuďová J, Makovníková J, Kizeková M (2016) Assessment of relationships between earthworms and soil abiotic and biotic factors as a tool in sustainable agricultural. Sustainability 8(9):906

Kibet LC, Blanco-Canqui H, Jasa P (2016) Long-term tillage impacts on soil organic matter components and related properties on a Typic Argiudoll. Soil Tillage Res 155:78–84

Kooch Y, Jalilvand H (2008) Earthworms as ecosystem engineers and the most important detritivors in forest soils. Pak J Biol Sci 11(6):819–825

Koubová A, Chroňáková A, Pižl V, Sánchez-Monedero MA, Elhottová D (2015) The effects of earthworms Eisenia spp. on microbial community are habitat dependent. Eur J Soil Biol 68:42–55

Lalander CH, Komakech AJ, Vinnerås B (2015) Vermicomposting as manure management strategy for urban small-holder animal farms–Kampala case study. Waste Manag 39:96–103

Lapied E, Nahmani J, Rousseau GX (2009) Influence of texture and amendments on soil properties and earthworm communities. Appl Soil Ecol 43(2–3):241–249

Lee KE (1985) Earthworms: their ecology and relationships with soil and land use. Academic Press, New York

Lipiec J, Frąc M, Brzezińska M, Turski M, Oszust K (2016) Linking microbial enzymatic activities and functional diversity of soil around earthworm burrows and casts. Front Microbiol 7:1361

Ljungstrom PO, De Orellana JA, Priano LJJ (1973) Influence of some edaphic factors on earthworm distribution in Santa Fe Province, Argentina. Pedobiologia 13:236–247

Marhan S, Scheu S (2005) Effects of sand and litter availability on organic matter decomposition in soil and in casts of Lumbricus terrestris L. Geoderma 128(1–2):155–166

Mariotte P, Le Bayon RC, Eisenhauer N, Guenat C, Buttler A (2016) Subordinate plant species moderate drought effects on earthworm communities in grasslands. Soil Biol Biochem 96:119–127

Martin A, Marinissen JCY (1993) Biological and physico-chemical processes in excrements of soil animal. Geoderma 56:331–347

Medina-Sauza RM, Álvarez-Jiménez M, Delhal A, Reverchon F, Blouin M, Guerrero-Analco JA, Cerdán CR, Guevara R, Villain L, Barois I (2019) Earthworms building up soil microbiota, a review. Front Environ Sci 7:81

Miller SA, Harley JB (1999) Zoology, 4th edn. McGraw-Hill, Boston, MA

Morris NL, Miller PCH, Orson JH, Froud-Williams RJ (2010) The adoption of non-inversion tillage systems in the United Kingdom and the agronomic impact on soil, crops and the environment—a review. Soil Tillage Res 108(1–2):1–15

Munnoli PM, Bhosle S (2011) Water-holding capacity of earthworms' vermicompost made of sugar industry waste (press mud) in mono-and polyculture vermireactors. Environmentalist 31 (4):394–400

Nachtergale L, Ghekiere K, De Schrijver A, Muys B, Luyssaert S, Lust N (2002) Earthworm biomass and species diversity in windthrow sites of a temperate lowland forest. Pedobiologia 46 (5):440–451

Nordstrom S, Rundgren S (1974) Environmental factors and lumbricid associations in southern Sweden. Pedobiologia 14:1–27

Perreault JM, Whalen JK (2006) Earthworm burrowing in laboratory microcosms as influenced by soil temperature and moisture. Pedobiologia 50(5):397–403

Piccoli I, Chiarini F, Carletti P, Furlan L, Lazzaro B, Nardi S et al (2016) Disentangling the effects of conservation agriculture practices on the vertical distribution of soil organic carbon. Evidence of poor carbon sequestration in North-Eastern Italy. Agric Ecosyst Environ 230:68–78

Pietz RI, Peterson JR, Prater JE, Zenz DR (1984) Metal concentrations in earthworms from sewage sludge-amended soils at a strip mine reclamation site 1. J Environ Qual 13(4):651–654

Postma-Blaauw MB, Bloem J, Faber JH, Van Groenigen JW, De Goede RG, Brussaard L (2006) Earthworm species composition affects the soil bacterial community and net nitrogen mineralization. Pedobiologia 50(3):243–256

Puga-Freitas R, Blouin M (2015) A review of the effects of soil organisms on plant hormone signalling pathways. Environ Exp Bot 114:104–116

Raja IA, Ehsan N, Rana N, Sarwar A (2017) Impact of soil pollutants on diversity and abundance of earthworms in cauliflower crop. Stud Zool Entomol J 5(4):1007–1012

Rice CW (2002) Organic matter and nutrient dynamics. In: Lal R (ed) Encyclopedia of soil science. Marcel Dekker Inc., New York, pp 925–928

Robinson SI, McLaughlin ÓB, Marteinsdóttir B, O'Gorman EJ (2018) Soil temperature effects on the structure and diversity of plant and invertebrate communities in a natural warming experiment. J Anim Ecol 87(3):634–646

Sankar AS, Patnaik A (2018) Impact of soil physico-chemical properties on distribution of earthworm populations across different land use patterns in southern India. J Basic Appl Zool 79(1):50

Satchell JE (1967) Lumbricidae. In: Burges A, Raw F (eds) Soil biology. Academic Press, London, pp 259–322

Sharma DK, Tomar S, Chakraborty D (2017) Role of earthworm in improving soil structure and functioning. Curr Sci 113:1064–1071

Shuster W, McDonald L, McCartney D, Parmelee R, Studer N, Stinner B (2002) Nitrogen source and earthworm abundance affected runoff volume and nutrient loss in a tilled-corn agroecosystem. Biol Fertil Soils 35(5):320–327

Singh S, Singh J, Vig AP (2016a) Earthworm as ecological engineers to change the physico-chemical properties of soil: soil vs vermicast. Ecol Eng 90:1–5

Singh S, Singh J, Vig AP (2016b) Effect of abiotic factors on the distribution of earthworms in different land use patterns. J Basic Appl Zool 74:41–50

Solomou AD, Sfougaris AI, Vavoulidou EM, Csuzdi C (2013) Species richness and density of earthworms in relation to soil factors in olive orchard production systems in Central Greece. Commun Soil Sci Plant Anal 44(1–4):301–311

Somero GN (2010) The physiology of climate change: how potentials for acclimatization and genetic adaptation will determine 'winners' and 'losers'. J Exp Biol 213(6):912–920

Spurgeon DJ, Keith AM, Schmidt O, Lammertsma DR, Faber JH (2013) Land-use and land-management change: relationships with earthworm and fungi communities and soil structural properties. BMC Ecol 13(1):46

Topoliantz S, Ponge JF, Viaux P (2000) Earthworm and enchytraeid activity under different arable farming systems, as exemplified by biogenic structures. Plant Soil 225(1–2):39–51

Tripathi G, Bhardwaj P (2004) Comparative studies on biomass production, life cycles and composting efficiency of Eisenia fetida (Savigny) and Lampito mauritii (Kinberg). Bioresour Technol 92(3):275–283

Tripathi G, Kachhwaha N, Dabi I, Bandooni N (2011) Temperature-dependent alterations in metabolic enzymes and proteins of three ecophysiologically different species of earthworms. Braz Arch Biol Technol 54(4):769–776

Van Groenigen JW, Lubbers IM, Vos HM, Brown GG, De Deyn GB, Van Groenigen KJ (2014) Earthworms increase plant production: a meta-analysis. Sci Rep 4:6365

Wever LA, Lysyk TJ, Clapperton MJ (2001) The influence of soil moisture and temperature on the survival, aestivation, growth and development of juvenile Aporrectodea tuberculata (Eisen) (Lumbricidae). Pedobiologia 45(2):121–133

Whalen JK (2004) Spatial and temporal distribution of earthworm patches in corn field, hayfield and forest systems of southwestern Quebec, Canada. Appl Soil Ecol 27(2):143–151

Whalen JK, Parmelee RW (2000) Earthworm secondary production and N flux in agroecosystems: a comparison of two approaches. Oecologia 124(4):561–573

Yvan C, Stéphane S, Stéphane C, Pierre B, Guy R, Hubert B (2012) Role of earthworms in regenerating soil structure after compaction in reduced tillage systems. Soil Biol Biochem 55:93–103

Zaller JG, Arnone JA III (1999) Earthworm and soil moisture effects on the productivity and structure of grassland communities. Soil Biol Biochem 31(4):517–523

Zaller JG, Saccani F, Frank T (2011) Effects of earthworms and mycorrhizal fungi on the growth of the medicinal herb Calendula officinalis (Asteraceae). Plant Soil Environ 57(11):499–504

Zhang GS, Chan KY, Oates A, Heenan DP, Huang GB (2007) Relationship between soil structure and runoff/soil loss after 24 years of conservation tillage. Soil Tillage Res 92(1–2):122–128

Chapter 20
Effect of Methyl Parathion on the Growth and Reproduction of *Eisenia fetida* in Natural Soil

Ankurita Nath and Subrata Hait

Abstract A laboratory study was carried out to assess the toxic effects of organophosphate insecticide methyl parathion (MP) 2% dust (active ingredient, a.i.) on epigeic earthworm *Eisenia fetida* at different lethal and sub-lethal concentrations. A series of toxicity tests starting from range finding acute to chronic were conducted in natural soil to determine the sensitive endpoints like survival rate, growth rate and reproduction rate of *E. fetida*. Results from the study revealed a reduction in the earthworm biomass and reproduction potential. The median lethal concentration (LC_{50}) determined from 14-day acute toxicity test was 36.63 mg a.i./kg, which confirmed the extremely toxic nature of MP 2% dust. The addition of septic tank sludge as a feeding material in the chronic toxicity test helped the earthworms in gaining biomass at lower doses of 2.5 and 4.0 mg a.i./kg of MP 2% dust. At a lower concentration in the range of (6.4–26.2 mg a.i./kg), MP 2% dust showed sub-lethal toxicity to *E. fetida*. Change in the earthworm biomass and reproduction rate during the toxicity test appeared to be the more sensitive endpoints compared to the LC_{50} value used for assessing the earthworm mortality from pesticide. This study provides an insight into the ecological risks associated with pesticides and helps to provide an understanding of the sustainable use of pesticides for ecological protection.

Keywords Methyl parathion · Organophosphate insecticide · *Eisenia fetida* · Eco-toxicology · Earthworm growth and reproduction

---

A. Nath · S. Hait (✉)
Department of Civil and Environmental Engineering, Indian Institute of Technology Patna, Patna, Bihar, India
e-mail: shait@iitp.ac.in

S. A. Bhat et al. (eds.), *Earthworm Assisted Remediation of Effluents and Wastes*,
https://doi.org/10.1007/978-981-15-4522-1_20

20.1 Introduction

India is an agrarian economy and the demand for higher agricultural growth is arising with an increasing population. The rising demand for food remains an unaccomplished target due to the inadequate knowledge about agricultural practices and services. In order to ensure higher agricultural growth, the need for use of pesticides has emerged. The usage of pesticides to sustain higher agricultural production affects the non-target organisms in soil and water (Abdullah et al. 1997; Nath et al. 1997; Suh et al. 2000; Reinecke and Reinecke 2007; Yasmin and D'Souza 2010) as well as decreases their growth rate and reproductive power (Takada et al. 2000; Yasmin and D'Souza 2010). The population of beneficial soil microorganisms declines due to heavy treatment with an increased amount of pesticide residue levels in the soil. The widespread lack of awareness among the Indian farmers regarding the recommended application dose of pesticide, method of application and unscientific disposal of packaging material has resulted in the application of multiple doses of pesticides into the fields causing environmental and health degradation. The indiscriminate use of pesticides to increase the productivity of crops leads to toxic hazards to beneficial soil organisms like earthworms. Earthworms are one of the non-target organisms of soil biota which play a crucial role in improving soil fertility and structure. Earthworms improve the soil organic matter physically and chemically by mixing leaf litter with soil and thereby facilitating the formation and stabilization of soil aggregates. Exposure of earthworms to pesticides leads to bioaccumulation of pesticides in them, which causes serious damages to higher trophic levels compared to the effects caused in the earthworms itself (Darling and Thomas 2005; Hobbelen et al. 2006; Van Gestel et al. 2011). Therefore, earthworms are used as bio-indicators of soil contamination to provide safety limit values for pesticides application (Suthar et al. 2008; Lourenço et al. 2011).

Methyl parathion (MP) 2% dust is a commonly used pesticide in India for paddy and wheat crops. According to the World Health Organization (WHO), MP 2% dust is an extremely hazardous Class 1A pesticide belonging to the class of the organophosphate group. Like other organophosphates, MP is also a cholinesterase inhibitor (Suthar 2014). India is considered as the second largest consumer of organophosphate insecticide (about 40–45% of the total consumption of pesticide) (Abhilash and Singh 2009). MP 2% dust is focused mainly on controlling chewing and sucking insects in a varied range of crops, vegetables and cotton crops (Suthar 2014). In an experiment by Suthar (2014), MP 2% dust was exposed to different species of tropical earthworms which caused less injury to the epigeic earthworms which dwelt on the surface but the damage caused to the endogeic earthworm was lethal. After the review of scientific literature, it is realized that the effect of MP 2% dust at lethal and sub-lethal doses on epigeic earthworm *Eisenia fetida* has not been well documented. In this context, a series of toxicity tests were conducted on epigeic earthworm *Eisenia fetida* using MP 2% dust (active ingredient, a.i.) to validate whether MP 2% dust is toxic or non-toxic to non-target organisms. The aims of this study were: (1) to determine the range of MP 2% dust a.i. causing no mortality to 100% mortality of epigeic earthworm *Eisenia fetida*, (2) to assess the effects of acute doses of MP 2% dust a.i. on the growth and survivability of epigeic earthworm *Eisenia fetida* and (3) to assess the effects of chronic doses of MP 2% dust a.i. on the growth, survivability and reproduction of epigeic earthworm *Eisenia fetida* in a natural soil with septic tank sludge added as a feeding material.

20.2 Materials and Methods

20.2.1 Earthworms

Adult earthworms *Eisenia fetida* with well-developed clitellum (about 400 mg average weight) were obtained from the vermiculture maintained in the Environmental Engineering Laboratory at Indian Institute of Technology (IIT) Patna, Bihar, India. They were acclimatized in the test substrate (natural soil) for 24 h after removing them from the synchronized culture in the laboratory. Earthworms were properly rinsed and dried using a filter paper before measuring their initial weight.

20.2.2 Pesticide

Pesticide methyl parathion (MP) 2% dust belonging to the group of organophosphate with 2% active ingredient (a.i.) was used for the purpose of the study. It was purchased from a local agro-chemical shop at Bihta, Patna, Bihar, India as MP is a widely used pesticide for agricultural purposes in India. MP is a highly toxic chemical that has short-term and long-term repercussions on living matter. MP mixed with tap water was in the form of a suspension and it was agitated properly before applying the solution in the reactors.

20.2.3 Test Soil Preparation and Apparatus

The natural soil for the laboratory experiment was collected from the nearby unconstructed areas of IIT Patna, Bihta campus which was earlier used as agricultural land. The use of natural soil from natural sites seems to be more effective in representing a particular area than the artificial soils and other test and/or culture soils like Organization for Economic Co-operation and Development (OECD) standard soil, German standard soil Lufa, which are used for eco-toxicity test (Fairbrother et al. 1999; Løkke and Van Gestel 1998). The soil was air-dried in a hot air oven for 24 h maintained at a temperature 55 ± 5 °C. The air-dried sample

Table 20.1 Physico-chemical characteristics of natural soil and septic tank sludge

Parameters	Unit	Natural soil	Septic tank sludge
pH (1:10 w/v)	–	6.90	6.30
Electrical conductivity	mS/cm	11.60	1.20
Moisture content	%	30–35	–
Total nitrogen	%	4.60	6.90
Total carbon	%	0.23	27.60
Total organic carbon	%	0.20	24.23
Volatile solids	%	1.08	6.92

was sieved using a 1.12-mm sieve. Prior to use in the experiment, the soil was analysed for its physico-chemical parameters as presented in Table 20.1.

One kilogram of soil was collected in a plastic container of capacity 3 L with dimension of 20 cm diameter and 10 cm height. The containers were provided with a lid which was perforated to provide aeration, but to prevent the earthworms from escaping. One kilogram of air-dried natural soil was hand-mixed with the suspended solution of MP 2% dust a.i. in tap water. For proper hand-mixing of the pesticide solution with natural soil, safety gloves and mask were worn. After the mixing of pesticide solution in the soil, a group of ten earthworms was weighed and placed in each reactor. Replicates and control reactors were maintained for each treatment. Control reactors were prepared without the addition of pesticide solution into it. The moisture content of the soil was maintained at 35–40% and was replenished with tap water whenever there was an alteration from the optimum condition.

20.2.4 Organic Admixture

Septic tank sludge as an organic admixture was used as feeding material for the earthworms. Five grams of air-dried septic tank sludge in finely grounded form was added every week in each reactor for 4 weeks in chronic toxicity test. Before using the septic tank sludge as a feeding material, its characterization was done in order to confirm whether it would be suitable as feeding material for earthworms. The physico-chemical characteristics of septic tank sludge as an organic admixture are presented in Table 20.1.

20.2.5 Range Finding Test

A preliminary range finding test in accordance with the procedure mentioned in the OECD 207 test (OECD 1984) was conducted initially in order to determine the range which is responsible for total mortality and for the absence of mortality of earthworms. Ten earthworms were added in each of the reactors. It is an important test as it gives the range to be used for the definitive acute toxicity test. Six concentration levels of 15, 23, 34, 50, 75 and 112 mg a.i./kg (dry weight of natural soil) along with control were used for determining 0% and 100% mortality at 7th and 14th day. Three replicates were used for each test.

20.2.6 Acute Toxicity Test

Four concentration levels of MP 2% dust a.i. along with control were selected for the acute toxicity test. The concentration levels were 25, 40, 55 and 70 mg a.i./kg of dried natural soil. Three replicates were used for each dose. The suspended solution of MP 2% dust a.i. was thoroughly mixed in the natural soil. For this test, adult earthworms (with clitellum) having a weight between 300 and 450 mg were used. Earthworms that were to be used were acclimatized for 24 h in natural soil before the start of the test. Ten earthworms were added in each reactor after one day. The earthworm mortality and biomass change were monitored at 7th and 14th day after proper cleaning and drying of earthworms on the Whatman filter paper. The reactors examined on the 7th day were replaced by the earthworms after the counting and weighing process was completed. The median lethal concentration (LC_{50}) was determined in this acute test as per the procedure mentioned in the ISO 11268-1 (ISO 2012).

20.2.7 Chronic Toxicity Test

Six concentration levels of MP of 2.5, 4.0, 6.4, 10.2, 16.3 and 26.2 mg a.i./kg of dried natural soil along with control were selected for the test. Ten earthworms were placed in each reactor. Three replicates were used for each dose and the test was administered for the duration of 28 and 56 days. According to the OECD 222 test (OECD 2016), the feeding material, i.e. powdered septic tank sludge was added on the day after the addition of the earthworms. Around 5 g of sludge as feed was spread over the soil surface in each reactor and was moistened by adding water. The production of cocoons from mature earthworms requires a time span of 28 days. During the period of 28 days, 5 g of sludge as feed was added once a week. The mature earthworms were removed on the 28th day from the reactors and the change in biomass was noted after cleaning and drying the earthworms on Whatman filter paper. The number of cocoons produced in each reactor was counted and left in the reactors for the remaining 4 weeks without any further addition of feeding materials. According to Shi et al. (2007), the weight of the earthworms was measured and used for computing growth inhibition. The growth inhibition is given by:

$$GI\% = \frac{(W_0 - W_t)}{W_0} \times 100\%$$

where

W_0: earthworm weight on day 0
W_t: earthworm weight after t days

20.2.8 *Statistical Analysis*

Statistical analyses of the data obtained were performed using the SPSS 26 software. The significant difference between the treatments was determined by one-way ANOVA using Tukey's test as a post-hoc pairwise multiple comparison process for the contrast of the differences among treatment means. Levene's test was used to ensure the homogeneity of variances. The level of probability used for statistical significance was $P < 0.05$.

20.3 Results and Discussion

20.3.1 *Range Finding Test*

The range finding test helps to determine the range which is responsible for total mortality and for the absence of mortality of earthworms. The mortality effect of different concentrations of MP 2% dust a.i. in the range finding test is presented in

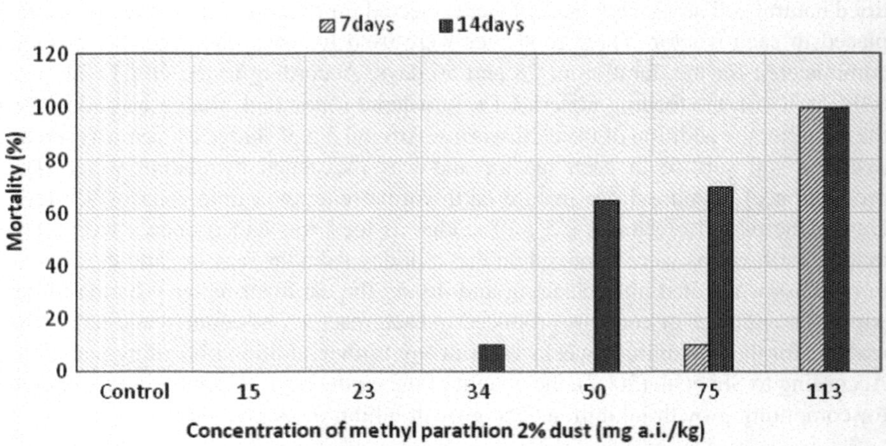

Fig. 20.1 Mortality of *Eisenia fetida* as observed in the range finding test using MP 2% dust a.i.

Fig. 20.1. In the range finding test, no mortality of earthworm was observed in the control reactors on the 7th and 14th day. The reactors with a concentration of 15 and 23 mg a.i./kg of MP 2% dust also showed no mortality. When the dosage level of MP was increased to 50 and 75 mg a.i./kg, a mortality of 65% and 70% was observed respectively at the 14th day. The observed earthworm mortality at 14th day was very high as compared to that of the 7th day on which no mortality was observed at 50 mg a.i./kg. Thus, it can be inferred that the toxicity of the insecticide affects the earthworms with an increase in time and results in their mortality. At the maximum dose level of MP, i.e. 113 mg a.i./kg, 100% earthworm mortality was observed. The range of concentration of MP which caused mortality between 0% and 100% as determined from this preliminary test was between 23 and 113 mg a.i./kg.

20.3.2 Acute Toxicity Test

20.3.2.1 Effect on Earthworm Mortality

The doses to be used for acute toxicity test were determined on the basis of range finding test. The doses chosen were between 0% and 100% mortality of range finding test. The mortality effect of different concentrations of MP 2% dust a.i. from the acute toxicity test is presented in Fig. 20.2. No mortality was found in the control reactors. From the experiment, it was found that there was no mortality

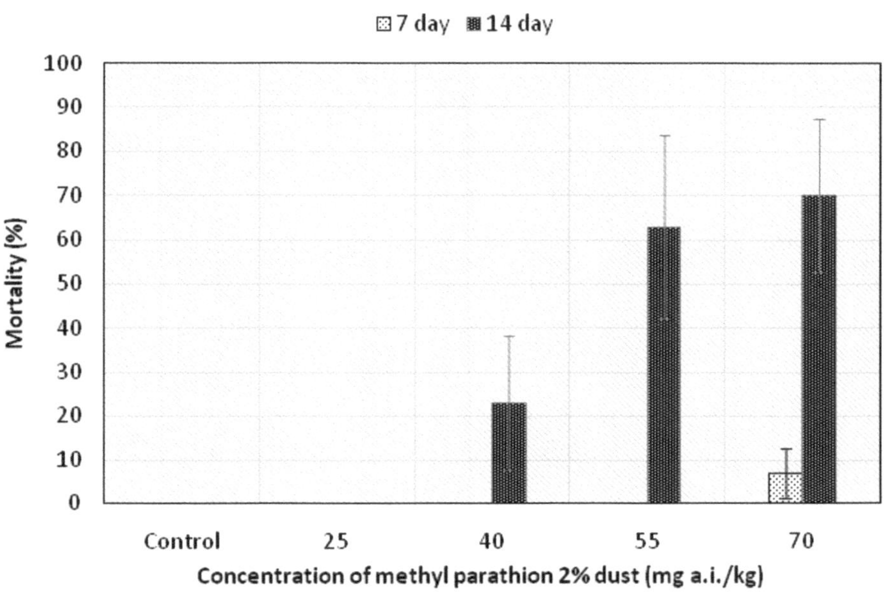

Fig. 20.2 Mortality rate of *Eisenia fetida* from acute exposure to MP 2% dust a.i.

up to an MP concentration of 25 mg a.i./kg. The range of doses considered below 55 mg a.i./kg did not cause death of earthworms till an exposure period of 7 days. However, the mortality rate of earthworms was observed to start increasing with an increase in the dose of MP 2% dust a.i. beyond an exposure period of 7 days. Further, the earthworm mortality was also observed to be very high at higher doses on the 14th day. At a dose of 40 mg a.i./kg, the earthworm mortality was 23% and further increased to 70% with an increase in the concentration of MP to 70 mg a.i./kg. At a high concentration, the earthworms were observed to become stiff with the coiling of their structure. Thus, it can be inferred that MP is extremely toxic to *Eisenia fetida.* The median lethal concentration (LC_{50}) of *Eisenia fetida* for a 14-day exposure to MP 2% dust a.i. was found to be 36.63 mg a.i./kg.

20.3.2.2 Earthworm Growth Inhibition

The growth inhibition of earthworms at 7th and 14th day as observed at different doses of MP 2% dust a.i. is shown in Fig. 20.3. A large extent of positive impact on the growth inhibition of *Eisenia fetida* was observed on the 14th day as compared to the 7th day. Growth inhibition at higher doses was notably higher than that of controls. The control reactors had also shown a bit of growth inhibition, i.e. around 7%. The earthworm biomass loss in control reactors might be due to the presence of less organic matter in the soil as evident from the total organic carbon (TOC) content of around 0.2%. The unavailability of proper amount of feeding material plausibly led to the loss of earthworm biomass in the control reactors although the condition of the earthworms was better than those in the test reactors with different doses of MP. According to Bustos-Obregón and Goicochea (2002),

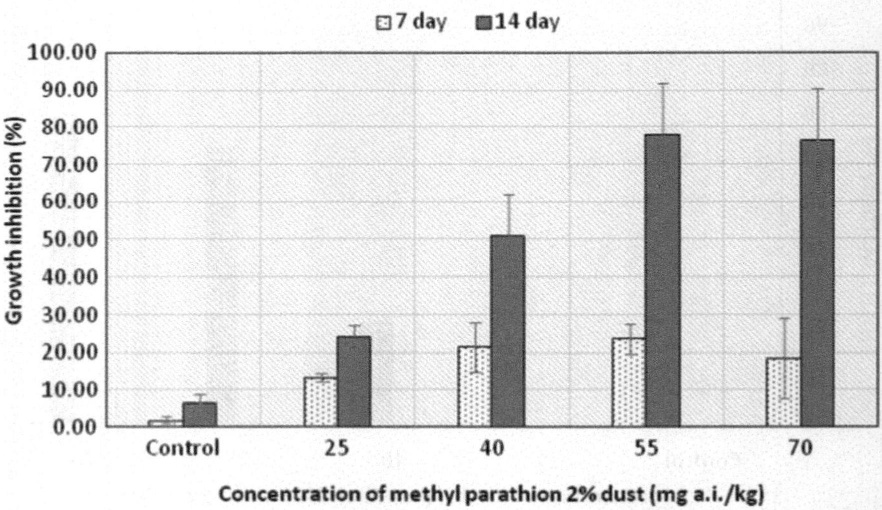

Fig. 20.3 Growth inhibition response in *Eisenia fetida* from acute exposure to MP 2% dust a.i.

the weight loss and alteration in muscular function were the reasons behind the coiling of earthworms in 100% of the cases treated with parathion. It has also been reported that earthworms treated in organophosphate pesticides find difficulty in locomotion after getting intoxicated and they become unable to feed themselves which causes a considerable weight loss in them (Bustos-Obregón and Goicochea 2002).

A dose-dependent trend showing the decrease in earthworm biomass over a 14-day exposure period was further confirmed by the one-way ANOVA test. After the 7-day exposure to MP, it was found that there was no significant difference in the growth inhibition in low and medium dosage (25, 40 and 55 mg a.i./kg) (Tukey's test, $P < 0.05$) whereas, there was a significant ($P < 0.05$) difference from the control at a dose of 70 mg a.i./kg. On the 14th day, there was positive growth inhibition in all the reactors. In dosage of 55 and 70 mg a.i./kg, there was more than 80% biomass loss of earthworms as compared to the control. From Tukey's test as a post-hoc analysis, it was observed that the earthworm growth inhibition at 14th day was significantly ($P < 0.05$) different from the control at higher doses of 55 and 70 mg a.i./kg. However, there was no significant difference compared to control ($P > 0.05$) at a low concentration of MP of 25 and 40 mg a.i./kg.

20.3.3 Chronic Toxicity Test

20.3.3.1 Effect on Earthworm Growth Inhibition

The growth inhibition of earthworms at 28th day in the chronic toxicity test was observed at different doses of MP 2% dust a.i. as presented in Fig. 20.4. A negative growth inhibition was observed in the control as well for doses of 2.5 and 4.0 mg a.i./kg. The earthworm growth rates observed as compared to control at MP doses of 2.5, 4.0, 6.4 and 10.2 mg a.i./kg were not significantly ($P > 0.05$) reduced with values around 3.87%, 8.45%, 16.04% and 7.20%, respectively. However, the earthworm growth rate was significantly ($P < 0.05$) reduced to 21% and 78% at MP doses of 16.3 and 23.0 mg a.i./kg, respectively. The negative growth inhibition at the lower dose was plausibly due to the weekly addition of 5 g of dried septic tank sludge as feeding material for 4 weeks. The high organic carbon content of the septic tank sludge provided the earthworms with the necessary food required for their growth. At lower doses, the earthworms were in the proper state but started to become stiff and structural coiling beyond an MP dose of 6.4 mg a.i./kg. Coiling has been regarded as a symptom of weight loss (Bustos-Obregón and Goicochea 2002). From ANOVA Tukey's test, it was observed that there was no statistically significant difference of lower and medium doses with control ($P > 0.05$) at 28 days indicating that MP 2% dust a.i. has little effect on the growth of earthworms. A statistically significant ($P < 0.05$) difference with the control was observed at the MP doses of 16.3 and 25 mg a.i./kg.

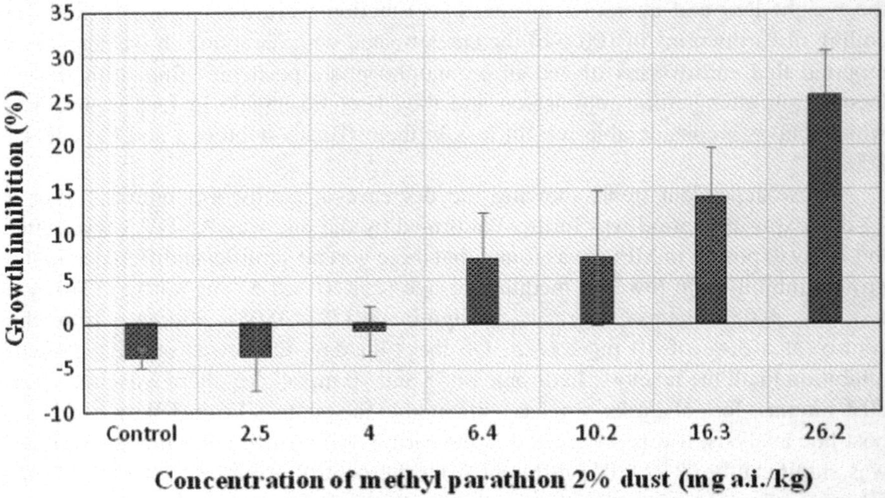

Fig. 20.4 Growth inhibition in *Eisenia fetida* on 28th day from chronic exposure to MP 2% dust a.i.

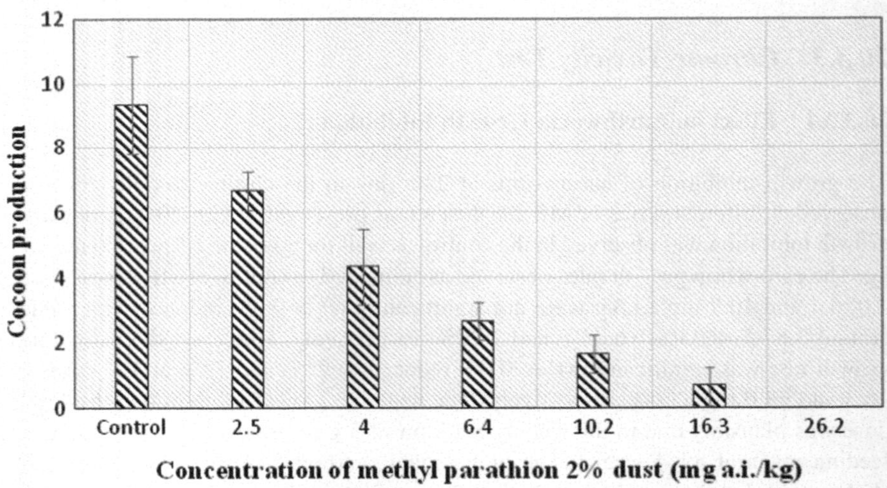

Fig. 20.5 Cocoon production by *Eisenia fetida* at different doses of MP 2% dust a.i.

20.3.3.2 Effect on the Earthworm Reproduction

The cocoon production by *Eisenia fetida* as a measure of the earthworm reproduction ability at different doses of MP 2% dust a.i. is shown in Fig. 20.5. The cocoon production in a period of 4 weeks was highest in the control reactors with an average production of 9.33 ± 1.53 cocoons in each reactor. The earthworms exposed to MP

2% dust a.i. showed lower cocoon production. The difference in the cocoon production was significantly ($P < 0.05$) different in all the doses of MP as compared to the control. Production of cocoons by the earthworms was found to be a sensitive parameter with the exposure to pesticides like paraquat, fentin, benomyl, phenmedipham, carbaryl, copper oxychloride and dieldrin (Cikutovic et al. 1993; Gupta and Saxena 2003; Reinecke and Reinecke 2007; Van Gestel 1992).

At MP doses of 2.5, 4.0, 6.4, 10.2 and 16.3 mg a.i./kg, the average weight per cocoon was not statistically ($P > 0.05$) different from the control. When the dose of MP was 26.2 mg a.i./kg, there was a statistically significant ($P < 0.05$) difference as compared to the control. The results obtained in this study were in good agreement with other studies (Xiao et al. 2006) showing the effect of pesticides like acetochlor on the reproduction ability of the earthworms.

20.4 Conclusions

The toxicological effects of organophosphate insecticide methyl parathion (MP) 2% dust (active ingredient, a.i.) on epigeic earthworm *Eisenia fetida* at different lethal and sub-lethal concentrations were assessed in this study. MP 2% dust a.i. was found to be extremely toxic to the epigeic earthworm *Eisenia fetida*. The median lethal concentration (LC_{50}) for *E. fetida* was found to be 36.63 mg a.i./kg, which confirms the extremely toxic nature MP 2% dust. A lower dose of 25 mg a.i./kg of MP 2% dust in acute toxicity test resulted in no mortality but showed a 24% loss in biomass as compared to the initial weight of the earthworms following a dose-dependent relationship. Septic tank sludge as an organic admixture was found to be a suitable feeding material for earthworms as it was observed to lead to a negative growth inhibition at a lower dose of MP 2% dust a.i. in the chronic toxicity test. This negative growth inhibition was essentially due to the high organic carbon content of septic tank sludge. In the chronic toxicity test, it was observed that the cocoon production by the earthworms was very less at higher sub-lethal doses. This can be attributed to the structural coiling of the earthworms as the earthworms plausibly find it difficult to copulate in terms of their posture of mating. It can be concluded from the study that the change in the earthworm biomass and the reproduction potential are the sensitive endpoints than the lethal concentration (LC_{50}) as they provide an early response to the sub-lethal doses of pesticides.

References

Abdullah AR, Bajet CM, Matin MA, Nhan DD, Sulaiman AH (1997) Ecotoxicology of pesticides in the tropical paddy field ecosystem. Environ Toxicol Chem 16(1):59–70

Abhilash PC, Singh N (2009) Pesticide use and application: an Indian scenario. J Hazard Mater 165 (1–3):1–12

Bustos-Obregón E, Goicochea RI (2002) Pesticide soil contamination mainly affects earthworm male reproductive parameters. Asian J Androl 4(3):195–200

Cikutovic MA, Fitzpatrick LC, Venables BJ, Goven AJ (1993) Sperm count in earthworms (Lumbricus terrestris) as a biomarker for environmental toxicology: effects of cadmium and chlordane. Environ Pollut 81(2):123–125

Darling CT, Thomas VG (2005) Lead bioaccumulation in earthworms, Lumbricus terrestris, from exposure to lead compounds of differing solubility. Sci Total Environ 346(1–3):70–80

Fairbrother A, Glazebrook PW, van Straalen NM, Tarazona JV (1999) Test methods for hazard determination of metals and sparingly soluble metal compounds in soils: summary of a SETAC Pellston workshop, 19–23 June, San Lorenzo de El Escorial, Spain

Gupta SK, Saxena PN (2003) Carbaryl-induced behavioural and reproductive abnormalities in the earthworm Metaphire posthuma: a sensitive model. Altern Lab Anim 31(6):587–593

Hobbelen PHF, Koolhaas JE, Van Gestel CAM (2006) Bioaccumulation of heavy metals in the earthworms Lumbricus rubellus and Aporrectodea caliginosa in relation to total and available metal concentrations in field soils. Environ Pollut 144(2):639–646

ISO (2012) ISO 11268-1: Soil-quality-effects of pollutants on earthworms (*Eisenia fetida*)—part 1: determination of acute toxicity using artificial soil substrate. International Organization for Standardization (ISO), Geneva

Løkke H, Van Gestel CAM (1998) Handbook of soil invertebrate toxicity tests. Wiley, Chichester

Lourenço JI, Pereira RO, Silva AC, Morgado JM, Carvalho FP, Oliveira JM, Malta MP, Paiva AA, Mendo SA, Gonçalves FJ (2011) Genotoxic endpoints in the earthworms sub-lethal assay to evaluate natural soils contaminated by metals and radionuclides. J Hazard Mater 186 (1):788–795

Nath BS, Suresh A, Varma BM, Kumar RS (1997) Changes in protein metabolism in hemolymph and fat body of the silkworm, Bombyx mori (Lepidoptera: Bombycidae) in response to organophosphorus insecticides toxicity. Ecotoxicol Environ Saf 36(2):169–173

OECD (1984) Test no. 207: Earthworm, acute toxicity tests, OECD Guidelines for the testing of chemicals, section 2. OECD Publishing, Paris

OECD (2016) Test no. 222: Earthworm reproduction test (*Eisenia fetida/Eisenia andrei*), OECD Guidelines for the testing of chemicals, section 2. OECD Publishing, Paris

Reinecke SA, Reinecke AJ (2007) The impact of organophosphate pesticides in orchards on earthworms in the Western Cape, South Africa. Ecotoxicol Environ Saf 66(2):244–251

Shi Y, Shi Y, Wang X, Lu Y, Yan S (2007) Comparative effects of lindane and deltamethrin on mortality, growth, and cellulase activity in earthworms (Eisenia fetida). Pestic Biochem Physiol 89(1):31–38

Suh CPC, Orr DB, Van Duyn JW (2000) Effect of insecticides on Trichogramma exiguum (Trichogrammatidae: Hymenoptera) preimaginal development and adult survival. J Econ Entomol 93(3):577–583

Suthar S (2014) Toxicity of methyl parathion on growth and reproduction of three ecologically different tropical earthworms. Int J Environ Sci Technol 11(1):191–198

Suthar S, Singh S, Dhawan S (2008) Earthworms as bioindicator of metals (Zn, Fe, Mn, Cu, Pb and Cd) in soils: is metal bioaccumulation affected by their ecological category? Ecol Eng 32 (2):99–107

Takada Y, Kawamura S, Tanaka T (2000) Biological characteristics: growth and development of the egg parasitoid Trichogramma dendrolimi (Hymenoptera: Trichogrammatidae) on the cabbage armyworm Mamestra brassicae (Lepidoptera: Noctuidae). Appl Entomol Zool 35 (3):369–379

Van Gestel CAM (1992) Validation of earthworm toxicity tests by comparison with field studies: a review of benomyl, carbendazim, carbofuran, and carbaryl. Ecotoxicol Environ Saf 23 (2):221–236

Van Gestel CAM, Ortiz MD, Borgman E, Verweij RA (2011) The bioaccumulation of molybdenum in the earthworm Eisenia andrei: influence of soil properties and ageing. Chemosphere 82 (11):1614–1619

Xiao N, Jing B, Ge F, Liu X (2006) The fate of herbicide acetochlor and its toxicity to Eisenia fetida under laboratory conditions. Chemosphere 62(8):1366–1373

Yasmin S, D'Souza D (2010) Effects of pesticides on the growth and reproduction of earthworm: a review. Appl Environ Soil Sci 2010:678360

The manufacturer's authorised representative in the EU is Springer Nature Customer Service Centre GmbH, Europaplatz 3, 69115 Heidelberg, Germany. If you have any concerns regarding our products, please contact ProductSafety@springernature.com

Printed and bound by CPI Group (UK) Ltd, Croydon, CR0 4YY
30/09/2026
02225557-0010